AF780791

Mechanism of Action of Mycotoxins

Mechanism of Action of Mycotoxins

Editor

Cristina Juan García

MDPI • Basel • Beijing • Wuhan • Barcelona • Belgrade • Manchester • Tokyo • Cluj • Tianjin

Editor
Cristina Juan García
Department of Preventive
Medicine and Public Health,
Food Sciences, Forensic
Medicine and Toxicology
University of Valencia
Burjassot
Spain

Editorial Office
MDPI
St. Alban-Anlage 66
4052 Basel, Switzerland

This is a reprint of articles from the Special Issue published online in the open access journal *Toxins* (ISSN 2072-6651) (available at: www.mdpi.com/journal/toxins/special_issues/mechanism_mycotoxins).

For citation purposes, cite each article independently as indicated on the article page online and as indicated below:

LastName, A.A.; LastName, B.B.; LastName, C.C. Article Title. *Journal Name* **Year**, *Volume Number*, Page Range.

ISBN 978-3-0365-7923-8 (Hbk)
ISBN 978-3-0365-7922-1 (PDF)

© 2023 by the authors. Articles in this book are Open Access and distributed under the Creative Commons Attribution (CC BY) license, which allows users to download, copy and build upon published articles, as long as the author and publisher are properly credited, which ensures maximum dissemination and a wider impact of our publications.
The book as a whole is distributed by MDPI under the terms and conditions of the Creative Commons license CC BY-NC-ND.

Contents

About the Editor . . . vii

Preface to "Mechanism of Action of Mycotoxins" . . . ix

Cheila Pereira, Sara C. Cunha and José O. Fernandes
Mycotoxins of Concern in Children and Infant Cereal Food at European Level: Incidence and Bioaccessibility
Reprinted from: *Toxins* **2022**, *14*, 488, doi:10.3390/toxins14070488 . . . 1

Marta Justyna Kozieł, Maksymilian Ziaja and Agnieszka Wanda Piastowska-Ciesielska
Intestinal Barrier, Claudins and Mycotoxins
Reprinted from: *Toxins* **2021**, *13*, 758, doi:10.3390/toxins13110758 . . . 23

Neda Alvarez-Ortega, Karina Caballero-Gallardo, Cristina Juan, Ana Juan-Garcia and Jesus Olivero-Verbel
Cytoprotective, Antiproliferative, and Anti-Oxidant Potential of the Hydroethanolic Extract of *Fridericia chica* Leaves on Human Cancer Cell Lines Exposed to α- and β-Zearalenol
Reprinted from: *Toxins* **2023**, *15*, 36, doi:10.3390/toxins15010036 . . . 49

Paula Llorens Castelló, Matteo Antonio Sacco, Isabella Aquila, Juan Carlos Moltó Cortés and Cristina Juan García
Evaluation of Zearalenones and Their Metabolites in Chicken, Pig and Lamb Liver Samples
Reprinted from: *Toxins* **2022**, *14*, 782, doi:10.3390/toxins14110782 . . . 61

Qun Cheng, Shuzhen Jiang, Libo Huang, Yuxi Wang and Weiren Yang
Zearalenone Exposure Affects the Keap1–Nrf2 Signaling Pathway and Glucose Nutrient Absorption Related Genes of Porcine Jejunal Epithelial Cells
Reprinted from: *Toxins* **2022**, *14*, 793, doi:10.3390/toxins14110793 . . . 73

Roua Gabriela Popescu, Andreea Luminița Rădulescu, Sergiu Emil Georgescu and Anca Dinischiotu
Aflatoxins in Feed: Types, Metabolism, Health Consequences in Swine and Mitigation Strategies
Reprinted from: *Toxins* **2022**, *14*, 853, doi:10.3390/toxins14120853 . . . 91

Silvia Iori, Marianna Pauletto, Irene Bassan, Federico Bonsembiante, Maria Elena Gelain and Anisa Bardhi et al.
Deepening the Whole Transcriptomics of Bovine Liver Cells Exposed to AFB1: A Spotlight on Toll-like Receptor 2
Reprinted from: *Toxins* **2022**, *14*, 504, doi:10.3390/toxins14070504 . . . 107

Yashodani Pillay, Savania Nagiah and Anil Chuturgoon
Patulin Alters Insulin Signaling and Metabolic Flexibility in HepG2 and HEK293 Cells
Reprinted from: *Toxins* **2023**, *15*, 244, doi:10.3390/toxins15040244 . . . 135

Mercedes Taroncher, Fiona Halbig, Yelko Rodríguez-Carrasco and María-José Ruiz
Stressful Effects of T-2 Metabolites and Defense Capability of HepG2 Cells
Reprinted from: *Toxins* **2022**, *14*, 841, doi:10.3390/toxins14120841 . . . 151

Philippe Guerre, Caroline Gilleron, Maria Matard-Mann and Pi Nyvall Collén
Targeted Sphingolipid Analysis in Heart, Gizzard, and Breast Muscle in Chickens Reveals Possible New Target Organs of Fumonisins
Reprinted from: *Toxins* **2022**, *14*, 828, doi:10.3390/toxins14120828 . . . 171

Yi Chen, Guanggang Qu, Hongkun Quan, Yihui Wang, Changjiang Wang and Md Atiqul Haque et al.
A Novel Cost-Effective Nanobody against Fumonisin B1 Contaminations: Efficacy Test in Dairy Milk and Chickens
Reprinted from: *Toxins* **2022**, *14*, 821, doi:10.3390/toxins14120821 . . . . **191**

About the Editor

Cristina Juan García

Cristina Juan García is Associate Professor of Food Science in the Faculty of Pharmacy at the University of Valencia (2008) and receieved her International PhD in Pharmacy at the UV. Her active research concerns are to develop and performance analytical methods to quantify and identify the presence of mycotoxins in different matrices (food, feed, biological matrices and culture cells) using different mass spectrometry equipment (QqQ, Qtrap and QTOF) coupled to or LC or GC. Her scientific production is documented by two books (editor/chapters), and she is author and co-author of more than 75 international publications (Research ID: F-3406-2013; ORCID: 0000-0002-8923-3219). She has collaborated with different research groups of food safety in mycotoxins projects, of Prof. Celeste Lino research group from University of Coimbra in Portugal; Prof. Alberto Ritieni from "Università degli Studi di Napoli " Federico II in Italy; and Prof. Doina Miere and Prof. Felicia Loghin research groups from University of Medicine and Pharmacy "Iuliu Haţieganu" of Cluj-Napoca in Rumania. She has been involved in several research projects (more than 10) supported by the Spanish Ministry of Education and Science and from the Valencian Community and participated as a member in the European MycoKey project H2020-EU.3.2.–SOCIETAL.

Preface to "Mechanism of Action of Mycotoxins"

Mycotoxins are a diverse group of chemicals that present wide toxicological responses in animals and humans. Their ingestion causes toxic effects that go from acute toxicity to long-term or chronic health disorders. Some mycotoxins have caused outbreaks of human toxicoses, and at least one mycotoxin, aflatoxin B1, is an assumed human hepatocarcinogen. As part of a comprehensive effort to curtail the adverse health effects posed by mycotoxins, substantial research has been conducted to determine the mechanism of action of mycotoxins. Although much information has been obtained regarding the action of several mycotoxins, future research topics should continue to address several areas of critical concern.

Detection of biomarkers in noninvasive samples, such as urine, requires the use of methods that are beginning to become an important tool in the measurement of human exposure to mycotoxins in populations that are particularly at risk. Among that, proteomics, metabolomics and genomics are actually goals in toxicological research not only for their innovative methodologies, but also because they can reveal changes in the levels of abundance of proteins and metabolites and their interactions.

In vitro studies in different cell lines could detail and explain many of these mechanisms, while in vivo can give a real scenario in the development of a toxic effect.

The focus of this Special Issue of *Toxin* was to gather the most recent reports on the mechanism of action of mycotoxins on single or combined mycotoxins studied in vivo or in vitro, the identification of known and unknown mycotoxin metabolites and other metabolites in different cell lines and animals or matrices (including organs, urine, or blood), and the development of analytical skills to study these mechanisms. Research papers and review articles are included in this reprint. A total of three review articles related to intestinal barrier, infants' and children's bioaccessibility and aflatoxin metabolism on swine have been consider (Kozieł et al., 2022; Pereira et al., 2022; Popescu et al., 2022). Furthermore, zearalenone and their metabolites have been studied in chicken, pig and lamb liver samples and in porcine jejunal epithelial cells to evaluate their presence and how their exposure affects the Keap1–Nrf2 signaling pathway and glucose nutrient absorption related genes (Castelló et al., 2022; Cheng et al., 2022). Zearalenone's effects have been evaluated by Alvarez-Ortega et al. (2022) on human cancer cell lines to study the cytoprotective, antiproliferative, and anti-oxidant potential of the hydroethanolic extract of Fridericia Chica leaves. Finally, five studies about aflatoxin B1, patuline, fumonisins and T-2 toxin have been included (Iori et al., 2022; Pillay et al., 2022; Guerrere et al., 2022; Chen et al., 2022; Taroncher et al., 2022). One study is related to their effect on liver (BFH12), hepatic (HepG2) and kidney (HEK293) cells.

Cristina Juan García
Editor

toxins

MDPI

Review

Mycotoxins of Concern in Children and Infant Cereal Food at European Level: Incidence and Bioaccessibility

Cheila Pereira, Sara C. Cunha * and José O. Fernandes

LAQV-REQUIMTE, Laboratory of Bromotology and Hidrology, Faculty of Pharmacy, University of Porto, Jorge de Viterbo Ferreira 228, 4050-313 Porto, Portugal; cheila.vpereira@gmail.com (C.P.); josefer@ff.up.pt (J.O.F.)
* Correspondence: sara.cunha@ff.up.pt

Abstract: Cereals are of utmost importance for the nutrition of infants and children, as they provide important nutrients for their growth and development and, in addition, they are easily digestible, being the best choice for the transition from breast milk/infant formula to solid foods. It is well known that children are more susceptible than adults to toxic food contaminants, such as mycotoxins, common contaminants in cereals. Many mycotoxins are already regulated and controlled according to strict quality control standards in Europe and around the world. There are, however, some mycotoxins about which the level of knowledge is lower: the so-called emerging mycotoxins, which are not yet regulated. The current review summarizes the recent information (since 2014) published in the scientific literature on the amounts of mycotoxins in infants' and children's cereal-based food in Europe, as well as their behaviour during digestion (bioaccessibility). Additionally, analytical methods used for mycotoxin determination and in vitro methods used to evaluate bioaccessibility are also reported. Some studies demonstrated the co-occurrence of regulated and emerging mycotoxins in cereal products used in children's food, which highlights the need to adopt guidelines on the simultaneous presence of more than one mycotoxin. Although very little research has been done on the bioaccessibility of mycotoxins in these food products, very interesting results correlating the fiber and lipid contents of such products with a higher or lower bioaccessibility of mycotoxins were reported. LC-MS/MS is the method of choice for the detection and quantification of mycotoxins due to its high sensibility and accuracy. In vitro static digestion models are the preferred ones for bioaccessibility evaluation due to their simplicity and accuracy.

Keywords: food toxins; mycotoxins; infancy/childhood nutrition; quality control; bioaccessibility; chromatography

Key Contribution: Summary of the latest scientific contributions on regulated and emerging mycotoxin incidence in cereal-based infant/children food with a review on their bioaccessibility.

Citation: Pereira, C.; Cunha, S.C.; Fernandes, J.O. Mycotoxins of Concern in Children and Infant Cereal Food at European Level: Incidence and Bioaccessibility. *Toxins* **2022**, *14*, 488. https://doi.org/10.3390/toxins14070488

Received: 23 May 2022
Accepted: 9 July 2022
Published: 15 July 2022

Publisher's Note: MDPI stays neutral with regard to jurisdictional claims in published maps and institutional affiliations.

Copyright: © 2022 by the authors. Licensee MDPI, Basel, Switzerland. This article is an open access article distributed under the terms and conditions of the Creative Commons Attribution (CC BY) license (https://creativecommons.org/licenses/by/4.0/).

1. Introduction

Nutrition in the first years of life is essential to provide optimal growth and development and to establish preferences and eating patterns [1–3]. At 6 months of age, breastfeeding should be gradually replaced by nutritionally adequate and safe complementary food, starting with small portions of food that gradually increase with child development [3,4]. From 6 to 23 months of age, children's nutritional needs require that they eat foods from at least four of the subsequent groups per day: grains, roots and tubers, legumes and nuts, dairy products, meat and fish, eggs, vitamin A-rich fruits and vegetables, and other fruits and vegetables. If their diet lacks this diversity of food, there is a risk of potential micronutrient deficiency [3].

Infant cereals are, usually, among the first foods introduced as complementary foods during weaning because they can complement breastfeeding, guaranteeing the nutritional requirements of the infant, providing a large number of proteins, vitamins (B1, B2, B3,

B6, B9, E, and K), minerals (iron, zinc, magnesium, sodium, potassium, and phosphorus), and bioactive compounds, and allowing iron fortification. Also, cereals provide non-digestible carbohydrates that are responsible for the development of microbiota (increasing the *Bacteroides* population). Due to their mild taste and semi-solid texture, cereals allow the transition from milk and the consequent acceptance of solid foods [5]. There are several different types of cereal-based products available for children, and they are composed of different base cereals, such as wheat, maize, barley, rice, and oats, which can be mixed with chocolate, honey, fruits, or nuts for the enhanced flavour and attractiveness of the product.

Cereals used in infant food are usually subjected to meticulous quality control processes before their release to the market; however, crops are susceptible to toxigenic fungi during both pre- and post-harvest steps, which can produce small toxic chemicals: mycotoxins. The main fungal producers of mycotoxins belong to the genera *Aspergillus* spp., *Penicillium* spp., and *Fusarium* spp. Generally, the identification of the fungi responsible for mycotoxin contamination is difficult since a single fungal species may be able to produce different mycotoxins, while a certain mycotoxin can be produced by more than one species. For instance, mycotoxins such as aflatoxins (AFs), ochratoxins (OTA), citrinin (CIT) [6], and sterigmatocystin (STE) [7], the last two being not regulated in this type of matrix, are produced by the *Aspergillus* species, while trichothecenes (TCs) (group A: HT-2, and T-2 toxins, and group B: deoxynivalenol (DON)), zearalenone (ZEA), fumonisins B1 (FB1) and B2 (FB2), and the emerging mycotoxins beauvericin (BEA) [8] and enniatins (ENs) [7] are often produced by the *Fusarium* species. Ergot alkaloids are produced by *Claviceps* [9].

Mycotoxins occur along all cereal food chains and can have a myriad of acute and/or chronic detrimental health effects, such as immunosuppressive, carcinogenic, estrogenic, gastrointestinal, and kidney events [10–12]. Therefore, several countries have adopted regulations to limit mycotoxin exposure through food and reduce their levels to as low as possible. In Europe, EC (European Community) set different maximum limits for mycotoxin exposure for adults and for children (Table 1)—Commission Regulation (EC) No 1881/2006, Commission Regulation (EC) No 165/2010, and Commission Recommendation (2013/165/EU). While the "classical mycotoxins" are already well-known and regulated, there are other mycotoxins, i.e., the so-called emerging mycotoxins, that are not routinely determined nor regulated, even though there is evidence of their rapidly increasing incidence [13].

Table 1. Maximum levels of mycotoxins in cereals and cereal-derived products according to the European Commission.

Mycotoxins	Processed Cereal-Based Foods and Baby Foods for Infants and Young Children (μg/kg)
Aflatoxin B1	0.1
Ochratoxin A	0.5
Patulin *	10
Deoxynivalenol	200
Sum T-2 and HT-2 toxin	15
Zearalenone	20
Sum Fumonisin B1 and Fumonisin B2	200

* Baby foods other than processed cereal-based foods for infants and children.

This manuscript aims to present for the first time a holistic review of mycotoxins in infants's and children's cereal-based foods found in Europe since 2014. To accomplish the objective, the amounts of mycotoxins, their bioaccessibility as well as the advantages and drawbacks of the analytical methods used for their determination will be provided. The impact of mycotoxins in infants and children and the protective effect of the ingestion of cereals against mycotoxicosis will be briefly summarized. Additionally, different factors that determine mycotoxins' bioaccessibility will be highlighted in order to understand whether the fraction of mycotoxins released from the food matrix into the gastrointestinal tract upon digestion could induce toxic health effects in the children. We provide an overview of the gaps in research on mycotoxins in cereal-based foods for infants and children, to stimulate and improve future research avenues.

2. Materials and Methods

A review of the literature was performed by using the databases Scopus® (www.scopus.com), Web of Science®, PubMed®, and Google Scholar. Several keywords were included to identify published works for mycotoxin incidence in cereal and cereal-based products for infants and children. These included "occurrence", "mycotoxin", "aflatoxin", "ochratoxin", "deoxynivalenol", "fumonisin", "patulin" "T-2", "HT-2", "enniatin", "beauvericin", "cereal" or "cereal-based products", "infant" or "children", and "Europe". The terms were searched across document titles, types, abstracts, and keywords lists across all years since 2014. Thirteen studies met the selected criteria and were included in the revision. The articles on bioavailability of mycotoxins in this matrix were selected using keywords such as "mycotoxin", "bioaccessibility" or "bioavailability", "cereal" or "cereal-based products", "infant" or "children"; since only four studies were found, no timeframe was established. Data from the studies were divided into different sections: major mycotoxins in cereals, bioaccessibility of mycotoxins in cereal-based infants' and children's food and strategies for its reduction, methods for mycotoxin analysis in food and bioaccessibility studies, and gaps in the research of mycotoxins in infants'/children's cereal-based food matrices (Figure 1).

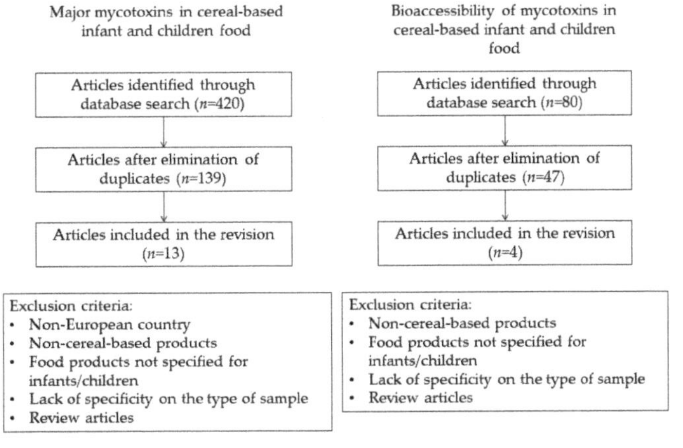

Figure 1. Methodology description diagram.

3. Mycotoxicosis in Infants and Children

Mycotoxin contamination is characterized by overtime exposure with the repeated consumption of the same kind of products. Children are more exposed to acute mycotoxicosis because they have stricter dietary patterns, consume some specific food products in larger amounts than adults (per kg of body weight), and their metabolism is not totally developed, so their system is more sensible [9,14–16]. However, as the levels of mycotoxins in food are usually low, the long-term effects are rarely seen in children [17,18].

Among children and infants, the most characterized symptoms are recurrent vomiting, caused by exposure to AFs, fumonisins, T-2, patulin (PAT), and DON, bone marrow failure, a well-recognized effect of contamination by TCs [15,19], in addition to recurrent apnoea, pneumonia and/or acute pulmonary haemorrhages, when the exposure route is inhalation [19]. Although unusual in children, there are some long-term consequences due to exposure to these toxins, such as hepatocellular carcinoma by AFs [10], oesophageal cancer and neural tube defects by fumonisins [10–12], renal cancer and Balkan nephropathy by OTA [10,20], estrogenic effects by ZEA [21,22], and ergotism by EAs [10] (Figure 2).

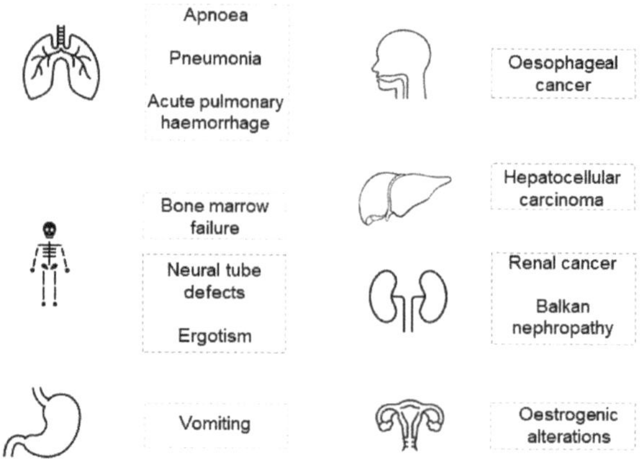

Figure 2. Most common mycotoxicosis health effects in children. Acute effects in green and chronic effects in orange.

Damage to the gastrointestinal system may also only be revealed in the long term. Mycotoxins affect the digestive system in two ways: first by altering the gut microbiota due to a direct toxic effect on the microorganisms and second by altering the structures of the intestine, e.g., TCs and PAT can affect the intestinal barrier, causing the weakening of the permeability and integrity of the intestinal epithelium, resulting in inflammation [17,23,24].

4. Protective Effects of Cereals against Mycotoxin Exposure

Cereals, especially whole-grain, provide several compounds with health-protective mechanisms that may diminish the effects caused by mycotoxins' toxicity. Particularly, wheat bran has shown consistent cancer-protective properties in human and animal experimental models [25]. The anti-carcinogenic properties of whole-grain foods are mainly due to antioxidant and anti-inflammatory compounds, such as phenolic acids [26–28], flavonoids, carotenoids [29], vitamin E [30], n-3 fatty acids [31], phytic acid [32,33], and selenium [34]. Whole-grain cereals are also an indirect source of short-chain fatty acids such as acetate, butyrate, and propionate, which have cancer-preventing properties by lowering intestinal pH and reducing the ability of bile acids to act as carcinogens [35,36]. Moreover, the presence of insoluble fibres in whole-grain cereals increases colon transit time and faecal bulking, which leads to the dilution of carcinogens and reduces their interaction with epithelial cells, resulting in cancer prevention [35]. Dietary fibres can also adsorb xenobiotics, resulting in diminished exposure and absorption by the digestive tract [37]. Each of these compounds acts complementary to each other, which enhances their protective action [38].

5. Major Mycotoxins in Cereals

There are several studies on the prevalence of mycotoxins in cereal-based products for young infants and children, as reported in some previous review manuscripts [39–41]. In Table 2, there is a summary of the data from those reviews with special focus on the incidence of mycotoxins in cereal and cereal-based foods for infants and children commercialized in Europe.

Table 2. Occurrence of mycotoxins in cereal-based infant and young children food in Europe (2014 –2021).

Country	Sample	Mycotoxin	Total Samples	Positive Samples			References
				%	Mean (μg/kg)	Range (μg/kg)	
Italy	Infant formulas and baby food	OTA	75	20	0.06	0.050–0.120	Juan et al., 2014 [42]
		DON		25.3	102.60	1–268	
		NIV		4	19.91	5.5–235	
		FUS-X		24	146.51	5.5–604	
		HT-2		2.7	12.65	2–151	
		β-ZOL		6.7	2.5	2–23.2	
		ENB		13.3	101.30	5–832	
		ENB1		1.3	7.80	5–117	
		ENB4		5.3	38.08	5–311	
		ENA1		4	6.58	5–125	
		BEA		1.3	1.18	5–21.3	
Portugal	Cereal baby food (maize, wheat, rice, barley, rye, oat, sorghum, millet, spelt)	DON	9	44	173.13	0.37–270.57	Pereira et al., 2015 [43]
		15AcDON		11	30.94	2.50–30.94	
		T2-Tetrol		11	112.18	10.48–112.18	
		NEO		11	87.21	1.28–87.21	
Portugal	Breakfast cereals for children (maize, wheat, rice, and multi-grain)	AFB1	26		0.028	0.040–0.400	Assunção et al., 2015 [44]
		AFB2			0.002	0.030–0.300	
		AFG1			0.006	0.045–0.450	
		AFM1			0.012	0.100–1.000	
		OTA			0.026	0.200–2.000	
		DON			59	15–360	
		NIV			6	25–360	
		FB1			13	2.5–8.0	
		FB2			3	2.5–8.0	
Turkey	Baby food (cereal based supplementary foods for infants and children)	OTA	50	68	0.034–0.374	0.042–0.380	Hampikyan et al., 2015 [45]
Portugal	Children cereal-based food	PAT	20	75 50 40	2.33 0.061	3.2–40.0 0.2–2.0	Assunção et al., 2016 [46]
Portugal	Breakfast cereals	AFB1	26	69	0.013	0.003–0.130	Martins et al., 2018 [47]
		AFB2		27	0.004	0.001–0.011	
		AFG1		4	0.013	0.006–0.014	
		AFM1		12	0.017	0.011–0.240	
		OTA		69	0.040	0.006–0.100	
		FB1		58	12.5	0.06–67.0	
		FB2		38	4.2	0.12–14.0	
		DON		62	91.5	0.4–207.8	
		NIV		4	27.1	5.6–27.1	
		ZEA		19	0.7	0.12–5.6	

Table 2. *Cont.*

Country	Sample	Mycotoxin	Total Samples	Positive Samples			References
				%	Mean (μg/kg)	Range (μg/kg)	
Portugal	Cereal-based children food	AFB1	26 breakfast cereals	73	0.036	NM	Assunção et al., 2018 [48]
		AFB2		46	0.07		
		AFG1		4	NA		
		AFM1		12	0.017		
		AFs		73	-		
		OTA		69	0.047		
		FB1		58	22.00		
		FB2		39	5.10		
		FMs		58	-		
		ZEA		73	1.20		
		DON		62	95.9		
		NIV		4	NA		
		AFB2	20 infant cereals (flours)	5	NA	NM	
		AFG1		10	0.014		
		AFM1		40	0.068		
		AFs		45	-		
		OTA		50	0.061		
		FB1		35	0.44		
		FMs		35	-		
		ZEA		30	0.48		
		DON		20	41.8		
		OTAOTA	6 biscuits	100	0.086	NM	
		DON		50	43.8		
Germany	Cereal-based baby food	AOH	19	36.8	0.89	4.73–7.13	Gotthardt et al., 2019 [49]
		AME		89.5	0.24	0.23–0.58	
		TEN		94.7	1	0.18–7.53	
		ATX I		15.8	0.17	NA	
		ATLP		5.3	0.24	NA	
		TA			50.2	5.66–221	
Spain	Cereal-based baby food	AFB1	60	11	0.03	0.02–0.23	Herrera et al., 2019 [50]
		AFB2		1	0.01	0.02–0.20	
		AFG1		6	0.02	0.02–0.16	
		AFG2		1	0.01	0.02–0.11	
		DON		12	37	33–245	

Table 2. *Cont.*

Country	Sample	Mycotoxin	Total Samples	Positive Samples			References
				%	Mean (μg/kg)	Range (μg/kg)	
Poland	Cereal-based infant and children food	DON	302	17	>LOD [a] <LOQ [b]	NM	Postupolski et al., 2019 [51]
		NIV		3			
		ZEA		14			
		OTA		4			
		HT-2		0			
		T-2		1			
		FB1		3			
		FB2		4			
Italy	Breakfast cereals Sweet cakes	OTA	84	2.38	1	NM	Capei et al., 2019 [52]
				35.7	1.34		
Austria and Czech Republic	Processed cereal-based infant foods	AFL	35	6	-	<LOQ–1.1	Braun et al., 2021 [53] *
		AFB1		-		-	
		STG		23	-	<LOQ–0.5	
		ZEA		3	0.24	1.2	
		DON		6	-	25–62	
		NIV		6	43	<LOQ–20	
		T-2		26	-	0.8–3.0	
		BEA		14	1.5	<LOQ–3.1	
		ENA		3	−1.9	<LOQ	
		ENB		11	0.7	<LOQ–2.1	
		ENA1		60	5.9	<LOQ–40	
		ENB1		26	3.9	<LOQ–10	
		FB1		20	4.8	<LOQ–8.3	
		AME		20	0.6	<LOQ–1.1	
		TA		31	48	<LOQ–124	
		TEN		34	0.9	<LOQ–1.5	
		ATPL		23	11	<LOQ–20	
Poland	Cereal-based baby foods	DON	110	9.09	107.8	62–148	Mruczyk et al., 2021 [54]

AFB1 (Aflatoxin B1), AFB2 (Aflatoxin B2), AFG1 (Aflatoxin G1), AFG2 (Aflatoxin G2), AFM1 (Aflatoxin M1), OTA (Ochratoxin A), DON (Deoxynivalenol), 15acDON (15-acetyldeoxynivelanol), NIV (Nivalenol), FUS-X (Fusarenon-x),T-2 (Mycotoxin T-2), HT-2 (Mycotoxin HT-2), T2-Tetrol (Mycotoxin T2-tetrol), β-ZOL (β-zearalenol), FB1 (Fumonisin B1), FB2 (Fumonisin B2), PAT (Patulin), ZEA (Zearalenone) ENB (Enniatin B), ENB1 (Enniatin B1), ENB2 (Enniatin B2), ENB4 (Enniatin B4), ENA (Enniatin A), ENA1 (Enniatin A1), ENA2 (Enniatin A2), BEA (Beauvericin), STG (Sterigmatocystin), NEO (Neosolaniol), AOH (Alternariol), AME (Alternariol monomethyl ether), TEN (Tentoxin), ATX I (Altertoxin 1), ATLP (Alterperylenol), TA (Tenuazonic acid) and AFL (Aflatoxicol). Maximum Limit (EC): AFB1—0.1 μg/kg; OTA—0.5 μg/kg; PAT—10.0 μg/kg; DON—200 μg/kg; ZEA—20 μg/kg; FB1 + FB2—200 μg/kg; T-2 + HT-2—15 μg/kg. Bold—above the maximum limit (EC). NA—not applicable; NM—not mentioned.[a] DON—2.0, NIV—18.6, ZEA—6.1, OTA—0.07, HT-2—1.1, T-2—0.1, FB1—1.4 and FB2—0.5 μg/kg. [b] DON—6.5, NIV—61.9, ZEA—20.5, OTA—0.24, HT-2—3.7, T-2—0.3, FB1—0.4 and FB2—1.5 μg/kg. * AFL—0.25; STG—0.1; NIV—16; BEA—0.4; ENA—0.4; ENA1—0.4; ENB—0.4; ENB1—0.1; FB1—7.0; AME—0.6; TA—24, TEN—1.0 and ATPL—10 μg/kg [53].

From 2014 to 2021, thirteen studies reported the presence of mycotoxins in cereal-based infants'/children's food products in Europe. The majority of the studies verified that TCs, particularly DON, are the most prevalent mycotoxins [42–44,47,48,50,51,54]. In some of these studies, DON levels were above the maximum limit established by EC (200 μg/kg) [42,43,48,50]. HT-2 was detected in two studies [42,51], one of which found two samples above the established maximum limit (15 μg/kg) [42], and T-2 was also found in two studies at low levels in the first in four samples [51], and in the last in nine samples [53]. Nivalenol (NIV), a TCs unregulated mycotoxin, was found in the cereal samples in five studies, namely in four samples in the studies by Juan et al., 2014 [42], and Martins et al., 2018 [47], in three samples in the study by Postupolski et al., 2019 [51], and in two samples in the study by Braun et al., 2021 [53].

OTA was also a very prevalent mycotoxin in cereal-based food for young people, being reported in eight out of thirteen studies [42,44–48,51,52], usually in levels below the maximum limit established by the EC (0.5 μg/kg). In the study by Juan et al., 2014 [42], this mycotoxin was present in 15 of 75 (20%) samples of cereal-based baby food. Assunção and team [46] found that 40% of the samples presented co-occurrence of PAT and OTA, which underlines the necessity of establishing a maximum limit of PAT in processed cereal-based foods. Martins et al., 2018 [47] reported the presence of OTA in 18 samples. Assunção et al., 2018 [48] analysed different types of samples and reported the presence of OTA in 18 samples of breakfast cereals, 10 samples of infant flours, and 6 samples of biscuits. Lastly, Postupolski et al., 2019 [51], detected the presence of this mycotoxin in four samples.

Regarding AFs, despite being one of the most studied mycotoxins, only four studies reported their presence in cereal-based food samples marketed for children [44,47,48,50]. AFB1 was above the maximum limit established by the EC of 0.1 μg/kg in the works of Martins et al., 2018 [47] and Herrera et al., 2019 [50], both in six samples. In the last study, AFB2 (n = 1), AFG1 (n = 6), and AFG2 (n = 1) were also found.

Fumonisins and ZEA were found less in cereal-based baby foods, and when detected, the levels were below the maximum limits indicated by the EC, 200 μg/kg and 20 μg/kg, respectively [44,47,48,51]. In the study by Postupolski et al., 2019 [51], only one sample presented positive results for fumonisins, and 14 samples were contaminated with ZEA, while in the studies by Assunção et al., 2018 [48] and Martins et al., 2018 [47], both fumonisins and ZEA had a similar incidence of 19 and 15 positive samples, respectively.

ENs were quantified in cereal-based samples marketed for children in two works (Juan et al., 2014 and Braun et al., 2021), but their incidence was lower when compared to other regulated mycotoxins. ENB was the most detected enniatin, mostly in wheat cereal samples (n = 7) in the study by Juan et al., 2014 [42] and in 21 samples in the study by Braun et al., 2021 [53]. Even so, co-occurrence with these emerging mycotoxins demonstrates the importance of the establishment of new guidelines and the necessity of more studies in these matrices.

Other emerging mycotoxins such as STG [53], Alternaria toxins (AOH, AME, TEN, TA, ATX I or ATPL) [49,53], T2-tetrol [43], and AFL [53], were also found in the studies reported here, however, with a much lower frequency.

6. Bioaccessibility of Mycotoxins in Cereal-Based Infants' and Children's Food and Strategies for Its Reduction

Despite the knowledge already acquired about the presence of mycotoxins in food products, the amounts that are effectively absorbed by the human body are not known, as little is known about their bioavailability and bioaccessibility once ingested [55–57] (Figure 3).

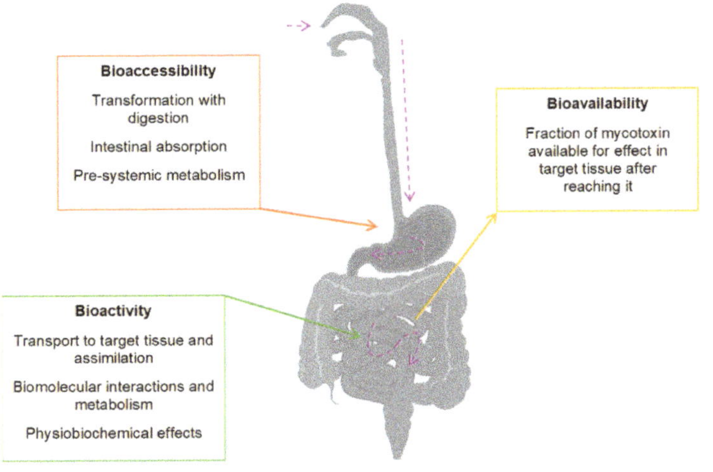

Figure 3. Bioaccessibility, bioactivity, and bioavailability definitions. Purple arrow—mycotoxin path after ingestion.

Mycotoxin bioavailability is the measurement of the amount of mycotoxin that can reach circulation after absorption by the intestinal epithelium (Figure 2). Different food matrices can affect the bioaccessibility of a mycotoxin; however, the metabolism and absorption depend on the mycotoxin itself [55,57,58].

There are few studies on the bioaccessibility of mycotoxins in children (Table 3), but it is known that a child may be more susceptible to significant damage to the intestinal enterocytes caused by the presence of these toxins in the intestinal fluid when compared to adults [15,58,59].

Table 3. Bioaccessibility of mycotoxins in cereal-based infants' and young children's food.

Country	Sample	Mycotoxin	Total Samples	Bioaccessibility (%)	References
The Netherlands	Infant formula (spaghetti Bolognese) supplemented with 2 mL sunflower oil per 100 g of food	AFB1 OTA	2	88 ± 16–94 ± 8 29 ± 6–32 ± 4	Kabak et al., 2009 [60]
Italy	Commercial pasta	DON	6	2.12–41.5	Raiola et al., 2012 [59]
Spain	Breakfast cereals Cookies Breads	ENA ENA1 ENB ENB1	14	50 ± 3–80 ± 3 40 ± 2–64 ± 2 43 ± 3–70 ± 3 46 ± 3–74 ± 2	Prosperini et al., 2013 [61]
Portugal	Cereal-based food	PAT OTA PAT+OTA PAT+OTA	6	30 ± 3–77 ± 2 95 ± 0–105 ± 2 33 ± 1–64 ± 2 (PAT) 103 ± 1–109 ± 0 (OTA)	Assunção et al., 2016 [46]

Kabak and colleagues evaluated the effects of probiotic bacteria on bioaccessibility of AFB1 and OTA alone and together in infant formula, with an in vitro static digestion model. The bioaccessibility levels of AFB1 and OTA differ from each other, 88–94% for AFB1 and 29–32% for OTA, which shows that the bioaccessibility depends on the mycotoxin [60]. The authors found that the effect on bioaccessibility depended on bacteria species and type of mycotoxin, with *Lactobacillus acidophilus* and *L. casei* Shirota being more effective in reducing the bioaccessibility of AFB1, while no effect was observed on the bioaccessibility of OTA.

In 2012, Raiola's team evaluated the bioaccessibility of DON in six Italian commercial pastas marketed for young children using an in vitro static digestion model that simulated child physiology. The percentage of DON in the gastric fluid ranged from 2.12 ± 0.11–38.41 ± 2.95%, and the bioaccessibility of DON after the duodenal process ranged from 1.11 ± 0.01–17.91 ± 0.80%. The authors concluded that the bioaccessibility of DON is high enough to produce toxicity and cause higher damage to children, especially due to the small dimension of the intestinal epithelium [59].

Prosperini and colleagues [61] used an in vitro static digestion model to study the bioaccessibility of ENA, ENA1, ENB, and ENB1 in 14 samples of breakfast cereals, cookies, and bread. The samples were spiked with the mycotoxins at 1.5 and 3.0 μmol/g. Their results showed no significant differences between spiked levels in the same type of samples for all mycotoxins except ENB1, where there was higher bioaccessibility in the samples spiked with 1.5 μmol/g (67.3 ± 2.7%). Among the mycotoxins, ENA has the highest bioaccessibility, followed by ENB1, ENB, and ENA1. There are, however, differences in bioaccessibility between similar types of samples that may be related to the chemical structure of the mycotoxins, the food product composition, or matrix. When comparing different cereals, the ones with higher fibre content showed lower bioaccessibility, as the fibres combined with the mycotoxins, leaving less available to produce toxic effects. The inclusion of dietary fibres offers some protection against mycotoxicosis, resulting in a cheap and effective method for the detoxification of feed and food products. Likewise, when comparing different samples with different lipid content, the authors noticed that food products with a higher lipid content show lower bioaccessibility. When the mycotoxins interact with the fat content of the food product, they are not released completely.

The most recent study on bioaccessibility of mycotoxins in cereal-based food for children is from Assunção et al. [46], who studied the bioaccessibility of PAT and OTA in processed cereal-based foods with an in vitro static digestion model. The authors analysed six samples (three with fruit and three without), which were spiked with 20 μg/kg of PAT and 1 μg/kg of OTA, and they verified that the bioaccessibility values of OTA were significantly higher than those of PAT, 106 ± 0.6%, and 56 ± 2.7%, respectively. When PAT was assayed alone or combined with OTA, there was no statistical difference in the values of bioaccessibility, with mean values of 52 ± 4.2% and 56 ± 2.7%, respectively. There was, however, a statistical difference when assaying OTA alone or combined with PAT, with mean values of 100 ± 1.1% and 106 ± 0.6%, which means that OTA demonstrated higher bioaccessibility when in a mixture with PAT; the values above 100% may be due to possible interactions between food matrix, mycotoxins, and digestive fluids.

In vitro biotransformation studies with cell lines offer deeper insight into the effect of the digestion process on mycotoxins and their bioavailability, and consequently, on the impact of these contaminants on the intestinal cells. The biotransformation process of mycotoxins is composed of two phases: in the first, mycotoxins may suffer oxidation, reduction, or hydrolyzation by cytochrome P (CYP) enzymes present in the endoplasmic reticulum and mitochondria, and in the second phase, the resulting metabolites may form conjugates with glutathione, glucuronic acid, and sulphate [62]. The Caco-2 (colorectal adenocarcinoma cell) cell line is the preferred in vitro model method to analyse the intestinal absorption of compounds because these cells form a continuous monolayer with tight junctions that mimic the intestinal wall [63]; also, the ability to express various phase I and II enzymes makes it a good option for biotransformation studies. Other models, such as IPEC-J2 (intestinal porcine epithelial cell), are also used [64,65]. Epithelial integrity and the organization of the tight junction in the cell monolayer are evaluated by monitoring the trans-epithelial electrical resistance (TEER) and the expression of tight junction proteins.

The biotransformation processes of the most studied mycotoxins are already known. AFB1, while not reactive by itself, after bioactivation by different CYP enzymes, forms different metabolites with different degrees of toxicity, such as AFB1-8,9-exposide (AFBO) [66], which is very toxic, with mutagenic and carcinogenic characteristics via the action of CYP450, aflatoxicol (AFL), which is mildly toxic via a ketoreduction, and AFM1, which is also mildly toxic via hydroxylation. The detoxification process of AFBO and AFM1 involves a phase II reaction, conjugation with glutathione-by-glutathione S-transferase (GST) [67,68]. Studies reported that AFB1 reduces TEER in a time-dependent manner in Caco-2 cell line culture [69–71] and significantly decreases mRNA expression of tight junction proteins, which are essential for maintenance of adhesive and selective permeability characteristics of the intestinal epithelium [70,71], while AFM1, which is less toxic than AFB1, damaged epithelial integrity to a lesser extent [71,72]. The metabolites resulting from OTA biotransformation are usually less toxic than the original compound [73]. The most common metabolites formed are OTα, formed by the action of carboxypeptidases, 4-hydroxy-ochratoxin A (4-OH-OTA), and 10-hydroxyochratoxin A (10-OH-OTA), formed by reaction with CYP450 enzymes [66,74,75]. A study by Huang et al., [72] showed that OTA has the capacity to reduce TEER in a dose-dependent manner, both individually and in conjugation with AFM1. A study by Alizadeh et al., 2019 [76] demonstrates that OTA reduced TEER in a dose- and time-dependent manner and also decreased the expression of tight junction proteins. The same effect was demonstrated in a different type of cellular model, porcine intestinal cell line IPEC-J2 [77]. DON metabolites are formed by phase II metabolism with glucuronide and sulphate conjugation. The most common metabolites are 3-acetyl-DON, 15-acetyl-DON, diepoxy-deoxynivalenol (DOM-1), DON-3-guccoside (D3G), DON-sulfonates and DON-3-glucuronides [78–80]. In studies with Caco-2 cell lines, there was no metabolization of DON, and TEER reduction was apparent only with high exposure times [81,82]. Kadota et al., 2013 compared DON and two metabolites, 3AcDON and 15AcDON, on their intestinal transport in Caco-2 cell line. The authors found that all compounds reduced TEER; however, 15AcDON induced a significantly higher loss of epithelial integrity than DON or 3AcDON [83]. In IPEC-J2 cell line, DON was capable of significantly reducing tight junction protein expression, reducing stability, and increasing the degradation of existing tight junction proteins, leading to increased epithelial permeability [65]. The effect of DON on proliferating and differentiated Caco-2 cell lines was evaluated by Luo et al., 2021. This mycotoxin induces a reduction of TEER in differentiated cells and delays the establishment of TEER in proliferating cells, which may lead to a reduction in the renewal and repair rates of the intestinal epithelium [84]. T-2 toxin can be metabolized by hydroxylation, hydrolysis, and deepoxidation reactions (phase I) and by conjugation (phase II). The most common metabolites formed are HT-2, neosolaniol (NEO), and T-2 triol. While in phase I, CYP450 enzymes are the major contributors, and in phase II, the conjugation is most common with glucuronide [79,85]. T-2 induces a reduction in TEER values in cytotoxic concentration in IPEC-J2 cells, while non-cytotoxic concentrations have little effect [64]. The study by Romero et al., 2016 demonstrated that T-2 was capable of decreasing TEER and the expression of tight junction proteins in a concentration-dependent manner [70]. FB1 is not metabolized by the action of CYP450 enzymes and can act as an inhibitor of specific CYP450 enzymes, such as CYP2C11 and CYP1A2. When FB1 is acetylated, the resulting N-acetylated metabolites are more toxic than the original FB1 [62,66,86,87]. This mycotoxin also affects epithelial integrity by downregulating tight junction protein expression and decreasing TEER. In a study evaluating four mycotoxins (AFB1, FB1, T-2, and OTA), FB1 caused the second-highest reduction of TEER, only surpassed by AFB1 [70]. In the case of ZEA, the most common metabolites are α-zearalenol (α-ZEA), which has a higher affinity for oestrogen receptors in humans, β-zearalenol (β-ZEA) via action of CYP450 enzymes, and zearalenone-14-glucoside (ZEA14Glc) via glycosylation [88,89]. The study by Videmann et al., 2008 demonstrated the metabolization of ZEA into mainly α-ZEA, followed by β-ZEA and ZEA glucuronide, leading to a higher susceptibility to the effects of this mycotoxin [90].

Besides good agricultural practices on crop cereals production, different types of food processing such as washing, roasting, grinding, cooking, radiation, and others [10,91] are strategies used to reduce the occurrence of mycotoxins in cereal food staples. For instance,

Meca and co-workers studied the influence of heat treatment on the degradation of BEA, at 160, 180, and 200 °C and at different incubation times, 3, 6, 10, 15, and 20, in crispy breads with different flours. They found a 43% reduction of BEA bioaccessibility at 160 °C, 69% reduction at 180 °C, and 87% reduction at 200 °C, with increasing reduction with longer time of incubation [92]. There are, also, strategies to reduce the bioaccessibility of mycotoxins during digestion, such as the use of probiotic bacteria [58], chemical reduction, and food processing [93]. In 2009, Kabak and colleagues reported a reduction of 37% and 73% for AFB1 and OTA, respectively, in the presence of Lactobacillus and Bifidobacterium bacteria in pistachios, buckwheat, and infant food [60]. Another study on loaf bread samples showed a reduction of AFB1 (15–98%) and AFB2 (11–98%) bioaccessibility in gastric and duodenal digestion, with the presence of probiotic bacteria [94]. Isothiocyanates are compounds present in plants from the Brassicaceae family that have strong antimicrobial properties. Luciano and co-workers studied the reduction of BEA in cereals with BITC (benzyl isothiocyanate) and PITC (phenyl isothiocyanate) fumigation and found a significant reduction, with the highest rate at 92.5% in rice kernels [95].

7. Methods for Mycotoxin Analysis in Food and Bioaccessibility Studies

7.1. Analysis of Mycotoxin in Food

Identification and quantification of mycotoxins in food are critical steps to support production quality control and health risk assessment. Selective and robust methods are needed for mycotoxin analysis because they are very complex molecules with vastly different chemical structures that are present in a range of matrices, usually in low concentrations [96]. Co-occurrence of different mycotoxins is also an issue to consider when analyzing a product.

Mycotoxin analytical methods include several steps such as sampling, homogenization, extraction followed by a clean-up step to decrease and/or eliminate matrix compounds, sample concentration, separation, and detection. The detection step is performed either via a chromatographic technique with different detectors or via an immuno-chemical method [97,98]. More details about the sampling, sample preparation and detection method see Supplemental Material File S1.

7.2. Bioavailability and Bioaccessibility Analysis

Bioaccessibility assessment can be carried out by in vivo or in vitro assays. In vivo approaches are very complicated due to analytical and ethical questions, are very time-consuming and require specific planning, resources, and a high level of experimental control. Although less realistic, in vitro approaches have a wider experimental scope and are simpler to perform while still producing useful results [57].

In vivo studies can be divided in two groups: overall balance studies and determination of substances (or metabolites) in relevant tissues, while in vitro studies, can be divided into static and dynamic processes [57]. Static models mimic the digestive transit by the sequential exposure of the food product to the conditions of the stomach and small intestine, whereas dynamic models simulate the gradual transit of the food product through the different processes of the digestive tract, ensuring a more realistic simulation of the digestive process. These last models simulate the gastric emptying patterns, food transit times, pH value modifications, different concentrations of enzymes, bile salts and electrolytes, water absorption percentage, and in some models, the microbial activity in the different compartments of the gastrointestinal tract [58]. Some models simulate the digestive process in the stomach and small intestine of monogastric animals while other models mimic the large intestine and include microbiota derived from the intestine. As mycotoxin absorption occurs in the small intestine, large intestine models are not used in these studies [57,58].

In vitro digestion models have been developed for the study of several different contaminants, based on human and animal physiology. Some requirements need to be considered to achieve a good methodology, such as the following: (1) representative of the physiological processes of the organism in the study; (2) biochemical reactions, hydrodynamics, and mechanical forces used must match with digestion kinetics; (3) simulation of fasted and fed conditions including proper an-aerobic conditions for the presence of gastrointestinal microorganisms; and (4) easy, robust, reproducible, and applicable methodology [55,57].

The main physiological components of in vitro digestion models are (1) saliva, because the digestion process starts in the mouth with a mechanical action of chewing aided by salivary digestion. This fluid is an exocrine secretion consisting of 99% water, electrolytes, and proteins. In most models simulated saliva is a simplified version of this fluid containing electrolytes, urea, and α-amylase. (2) Gastric juices, which are secreted by the gastric glands in the stomach, are composed by mucus, enzymes, and aqueous component, and in models, they are simulated by a strong decrease of pH and addition of pepsin and electrolytes. (3) Intestinal juices, because after gastric digestion, the food is released into the duodenum, where enzymes from the pancreas and bile from the liver continue the digestion process, and the simulated fluid is composed of electrolytes, pancreatin, and bile salts. Other important components in the digestion models are temperature, peristalsis, incubation time, and microbial interactions, the last not usual component in static models for mycotoxins because these compounds are absorbed in the small intestine [55].

All studies on bioaccessibility reported here (Table 4) used the in vitro static digestive model, with three steps (mouth, stomach, and small intestine digestion) because it is an easier and cheaper model. Kabak et al., 2009 [60] used a method developed by Versantvoort et al., 2005 [99]. The process starts with adding simulated saliva, pH 6.8, to a food sample and incubation for 5 min at 37 °C. Then, artificial gastric juice, pH 1.3, was added, following 2 h incubation at 37 °C. Finally, artificial duodenal fluid at pH 8.1 is added and the mixture is incubated for another 2 h at 37 °C. After digestion, the mycotoxin levels were evaluated in the extract. Raiola et al., 2012 [59] and Prosperini et al., 2013 [61] used a similar method adapted from Gil-Izquierdo et al., 2002 [100], with small alterations to components of the artificial fluids and the pH in each phase of digestion. The first step combines oral and gastric digestion, where the pH starts at 6.5 and is then adjusted to 2, and in the duodenal digestion, the pH is 6.5. After each digestion, the extract is obtained for mycotoxin extraction and detection. For the child digestion model, in the study of Raiola and team [59], there are slight modifications, where the pH of the stomach is 3 and the amount of pepsin, pancreatin, and bile salts were reduced. Assunção and team [46] used a method adapted from Minekus et al., 2014 [101], a standardized method, very similar to the one used by Kabak's team, with a pH of 7 in the oral digestion, 3 in the gastric digestion, 7 in the intestinal digestion, and some alterations to the components of the artificial digestion fluids, as reported in Table 4. After the intestinal digestion, the mycotoxin levels were quantified in the extract.

Table 4. Methods for analysis of mycotoxins bioaccessibility in cereal-based food for infants and children.

Matrix	Mycotoxin	Digestion Model	Extraction	Detection Method	References
Infant formula	AFB1 OTA	Static in vitro digestion model: Oral phase (KCl/KSCN/NaH_2PO_4/$NaSO_4$/NaCl/$NaHCO_3$/urea/a-amylase/uric acid/mucin) Gastric phase (NaCl/NaH2PO4/KCl/$CaCl_2$/NH_4Cl/HCl/glucose/glucuronic acid/urea/glucosamine hydrochloride/BSA/pepsin/mucin) Intestinal phase (NaCl/$NaHCO_3$/KCl/HCl/urea/$CaCl_2(2H_2O)$/BSA/bile)	Phosphoric acid/chloroform + IAC AflaOchra HPLC™	HPLC-FD	Kabak et al., 2009 [60]
Commercial pasta	DON	Static in vitro digestion model: Oral phase (KCl/KSCN/NaH_2PO_4/$NaSO_4$/NaCl/$NaHCO_3$/urea/a-amylase)Gastric phase (HCl/pepsin) Intestinal phase ($NaHCO_3$/pancreatin/bile salts/H_2O)	ACN:water (84:16; *v/v*)	LC-MS/MS	Raiola et al., 2012 [59]
Breakfast cereals Cookies Breads	ENA ENA1 ENB ENB1	Static in vitro digestion model: Oral phase (KCl/KSCN/NaH_2PO_4/$NaSO_4$/NaCl/$NaHCO_3$/urea/a-amylase) Gastric phase (HCl/pepsin) Intestinal phase ($NaHCO_3$/pancreatin/bile salts/H_2O)	Ethyl acetate	LC-DAD	Prosperini et al., 2013 [61]
Cereal-based food	PAT	Static in vitro digestion model: Oral phase (KCl/KH_2PO_4/$NaHCO_3$/$MgCl_2(H_2O)_6$/$(NH_4)_2CO_3$) Gastric phase (KCl/KH_2PO_4/$NaHCO_3$/NaCl/$MgCl_2(H_2O)_6$/$(NH_4)_2CO_3$/pepsin) Intestinal phase (KCl/KH_2PO_4/$NaHCO_3$/NaCl/$MgCl_2(H_2O)_6$/pancreatin/bile)	Ethyl acetate + sodium sulphate + sodium hydrogenocarbonate + SPE column	RP-HPLC-UV	Assunção et al., 2016 [46]
	OTA		MeOH:water (80:20) + IAC AflaOchra	RP-HPLC-FD	

AFB1—Aflatoxin B1; DON—Deoxynivalenol; ENA—Enniatin A; ENA1—Enniatin A1; ENB—Enniatin B; ENB1—Enniatin B1; OTA—Ochratoxin; PAT—Patulin; ACN—Acetonitrile; HPLC-FD —High-performance liquid chromatography coupled to fluorescence detector; HPLC-UV—High-performance liquid chromatography coupled to ultraviolet detector; IAC—Immunoaffinity columns; LC-DAD—Liquid chromatography coupled to diode array detector; LC-MS/MS—Liquid chromatography coupled to tandem mass spectrometry; RP—Reverse-phase; SPE—Solid-phase extraction.

8. Gaps in the Research of Mycotoxins in Infant/Children Cereal-Based Food Matrices

This review of the available research on mycotoxin quantification and bioaccessibility in cereal-based children and infant food products makes it clear that this is a topic that requires much more attention.

The majority of the studies on mycotoxin determination and quantification in these types of food products reported the presence of regulated mycotoxins, in many cases presenting values above the EU legal limits, and the co-occurrence of several compounds. Although only three studies have evaluated the presence of emerging non-regulated mycotoxins, positive results were always found, which leads to the possibility of a higher mycotoxin presence in cereal-based food products than the one reported in the studies based on the assessment of only regulated mycotoxins.

The type of samples used was quite diversified over the years, with almost all studies differentiating the samples by cereal composition, such as wheat, rice, barley, or maize, or by the presence of fruits and/or cocoa and presenting results in relation to that differentiation. However, only two studies (Herrera and Braun) mentioned products with cereals from organic/biologic cultures [50,53], and only Herrera et al., 2019 [50] presented the results of these specific products. There is an ever-increasing search for organic and biological products for their assumed health benefits, and more families are starting to feed their infants and children with homemade products and formulas, leading to the necessity of more research on these types of products. Braun et al., 2021 [53], diversified their sample pool by also quantifying mycotoxins in raw flours samples intended for production of infant foods. Some studies present other types of cereal-based products for children such as sweet cakes [52] and biscuits [48], highlighting the necessity to add other cereal-based products for children such as cereal bars and cookies. It is also possible to see some distinctions between cereal-based food products for babies, for infants, and for children, as with different age groups there are different nutritional needs, different eating habits, and different susceptibilities. Some studies [48,50,54] clearly present their results with a separation of these types of products, where the first team divided their samples into infant cereals and breakfast cereals [48], the second team considered two types of samples, gluten-free samples for babies from 4 to 6 months of age and multi-cereals for infants aged 7 to 12 months [50], and the last team separated their samples by brand, type of cereal, and consumption age [54].

Bioaccessibility studies on cereal-based infant food are scarce (only four in total as far as we know) and cover a small number of mycotoxins. For instance, ZEA, T-2, HT-2, AME, and TA, which were found in several studies [42,44,47–49,53], were never accessed for their bioaccessibility. Despite these types of studies being time-consuming, they do provide valuable information on the potentially harmful effect of mycotoxins on infants and children. In the study by Prosperini and colleagues [61], they reported that in the same type of food, a different composition can change the bioaccessibility of a mycotoxin; for example, a higher percentage of fibre or fat can result in lower bioaccessibility. This highlights the importance of carefully choosing the type of matrix when analysing the effect of a mycotoxin.

The co-occurrence of different mycotoxins [42,46,47,50,53] and whether they act as antagonists or synergists of each other needs to be further researched. More bioavailability studies, with more and different matrices, and other mycotoxins, regulated and non-regulated, will be of extreme importance for the evaluation of the cumulative effect of the compounds in the organism. It is also important to note the lack of studies that correlate the health benefits of cereal ingestion and their protective action against mycotoxins.

9. Final Considerations

Child nutrition is of extreme importance to guarantee the nutritional and energy requirements for proper growth and development. Cereals are one of the best types of food for the weening period in children, as they offer the most important nutrients, are easy to digest, and have a tolerable texture to allow an easier transition to solid foods.

Mycotoxin occurrence in food and processed food products is a worldwide issue due to its high prevalence, particularly in cereals. Nowadays, there are several EU regulations on maximum limits allowed for the most recognized mycotoxins, such as AFs, OTA, ZEA, FMs, and PAT. These regulations are especially severe regarding the limits required for food and food products destined for children, as they are more susceptible to the effects of mycotoxin contamination.

This review about the presence of mycotoxins in cereal-based infant foods commercialized in Europe shows that besides the regular presence of "classical" mycotoxins, which in some cases are above the legal limits, the co-occurrence of the so-called emerging mycotoxins that are not yet regulated is quite common. This finding underlines the necessity of a careful re-evaluation of current guidelines, as these only take into consideration the individual effect of each mycotoxin. Moreover, regulated mycotoxins are the only ones regularly tested for quality control purposes, which may lead to a misguided belief that infant food products are within the safety limits.

Few studies were reported on the bioavailability and bioaccessibility of mycotoxins from cereal-based food matrices, which leads to a diminished understanding of the full effects of the presence of these compounds in infant foods. More studies are needed in this area to gather a consensus on the quality parameters required for the products being marketed for and consumed by children in Europe. The favored method for bioaccessibility analysis is the static in vitro digestion model, due to its cost-effectiveness ratio. It is predictable that in the future, there may be wider use of dynamic digestion models because they offer a more realistic approach to the digestion process.

Supplementary Materials: The following supporting information can be downloaded at: https://www.mdpi.com/article/10.3390/toxins14070488/s1, Table S1: Methods for analysis of mycotoxins in cereal-based food for infants and children, in Europe (2014–2021), File S1: Methods for Mycotoxin Analysis in Food. References [102–118] are cited in the Supplementary Materials.

Author Contributions: Conceptualization, S.C.C. and J.O.F.; Writing—Original Draft, C.P. Writing—Review and Editing, S.C.C. and J.O.F. All authors have read and agreed to the published version of the manuscript.

Funding: This work is co-financed by the European Regional Development Fund (ERDF) through the Intarreg VA Spain-Portugal (POCTEP) 2014-2020 Programme under Grant Agreement 0591_FOODSENS_1_E. This output reflects the views only of the authors, and the programme authorities cannot be held responsible for any use which may be made of the information contained therein. Sara C. Cunha acknowledges FCT for IF/01616/2015 contract. This research was partially supported by national funds through FCT within the scope of UIDB/04423/2020 and AgriFood XXI R&D&I project, operation No. NORTE-01-0145-FEDER-000041, co-financed by the European Regional Development Fund (ERDF) through NORTH 2020 (Northern Regional Operational Program 2014/2020).

Institutional Review Board Statement: Not applicable.

Informed Consent Statement: Not applicable.

Data Availability Statement: Not applicable.

Conflicts of Interest: The authors declare no conflict of interest.

References

1. Cascant, M.M.; Garrigues, S.; de la Guardia, M. Direct determination of major components in human diets and baby foods. *Anal. Bioanal. Chem.* **2015**, *407*, 1961–1972. [CrossRef] [PubMed]
2. Amezdroz, E.; Carpenter, L.; O'Callaghan, E.; Johnson, S.; Waters, E. Transition from milks to the introduction of solid foods across the first 2 years of life: Findings from an Australian birth cohort study. *J. Hum. Nutr. Diet.* **2015**, *28*, 375–383. [CrossRef] [PubMed]
3. Bentley, M.E.; Nulty, A.K. When Does It All Begin: What, When, and How Young Children Are Fed. *Nestle Nutr. Inst. Workshop Ser.* **2020**, *93*, 15–24. [PubMed]
4. UNICEF. *From the First Hour of Life*; Data and Analytics, Division of Data, Research and Policy and Nutrition Section, Programe Division—UNICEF: New York, USA, 2016; ISBN 978-92-806-4852-2.

5. Klerks, M.; Bernal, M.J.; Roman, S.; Bodenstab, S.; Gil, A.; Sanchez-Siles, L.M. Infant Cereals: Current Status, Challenges, and Future Opportunities for Whole Grains. *Nutrients* **2019**, *11*, 473. [CrossRef]
6. Doughari, J.H. The Occurrence, Properties and Significance of Citrinin Mycotoxin. *J. Plant Pathol. Microbiol.* **2015**, *6*, 11.
7. Gruber-Dorninger, C.; Novak, B.; Nagl, V.; Berthiller, F. Emerging Mycotoxins: Beyond Traditionally Determined Food Contaminants. *J. Agric. Food Chem.* **2017**, *65*, 7052–7070. [CrossRef]
8. Mallebrera, B.; Prosperini, A.; Font, G.; Ruiz, M.J. In vitro mechanisms of Beauvericin toxicity: A review. *Food Chem. Toxicol.* **2018**, *111*, 537–545. [CrossRef]
9. Peraica, M.; Richter, D.; Rašić, D. Mycotoxicoses in children. *Arch. Ind. Hyg. Toxicol.* **2014**, *65*, 347–363. [CrossRef]
10. Bennett, J.W.; Klich, M. Mycotoxins. *Clin. Microbiol. Rev.* **2003**, *16*, 497–516. [CrossRef]
11. Gelderblom, W.C.; Marasas, W.F. Controversies in fumonisin mycotoxicology and risk assessment. *Hum. Exp. Toxicol.* **2012**, *31*, 215–235. [CrossRef]
12. Alizadeh, A.M.; Rohandel, G.; Roudbarmohammadi, S.; Roudbary, M.; Sohanaki, H.; Ghiasian, S.A.; Taherkhani, A.; Semnani, S.; Aghasi, M. Fumonisin B1 contamination of cereals and risk of esophageal cancer in a high risk area in northeastern Iran. *Asian Pac. J. Cancer Prev.* **2012**, *13*, 2625–2628. [CrossRef]
13. Vaclavikova, M.; Malachova, A.; Veprikova, Z.; Dzuman, Z.; Zachariasova, M.; Hajslova, J. 'Emerging' mycotoxins in cereals processing chains: Changes of enniatins during beer and bread making. *Food Chem.* **2013**, *136*, 750–757. [CrossRef]
14. Sherif, S.O.; Salama, E.E.; Abdel-Wahhab, M.A. Mycotoxins and child health: The need for health risk assessment. *Int. J. Hyg. Environ. Health* **2009**, *212*, 347–368. [CrossRef]
15. Raiola, A.; Tenore, G.C.; Manyes, L.; Meca, G.; Ritieni, A. Risk analysis of main mycotoxins occurring in food for children: An overview. *Food Chem. Toxicol.* **2015**, *84*, 169–180. [CrossRef]
16. Hulin, M.; Bemrah, N.; Nougadère, A.; Volatier, J.L.; Sirot, V.; Leblanc, J.C. Assessment of infant exposure to food chemicals: The French Total Diet Study design. *Food Addit. Contam. Part A* **2014**, *31*, 1226–1239. [CrossRef]
17. Arce-López, B.; Lizarraga, E.; López de Mesa, R.; González-Peñas, E. Assessment of Exposure to Mycotoxins in Spanish Children through the Analysis of Their Levels in Plasma Samples. *Toxins* **2021**, *13*, 150. [CrossRef]
18. Kushnir-Sukhov, N.M. A Novel Link between Early Life Allergen Exposure and Neuroimmune Development in Children. *J. Clin. Exp. Immunol.* **2020**, *5*, 188–195.
19. Etzel, R.A. What the primary care pediatrician should know about syndromes associated with exposures to mycotoxins. *Curr. Probl. Pediatr. Adolesc. Health Care* **2006**, *36*, 282–305. [CrossRef]
20. Malir, F.; Ostry, V.; Pfohl-Leszkowicz, A.; Malir, J.; Toman, J. Ochratoxin A: 50 Years of Research. *Toxins* **2016**, *8*, 191. [CrossRef]
21. Ferrigo, D.; Raiola, A.; Causin, R. Fusarium Toxins in Cereals: Occurrence, Legislation, Factors Promoting the Appearance and Their Management. *Molecules* **2016**, *21*, 627. [CrossRef]
22. Lee, H.J.; Ryu, D. Advances in Mycotoxin Research: Public Health Perspectives. *J. Food Sci.* **2015**, *80*, T2970–T2983. [CrossRef]
23. Guerre, P. Mycotoxin and Gut Microbiota Interactions. *Toxins* **2020**, *12*, 769. [CrossRef]
24. Akbari, P.; Braber, S.; Varasteh, S.; Alizadeh, A.; Garssen, J.; Fink-Gremmels, J. The intestinal barrier as an emerging target in the toxicological assessment of mycotoxins. *Arch. Toxicol.* **2017**, *91*, 1007–1029. [CrossRef]
25. Zhu, Y.; Sang, S. Phytochemicals in whole grain wheat and their health-promoting effects. *Mol. Nutr. Food Res.* **2017**, *61*, 1600852. [CrossRef]
26. Schöneberg, T.; Kibler, K.; Sulyok, M.; Musa, T.; Bucheli, T.D.; Mascher, F.; Bertossa, M.; Voegele, R.T.; Vogelgsang, S. Can plant phenolic compounds reduce Fusarium growth and mycotoxin production in cereals? *Food Addit. Contam. Part A* **2018**, *35*, 2455–2470. [CrossRef]
27. Stuper-Szablewska, K.; Kurasiak-Popowska, D.; Nawracała, J.; Perkowski, J. Quantitative profile of phenolic acids and antioxidant activity of wheat grain exposed to stress. *Eur. Food Res. Technol.* **2019**, *245*, 1595–1603. [CrossRef]
28. Ferruz, E.; Loran, S.; Herrera, M.; Gimenez, I.; Bervis, N.; Barcena, C.; Carramiñana, J.J.; Juan, T.; Herrera, A.; Ariño, A. Inhibition of Fusarium Growth and Mycotoxin Production in Culture Medium and in Maize Kernels by Natural Phenolic Acids. *J. Food Prot.* **2016**, *79*, 1753–1758. [CrossRef]
29. Atanasova-Penichon, V.; Barreau, C.; Richard-Forget, F. Antioxidant Secondary Metabolites in Cereals: Potential Involvement in Resistance to Fusarium and Mycotoxin Accumulation. *Front. Microbiol.* **2016**, *7*, 566. [CrossRef] [PubMed]
30. Fardet, A.; Rock, E.; Rémésy, C. Is the in vitro antioxidant potential of whole-grain cereals and cereal products well reflected in vivo? *J. Cereal Sci.* **2008**, *48*, 258–276. [CrossRef]
31. Slavin, J. Why whole grains are protective: Biological mechanisms. *Proc. Nutr. Soc.* **2003**, *62*, 129–134. [CrossRef] [PubMed]
32. Graf, E.; Eaton, J.W. Antioxidant functions of phytic acid. *Free Radic. Biol. Med.* **1990**, *8*, 61–69. [CrossRef]
33. Feizollahi, E.; Mirmahdi, R.S.; Zoghi, A.; Zijlstra, R.T.; Roopesh, M.S.; Vasanthan, T. Review of the beneficial and anti-nutritional qualities of phytic acid, and procedures for removing it from food products. *Food Res. Int.* **2021**, *143*, 110284. [CrossRef]
34. Liu, Y.; Dong, R.; Yang, Y.; Xie, H.; Huang, Y.; Chen, X.; Wang, D.; Zhang, Z. Protective Effect of Organic Selenium on Oxidative Damage and Inflammatory Reaction of Rabbit Kidney Induced by T-2 Toxin. *Biol. Trace Elem. Res.* **2021**, *199*, 1833–1842. [CrossRef]
35. Fardet, A. New hypotheses for the health-protective mechanisms of whole-grain cereals: What is beyond fibre? *Nutr. Res. Rev.* **2010**, *23*, 65–134. [CrossRef]
36. Slavin, J.; Jacobs, D.; Marquart, L. Whole-grain consumption and chronic disease: Protective mechanisms. *Nutr. Cancer* **1997**, *27*, 14–21. [CrossRef]

37. Aoudia, N.; Callu, P.; Grosjean, F.; Larondelle, Y. Effectiveness of mycotoxin sequestration activity of micronized wheat fibres on distribution of ochratoxin A in plasma, liver and kidney of piglets fed a naturally contaminated diet. *Food Chem. Toxicol.* **2009**, *47*, 1485–1489. [CrossRef]
38. Fardet, A. 16—New Concepts and Paradigms for the Protective Effects of Plant-Based Food Components in Relation to Food Complexity. In *Vegetarian and Plant-Based Diets in Health and Disease Prevention*; Mariotti, F., Ed.; Academic Press: London, UK, 2017; pp. 293–312.
39. Hernández, M.; Juan-García, A.; Moltó, J.C.; Mañes, J.; Juan, C. Evaluation of Mycotoxins in Infant Breast Milk and Infant Food, Reviewing the Literature Data. *Toxins* **2021**, *13*, 535. [CrossRef]
40. Coppa, C.F.S.C.; Khaneghah, A.M.; Alvito, P.; Assunção, R.; Martins, C.; Eş, I.; Gonçalves, B.L.; Valganon de Neeff, D.; Sant'Ana, A.S.; Corassin, C.H.; et al. The occurrence of mycotoxins in breast milk, fruit products and cereal-based infant formula: A review. *Trends Food Sci. Technol.* **2019**, *92*, 81–93. [CrossRef]
41. De Sá, S.V.M.; Monteiro, C.; Fernandes, J.O.; Pinto, E.; Faria, M.A.; Cunha, S.C. Emerging mycotoxins in infant and children foods: A review. *Crit. Rev. Food Sci. Nutr.* **2021**, 1–15. [CrossRef]
42. Juan, C.; Raiola, A.; Mañes, J.; Ritieni, A. Presence of mycotoxin in commercial infant formulas and baby foods from Italian market. *Food Control* **2014**, *39*, 227–236. [CrossRef]
43. Pereira, V.L.; Fernandes, J.O.; Cunha, S.C. Comparative assessment of three cleanup procedures after QuEChERS extraction for determination of trichothecenes (type A and type B) in processed cereal-based baby foods by GC–MS. *Food Chem.* **2015**, *182*, 143–149. [CrossRef] [PubMed]
44. Assunção, R.; Vasco, E.; Nunes, B.; Loureiro, S.; Martins, C.; Alvito, P. Single-compound and cumulative risk assessment of mycotoxins present in breakfast cereals consumed by children from Lisbon region, Portugal. *Food Chem. Toxicol.* **2015**, *86*, 274–281. [CrossRef] [PubMed]
45. Hampikyan, H.; Bingol, E.B.; Colak, H.; Cetin, O.; Bingol, B. Determination of ochratoxin a in baby foods by ELISA and HPLC. *Acta Aliment.* **2015**, *44*, 578–584. [CrossRef]
46. Assunção, R.; Martins, C.; Dupont, D.; Alvito, P. Patulin and ochratoxin A co-occurrence and their bioaccessibility in processed cereal-based foods: A contribution for Portuguese children risk assessment. *Food Chem. Toxicol.* **2016**, *96*, 205–214. [CrossRef]
47. Martins, C.; Assunção, R.; Cunha, S.C.; Fernandes, J.O.; Jager, A.; Petta, T.; Oliveira, C.A.; Alvito, P. Assessment of multiple mycotoxins in breakfast cereals available in the Portuguese market. *Food Chem.* **2018**, *239*, 132–140. [CrossRef]
48. Assunção, R.; Martins, C.; Vasco, E.; Jager, A.; Oliveira, C.; Cunha, S.C.; José, O.F.; Nunes, B.; Loureiro, S.; Alvito, P. Portuguese children dietary exposure to multiple mycotoxins—An overview of risk assessment under MYCOMIX project. *Food Chem. Toxicol.* **2018**, *118*, 399–408. [CrossRef]
49. Gotthardt, M.; Asam, S.; Gunkel, K.; Moghaddam, A.F.; Baumann, E.; Kietz, R.; Rychlik, M. Quantitation of Six Alternaria Toxins in Infant Foods Applying Stable Isotope Labeled Standards. *Front. Microbiol.* **2019**, *10*, 109. [CrossRef]
50. Herrera, M.; Bervis, N.; Carraminana, J.J.; Juan, T.; Herrera, A.; Arino, A.; Loran, S. Occurrence and Exposure Assessment of Aflatoxins and Deoxynivalenol in Cereal-Based Baby Foods for Infants. *Toxins* **2019**, *11*, 150. [CrossRef]
51. Postupolski, J.; Starski, A.; Ledzion, E.; Kurpinska-Jaworska, J.; Szczesna, M. Exposure assessment of infants and young children on selected Fusarium toxins. *Rocz. Panstw. Zakl. Hig.* **2019**, *70*, 5–14. [CrossRef]
52. Capei, R.; Pettini, L.; Mando Tacconi, F. Occurrence of Ochratoxin A in breakfast cereals and sweet snacks in Italy: Dietary exposure assessment. *Ann Ig* **2019**, *31*, 130–139. [PubMed]
53. Braun, D.; Eiser, M.; Puntscher, H.; Marko, D.; Warth, B. Natural contaminants in infant food: The case of regulated and emerging mycotoxins. *Food Control* **2021**, *123*, 107676. [CrossRef]
54. Mruczyk, K.; Cisek-Woźniak, A.; Mizgier, M.; Wójciak, R.W. Natural Occurrence of Deoxynivalenol in Cereal-Based Baby Foods for Infants from Western Poland. *Toxins* **2021**, *13*, 777. [CrossRef] [PubMed]
55. González-Arias, C.A.; Marín, S.; Sanchis, V.; Ramos, A.J. Mycotoxin bioaccessibility/absorption assessment using in vitro digestion models: A review. *World Mycotoxin J.* **2013**, *6*, 167–184. [CrossRef]
56. Arce-López, B.; Lizarraga, E.; Vettorazzi, A.; González-Peñas, E. Human Biomonitoring of Mycotoxins in Blood, Plasma and Serum in Recent Years: A Review. *Toxins* **2020**, *12*, 147. [CrossRef] [PubMed]
57. Cardoso, C.; Afonso, C.; Lourenço, H.; Costa, S.; Nunes, M.L. Bioaccessibility assessment methodologies and their consequences for the risk–benefit evaluation of food. *Trends Food Sci. Technol.* **2015**, *41*, 5–23. [CrossRef]
58. Rebellato, A.P.; dos Santos Caramês, E.T.; Pallone, J.A.L.; de Oliveira Rocha, L. Mycotoxin bioaccessibility in baby food through in vitro digestion: An overview focusing on risk assessment. *Curr. Opin. Food Sci.* **2021**, *41*, 107–115. [CrossRef]
59. Raiola, A.; Meca, G.; Mañes, J.; Ritieni, A. Bioaccessibility of Deoxynivalenol and its natural co-occurrence with Ochratoxin A and Aflatoxin B1 in Italian commercial pasta. *Food Chem. Toxicol.* **2012**, *50*, 280–287. [CrossRef]
60. Kabak, B.; Brandon, E.F.A.; Var, I.; Blokland, M.; Sips, A.J.A.M. Effects of probiotic bacteria on the bioaccessibility of aflatoxin B1 and ochratoxin A using an in vitro digestion model under fed conditions. *J. Environ. Sci. Health Part B* **2009**, *44*, 472–480. [CrossRef]
61. Prosperini, A.; Meca, G.; Font, G.; Ruiz, M.-J. Bioaccessibility of Enniatins A, A1, B, and B1 in Different Commercial Breakfast Cereals, Cookies, and Breads of Spain. *J. Agric. Food Chem.* **2013**, *61*, 456–461. [CrossRef]
62. Tran, V.N.; Viktorová, J.; Ruml, T. Mycotoxins: Biotransformation and Bioavailability Assessment Using Caco-2 Cell Monolayer. *Toxins* **2020**, *12*, 628. [CrossRef]

63. Assunção, R.; Ferreira, M.; Martins, C.; Diaz, I.; Padilla, B.; Dupont, D.; Bragança, M.; Alvito, P. Applicability of In Vitro Methods to Study Patulin Bioaccessibility and Its Effects on Intestinal Membrane Integrity. *J. Toxicol. Environ. Health Part A* **2014**, *77*, 983–992. [CrossRef]
64. Goossens, J.; Pasmans, F.; Verbrugghe, E.; Vandenbroucke, V.; De Baere, S.; Meyer, E.; Haesebrouck, F.; De Backer, P.; Croubels, S. Porcine intestinal epithelial barrier disruption by the Fusariummycotoxins deoxynivalenol and T-2 toxin promotes transepithelial passage of doxycycline and paromomycin. *BMC Vet. Res.* **2012**, *8*, 245. [CrossRef]
65. Li, E.; Horn, N.; Ajuwon, K.M. Mechanisms of deoxynivalenol-induced endocytosis and degradation of tight junction proteins in jejunal IPEC-J2 cells involve selective activation of the MAPK pathways. *Arch. Toxicol.* **2021**, *95*, 2065–2079. [CrossRef]
66. Wen, J.; Mu, P.; Deng, Y. Mycotoxins: Cytotoxicity and biotransformation in animal cells. *Toxicol. Res.* **2016**, *5*, 377–387. [CrossRef]
67. Gross-Steinmeyer, K.; Eaton, D.L. Dietary modulation of the biotransformation and genotoxicity of aflatoxin B1. *Toxicology* **2012**, *299*, 69–79. [CrossRef]
68. Deng, J.; Zhao, L.; Zhang, N.-Y.; Karrow, N.A.; Krumm, C.S.; Qi, D.-S.; Sun, L.-H. Aflatoxin B1 metabolism: Regulation by phase I and II metabolizing enzymes and chemoprotective agents. *Mutat. Res./Rev. Mutat. Res.* **2018**, *778*, 79–89. [CrossRef]
69. Gratz, S.; Wu, Q.K.; El-Nezami, H.; Juvonen, R.O.; Mykkänen, H.; Turner, P.C. *Lactobacillus rhamnosus* strain GG reduces aflatoxin B1 transport, metabolism, and toxicity in Caco-2 Cells. *Appl. Environ. Microbiol.* **2007**, *73*, 3958–3964. [CrossRef]
70. Romero, A.; Ares, I.; Ramos, E.; Castellano, V.; Martínez, M.; Martínez-Larrañaga, M.-R.; Anadón, A.; Martínez, M.-A. Mycotoxins modify the barrier function of Caco-2 cells through differential gene expression of specific claudin isoforms: Protective effect of illite mineral clay. *Toxicology* **2016**, *353–354*, 21–33. [CrossRef]
71. Gao, Y.; Bao, X.; Meng, L.; Liu, H.; Wang, J.; Zheng, N. Aflatoxin B1 and Aflatoxin M1 Induce Compromised Intestinal Integrity through Clathrin-Mediated Endocytosis. *Toxins* **2021**, *13*, 184. [CrossRef]
72. Huang, X.; Gao, Y.; Li, S.; Wu, C.; Wang, J.; Zheng, N. Modulation of Mucin (MUC2, MUC5AC and MUC5B) mRNA Expression and Protein Production and Secretion in Caco-2/HT29-MTX Co-Cultures. Following Exposure to Individual and Combined Aflatoxin M1 and Ochratoxin A. *Toxins* **2019**, *11*, 132. [CrossRef]
73. Kőszegi, T.; Poór, M. Ochratoxin A: Molecular Interactions, Mechanisms of Toxicity and Prevention at the Molecular Level. *Toxins* **2016**, *8*, 111. [CrossRef]
74. Ringot, D.; Chango, A.; Schneider, Y.-J.; Larondelle, Y. Toxicokinetics and toxicodynamics of ochratoxin A, an update. *Chem.-Biol. Interact.* **2006**, *159*, 18–46. [CrossRef]
75. Tao, Y.; Xie, S.; Xu, F.; Liu, A.; Wang, Y.; Chen, D.; Pan, Y.; Huang, L.; Peng, D.; Wang, X.; et al. Ochratoxin A: Toxicity, oxidative stress and metabolism. *Food Chem. Toxicol.* **2018**, *112*, 320–331. [CrossRef]
76. Alizadeh, A.; Akbari, P.; Varasteh, S.; Braber, S.; Malekinejad, H.; Fink-Gremmels, J. Ochratoxin A challenges the intestinal epithelial cell integrity: Results obtained in model experiments with Caco-2 cells. *World Mycotoxin J.* **2019**, *12*, 399–407. [CrossRef]
77. Wang, H.; Zhai, N.; Chen, Y.; Fu, C.; Huang, K. OTA induces intestinal epithelial barrier dysfunction and tight junction disruption in IPEC-J2 cells through ROS/Ca^{2+}-mediated MLCK activation. *Environ. Pollut.* **2018**, *242*, 106–112. [CrossRef]
78. Maresca, M. From the Gut to the Brain: Journey and Pathophysiological Effects of the Food-Associated Trichothecene Mycotoxin Deoxynivalenol. *Toxins* **2013**, *5*, 784–820. [CrossRef]
79. Wu, Q.-H.; Wang, X.; Yang, W.; Nüssler, A.K.; Xiong, L.-Y.; Kuča, K.; Dohnal, V.; Zhang, X.-J.; Yuan, Z.-H. Oxidative stress-mediated cytotoxicity and metabolism of T-2 toxin and deoxynivalenol in animals and humans: An update. *Arch. Toxicol.* **2014**, *88*, 1309–1326. [CrossRef]
80. Payros, D.; Alassane-Kpembi, I.; Pierron, A.; Loiseau, N.; Pinton, P.; Oswald, I.P. Toxicology of deoxynivalenol and its acetylated and modified forms. *Arch. Toxicol.* **2016**, *90*, 2931–2957. [CrossRef]
81. Sergent, T.; Parys, M.; Garsou, S.; Pussemier, L.; Schneider, Y.-J.; Larondelle, Y. Deoxynivalenol transport across human intestinal Caco-2 cells and its effects on cellular metabolism at realistic intestinal concentrations. *Toxicol. Lett.* **2006**, *164*, 167–176. [CrossRef]
82. Akbari, P.; Braber, S.; Gremmels, H.; Koelink, P.J.; Verheijden, K.A.T.; Garssen, J.; Fink-Gremmels, J. Deoxynivalenol: A trigger for intestinal integrity breakdown. *FASEB J.* **2014**, *28*, 2414–2429. [CrossRef]
83. Kadota, T.; Furusawa, H.; Hirano, S.; Tajima, O.; Kamata, Y.; Sugita-Konishi, Y. Comparative study of deoxynivalenol, 3-acetyldeoxynivalenol, and 15-acetyldeoxynivalenol on intestinal transport and IL-8 secretion in the human cell line Caco-2. *Toxicol. In Vitro* **2013**, *27*, 1888–1895. [CrossRef] [PubMed]
84. Luo, S.; Terciolo, C.; Neves, M.; Puel, S.; Naylies, C.; Lippi, Y.; Pinton, P.; Oswald, I.P. Comparative sensitivity of proliferative and differentiated intestinal epithelial cells to the food contaminant, deoxynivalenol. *Environ. Pollut.* **2021**, *277*, 116818. [CrossRef] [PubMed]
85. Li, Y.; Wang, Z.; Beier, R.C.; Shen, J.; De Smet, D.; De Saeger, S.; Zhang, S. T-2 toxin, a trichothecene mycotoxin: Review of toxicity, metabolism, and analytical methods. *J. Agric. Food Chem.* **2011**, *59*, 3441–3453. [CrossRef] [PubMed]
86. Voss, K.A.; Smith, G.W.; Haschek, W.M. Fumonisins: Toxicokinetics, mechanism of action and toxicity. *Anim. Feed Sci. Technol.* **2007**, *137*, 299–325. [CrossRef]
87. EFSA Panel on Contaminants in the Food Chain (CONTAM); Knutsen, H.-K.; Alexander, J.; Barregård, L.; Bignami, M.; Brüschweiler, B.; Ceccatelli, S.; Cottrill, B.; Dinovi, M.; Edler, L.; et al. Risks for animal health related to the presence of fumonisins, their modified forms and hidden forms in feed. *EFSA J.* **2018**, *16*, e05242.
88. Malekinejad, H.; Maas-Bakker, R.F.; Fink-Gremmels, J. Bioactivation of zearalenone by porcine hepatic biotransformation. *Vet. Res.* **2005**, *36*, 799–810. [CrossRef]

89. Malekinejad, H.; Maas-Bakker, R.; Fink-Gremmels, J. Species differences in the hepatic biotransformation of zearalenone. *Vet. J.* **2006**, *172*, 96–102. [CrossRef]
90. Videmann, B.; Mazallon, M.; Tep, J.; Lecoeur, S. Metabolism and transfer of the mycotoxin zearalenone in human intestinal Caco-2 cells. *Food Chem. Toxicol.* **2008**, *46*, 3279–3286. [CrossRef]
91. Sforza, S.; Dall'asta, C.; Marchelli, R. Recent advances in mycotoxin determination in food and feed by hyphenated chromatographic techniques/mass spectrometry. *Mass Spectrom. Rev.* **2006**, *25*, 54–76. [CrossRef]
92. Meca, G.; Ritieni, A.; Mañes, J. Influence of the heat treatment on the degradation of the minor Fusarium mycotoxin beauvericin. *Food Control* **2012**, *28*, 13–18. [CrossRef]
93. Luz, C.; Saladino, F.; Luciano, F.B.; Mañes, J.; Meca, G. Occurrence, toxicity, bioaccessibility and mitigation strategies of beauvericin, a minor Fusarium mycotoxin. *Food Chem. Toxicol.* **2017**, *107*, 430–439. [CrossRef]
94. Saladino, F.; Posarelli, E.; Luz, C.; Luciano, F.B.; Rodriguez-Estrada, M.T.; Mañes, J.; Meca, G. Influence of probiotic microorganisms on aflatoxins B1 and B2 bioaccessibility evaluated with a simulated gastrointestinal digestion. *J. Food Compos. Anal.* **2018**, *68*, 128–132. [CrossRef]
95. Luciano, F.B.; Meca, G.; Manyes, L.; Mañes, J. A chemical approach for the reduction of beauvericin in a solution model and in food systems. *Food Chem. Toxicol.* **2014**, *64*, 270–274. [CrossRef]
96. Zheng, M.Z.; Richard, J.L.; Binder, J. A Review of Rapid Methods for the Analysis of Mycotoxins. *Mycopathologia* **2006**, *161*, 261–273. [CrossRef]
97. Pereira, V.L.; Fernandes, J.O.; Cunha, S.C. Mycotoxins in cereals and related foodstuffs: A review on occurrence and recent methods of analysis. *Trends Food Sci. Technol.* **2014**, *36*, 96–136. [CrossRef]
98. Alshannaq, A.; Yu, J.H. Occurrence, Toxicity, and Analysis of Major Mycotoxins in Food. *Int. J. Environ. Res. Public Health* **2017**, *14*, 632. [CrossRef]
99. Versantvoort, C.H.M.; Oomen, A.G.; Van de Kamp, E.; Rompelberg, C.J.M.; Sips, A.J.A.M. Applicability of an in vitro digestion model in assessing the bioaccessibility of mycotoxins from food. *Food Chem. Toxicol.* **2005**, *43*, 31–40. [CrossRef]
100. Gil-Izquierdo, A.; Zafrilla, P.; Tomás-Barberán, F.A. An in vitro method to simulate phenolic compound release from the food matrix in the gastrointestinal tract. *Eur. Food Res. Technol.* **2002**, *214*, 155–159. [CrossRef]
101. Minekus, M.; Alminger, M.; Alvito, P.; Ballance, S.; Bohn, T.; Bourlieu, C.; Carrière, F.; Boutrou, R.; Corredig, M.; Dupont, D.; et al. A standardised static in vitro digestion method suitable for food—An international consensus. *Food Funct.* **2014**, *5*, 1113–1124. [CrossRef]
102. Rahmani, A.; Jinap, S.; Soleimany, F. Qualitative and Quantitative Analysis of Mycotoxins. *Compr. Rev. Food Sci. Food Saf.* **2009**, *8*, 202–251. [CrossRef]
103. Shanakhat, H.; Sorrentino, A.; Raiola, A.; Romano, A.; Masi, P.; Cavella, S. Current methods for mycotoxins analysis and innovative strategies for their reduction in cereals: An overview. *J. Sci. Food Agric.* **2018**, *98*, 4003–4013. [CrossRef]
104. Xie, L.; Chen, M.; Ying, Y. Development of Methods for Determination of Aflatoxins. *Crit. Rev. Food Sci. Nutr.* **2016**, *56*, 2642–2664. [CrossRef]
105. Turner, N.W.; Bramhmbhatt, H.; Szabo-Vezse, M.; Poma, A.; Coker, R.; Piletsky, S.A. Analytical methods for determination of mycotoxins: An update (2009–2014). *Anal. Chim. Acta* **2015**, *901*, 12–33. [CrossRef]
106. Bueno, D.; Istamboulie, G.; Muñoz, R.; Marty, J. Determination of Mycotoxins in Food: A Review of Bioanalytical to Analytical Methods. *Appl. Spectrosc. Rev.* **2015**, *50*, 728–774. [CrossRef]
107. Turner, N.W.; Subrahmanyam, S.; Piletsky, S.A. Analytical methods for determination of mycotoxins: A review. *Anal. Chim. Acta* **2009**, *632*, 168–180. [CrossRef]
108. Krska, R.; Schubert-Ullrich, P.; Molinelli, A.; Sulyok, M.; MacDonald, S.; Crews, C. Mycotoxin analysis: An update. *Food Addit. Contam. Part A* **2008**, *25*, 152–163. [CrossRef]
109. Wilson, T.J.; Romer, T.R. Use of the Mycosep Multifunctional Cleanup Column for Liquid Chromatographic Determination of Aflatoxins in Agricultural Products. *J. Assoc. Off. Anal. Chem.* **1991**, *74*, 951–956. [CrossRef]
110. Cunha, S.C.; Fernandes, J.O. Development and validation of a method based on a QuEChERS procedure and heart-cutting GC-MS for determination of five mycotoxins in cereal products. *J. Sep. Sci.* **2010**, *33*, 600–609. [CrossRef] [PubMed]
111. Cunha, S.C.; Fernandes, J.O. Chapter 21—Application in Food Analysis. In *Liquid-Phase Extraction*; Poole, C.F., Ed.; Elsevier: Amsterdam, The Netherlands, 2020; pp. 643–665.
112. Tamura, M.; Uyama, A.; Mochizuki, N. Development of a Multi-mycotoxin Analysis in Beer-based Drinks by a Modified QuEChERS Method and Ultra-High-Performance Liquid Chromatography Coupled with Tandem Mass Spectrometry. *Anal. Sci.* **2011**, *27*, 629. [CrossRef] [PubMed]
113. Shephard, G.S. Current Status of Mycotoxin Analysis: A Critical Review. *J. AOAC Int.* **2016**, *99*, 842–848. [CrossRef] [PubMed]
114. Boyer, R.F. *Modern Expeimental Biochemestry*, 3rd ed.; Benjamin Cummings: San Francisco, CA, USA, 2000.
115. Bessaire, T.; Mujahid, C.; Mottier, P.; Desmarchelier, A. Multiple Mycotoxins Determination in Food by LC-MS/MS: An International Collaborative Study. *Toxins* **2019**, *11*, 658. [CrossRef]
116. Korfmacher, W.A. Foundation review: Principles and applications of LC-MS in new drug discovery. *Drug Discov. Today* **2005**, *10*, 1357–1367. [CrossRef]

117. Wang, X.; Wang, S.; Cai, Z. The latest developments and applications of mass spectrometry in food-safety and quality analysis. *TrAC Trends Anal. Chem.* **2013**, *52*, 170–185. [CrossRef]
118. Santos Pereira, C.; Cunha, S.C.; Fernandes, J.O. Prevalent Mycotoxins in Animal Feed: Occurrence and Analytical Methods. *Toxins* **2019**, *11*, 290. [CrossRef]

toxins

MDPI

Review

Intestinal Barrier, Claudins and Mycotoxins

Marta Justyna Kozieł, Maksymilian Ziaja and Agnieszka Wanda Piastowska-Ciesielska *

Department of Cell Culture and Genomic Analysis, Medical University of Lodz, Zeligowskiego 7/9, 90-752 Lodz, Poland; marta.koziel@umed.lodz.pl (M.J.K.); ziaja.ziaja0@gmail.com (M.Z.)
* Correspondence: agnieszka.piastowska@umed.lodz.pl

Abstract: The intestinal barrier is the main barrier against all of the substances that enter the body. Proper functioning of this barrier guarantees maintained balance in the organism. Mycotoxins are toxic, secondary fungi metabolites, that have a negative impact both on human and animal health. It was postulated that various mycotoxins may affect homeostasis by disturbing the intestinal barrier. Claudins are proteins that are involved in creating tight junctions between epithelial cells. A growing body of evidence underlines their role in molecular response to mycotoxin-induced cytotoxicity. This review summarizes the information connected with claudins, their association with an intestinal barrier, physiological conditions in general, and with gastrointestinal cancers. Moreover, this review also includes information about the changes in claudin expression upon exposition to various mycotoxins.

Keywords: claudins; mycotoxins; intestinal barrier; tight junctions

Key Contribution: The literature survey conducted by us suggests that the most common mycotoxins affect the intestinal barrier and the main component of tight junctions—claudins. Moreover, mycotoxins might affect the process of inflammation, cause mucus layer dysfunction, disturbances in commensal microbiota as well as morphological and permeability changes which might associate with gastrointestinal cancers.

Citation: Kozieł, M.J.; Ziaja, M.; Piastowska-Ciesielska, A.W. Intestinal Barrier, Claudins and Mycotoxins. *Toxins* **2021**, *13*, 758. https://doi.org/10.3390/toxins13110758

Received: 31 August 2021
Accepted: 22 October 2021
Published: 26 October 2021

Publisher's Note: MDPI stays neutral with regard to jurisdictional claims in published maps and institutional affiliations.

cc BY

Copyright: © 2021 by the authors. Licensee MDPI, Basel, Switzerland. This article is an open access article distributed under the terms and conditions of the Creative Commons Attribution (CC BY) license (https://creativecommons.org/licenses/by/4.0/).

1. Introduction

In the last few decades, a growing body of evidence underlines the role of the intestinal barrier in mycotoxin toxicity. Intestinal barrier dysfunction contributes to many gastrointestinal diseases, e.g., inflammatory bowel disease or celiac disease [1]. The functioning of the intestinal barrier might be influenced by many factors including diet or lifestyle. At the molecular level, these factors may affect claudin expression and, by disturbing the connection and permeability of the epithelium, leading to many pathological conditions such as cancer development or its progression [2]. Claudins are a family of proteins that take part in the formation of tight junction connections between cells. Substances present in our everyday diet, that directly affect expression of claudins, are mycotoxins produced as secondary metabolites of various types of fungi. Mycotoxin exposure is a global health problem due to their abundance. Nevertheless, the evidence on the role of mycotoxins in carcinogenesis is still limited. This review presents a literature survey conducted to assess the role of the intestinal barrier and claudins in mycotoxin exposure. Moreover, we have evaluated the potential role of the most common mycotoxins in the regulation of the expression of claudins in gastrointestinal cancers.

2. Intestinal Barrier: In Health and Disease

The intestinal barrier plays a critical role in human health. First of all, it should be emphasized that the intestine is called the second brain for a significant reason. The influence of the gut on the functioning of the human body is not very well understood yet. Apart from the obvious role of participation in the absorption of water and nutrients (products of digestion), as a highly dynamic barrier between the external environment (intestinal lumen) and human tissues, it plays also a modulatory role between the gut microbiota and the central nervous system (CNS). The presence of the continuous and integral barrier allows the body to maintain homeostasis [3], and dead cells are replaced with new ones derived from intestinal stem cell niches (ISCs) [4].

The first element of the barrier is mucus. The intestinal mucus layer has a primary role in intestinal protection against mechanical, chemical, and biological attacks. Goblet cells are responsible for mucus production, and there are many types of goblet cells. The type of individual cells depends on their particular location in sections of the digestive system. Moreover, cell origin determines also the thickness of the mucosa [5,6]. Mucus consists of α-defensins, human α-defensin 5 (HD-5), and human α-defensin 6 (HD-6) (secreted by Paneth cells located in the small intestinal crypts) or secretory immunoglobulin type A (sIgA) [7–12]. The presence of these components helps to maintain the balance in the composition of the gut microbiota. Dysfunction in the production/secretion of individual proteins can lead to disturbances in the proportions of individual bacteria.

The second barrier is formed by epithelial cells and inter-epithelial tight junctions (TJ) that are primarily responsible for cellular integrity (Figure 1). These junctions are made by a number of the following proteins; claudins, occludin, junctional adhesion molecules (JAM), and tricellulin as well as cytoplasmic plaque proteins such as three zonula occludens (ZO) proteins [13]. Tight junctions regulate epithelial polarity and vectorial movement of solutes and fluids in the intercellular space [14–16]. Disturbances in the expression of individual proteins included in tight junctions are associated with many pathophysiological conditions, tumors in particular [17–20]. The intestinal epithelium is also the first barrier against food contaminants and is highly sensitive to *Fusarium* toxins, especially deoxynivalenol (DON) and zearalenone (ZEA). All chemical substances and biological factors, influencing the expression of genes encoding tight junction elements, affect the integrity and permeability of the digestive system barrier. Thus, in the case of mycotoxins, chronic exposure to low doses of mycotoxins leads to many pathophysiological conditions, including cancer.

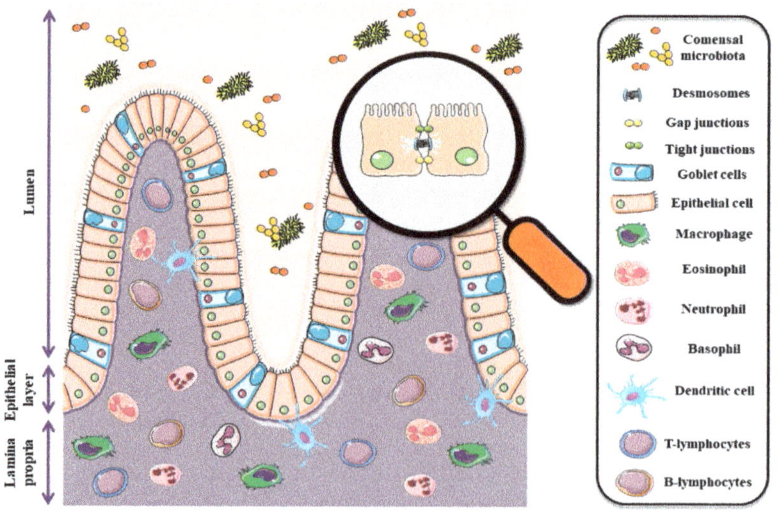

Figure 1. Diagram showing the intestinal barrier components. The graphical illustration was prepared by using the images from Servier Medical Art by Servier. Minor modifications were made (e.g., color of the stock images, some shapes) (https://smart.servier.com/smart_image/, accessed on 19 July 2021).

The third barrier is the cells of the immune system. Intraepithelial lymphocytes (IETs) and dendritic cells (DC) deserve special attention due to their contribution to the immune response. These cells are the first line of defense against enteric pathogens [21,22]. The gut-associated lymphoid tissue (GALT), literally the largest peripheral lymphoid tissue in the body, is definitely more complex. The role of this tissue is primarily to act in the context of flora selection in order to maintain homeostasis.

3. Claudins in Intestines-Schedule and Function in Health

Claudins are a family of proteins responsible for the formation of tight junctions between cells. Their mass is about 20–34 kDa. The structure of the various claudins is very similar and the major components that can be distinguished are: four transmembrane domains, two extracellular loops and amino- and carboxyl-terminal tails, which are located in the cytoplasm [23]. The first extracellular loop is involved in the regulation of paracellular charge and ion selectivity due to the presence of charged amino acids. The second one is associated with interactions between adjacent claudins. The amino tail is about seven amino acids and its length and sequence is generally similar among the claudin family. On the contrary, the carboxyl-tail is much more heterogeneous with 22 to 55 amino acids. The carboxyl end contains PDZ-motif that allows claudins to interact with other TJs proteins such as zonula occludens-1 (ZO-1). Moreover, post-translational modifications (e.g., phosphorylation or SUMOylation) take place in this region that may affect the functionality of the protein [23].

The tetraspan claudin family of proteins includes 26 family members in humans. However, the occurrence of particular claudins is determined by tissue and may differ depending on the type of the organism (Table 1). Their expression and location may be modulated by many substances including hormones [2]. For example, the estrogenic-like mycotoxin- zearalenone stimulated inflammation, disrupted the intestinal microflora and decreased the expression of claudin-4 in piglet intestine [24]. It should be underlined that pathological conditions (e.g., mucosal inflammation or cancer) are mainly associated with claudins disturbances what underlines their importance [25]. Claudins can be divided into two broad categories, pore-sealing and pore-forming claudins [26–28]. Alternations in the expression profile of individual claudin lead to changes in the paracellular transport/absorption of ions, fluids and substances such as drugs. Individual claudins can be regulated at several steps, including; transcription, microRNA repression, trafficking and phosphorylation. Claudins classified in the "pore-sealing" group, as the name suggests, lead to the sealing of junctions, thus reducing the permeability for various solutes and compounds. This group includes claudin-1, -3, -4, -5, -7, and -19. A group with opposite properties are claudins, classified as "pore-forming" (e.g., claudin-2 and -15), which are responsible for decreasing the tightness of the epithelium and increasing the permeability for different solutes [29,30]. Pore-forming claudins are responsible for size and charge selectivity of paracellular transport via tight junctions in a large variety of epithelia [25,31]. Apart from the obvious role of creating scaffolds, the purpose of which is to mechanically support epithelial cells, tight junctions (including claudins) are involved in signal transduction within the cell. An example that clearly shows the critical role of claudins as elements of tight connections is the claudin-1 knockout mice model- loss of claudin-1 leads to severe dehydration and postnatal death in mice [32]. The presence of claudins and other tight junction elements also plays a role in the intercellular communication process, but more importantly, it maintains a balance between proliferation, differentiation, and migration [33]. As different parts of the gastrointestinal tract vary in physiological properties, and therefore also in mucosal barrier permeability, they can have different claudin expression profiles. In mice, claudin-18 expression was observed in duodenum and jejunum [34,35]. However, there are contrary statements about its occurrence in the taste tissue [35,36]. In humans, claudin-18 was observed in gastric mucosa [37]. In the esophagus, the expression of claudin-4 and -7 was observed. The expression of *CLDN1* and *CLDN5* in the stomach was restricted to the glandular epithelium in mouse tissues [35]. The differences in claudin profiles in the human tissues were shown between the fundus and antrum of the stomach [38]. Claudin-3 and -4 were reported to be expressed at the highest levels in colon and rectum in rats and human samples [38,39]. Claudin-7 in rats and mice is observed in the ileocecal region [34,39], whereas in humans it was observed in the colon and rectum [38]. Expression of -8 in mice and human intestines was found to increase along the colon toward the rectum, contrary to claudin-15, which had its expression's peak in duodenum and jejunum [34,38,40]. Claudin-2 was predominantly expressed in the proximal intestine and maximal expression of claudin-12 was observed in mouse ileum [40] or jejunum in rat [39].

Table 1. Claudins expression in various parts of the gastrointestinal tract (GI) in mammals. * Distinguished parts of the colon.

Part of GI Tract	Claudins				References
	Human	Mouse	Rat	Pig	
Mouth	1, 4, 7, 8, 17	1, 2, 3, 4, 6, 7, 10, 11, 12, 17, 18, 23	-	4, 7	[35,36,41]
Esophagus	2, 3, 4, 7, 8, 12, 15, 18	-	-	-	[38,42,43]
Stomach	10, 11, 14, 17, 18, 23	1, 3, 5, 6, 11, 18	-	1	[35,37,38,44–46]
Duodenum	1, 2, 3, 4, 7, 8, 12, 15, 18	1, 2, 3, 4, 5, 7, 8, 9, 10, 11, 12, 14, 15, 18	1, 2, 3, 4, 5, 7, 8, 12	1, 3, 4, 5	[34,38–40,47–49]
Jejunum	2	1, 2, 3, 4, 5, 7, 8, 9, 10, 11, 12, 13, 14, 15, 18	1, 2, 3, 5, 7, 12	1, 3, 4, 5	[34,38–40,46–51]
Ileum	2, 3, 4, 7, 8, 12, 15, 18	1, 2, 3, 4, 5, 7, 8, 9, 10, 11, 12, 13, 14, 15	1, 2, 3, 5, 7, 8, 12	1, 3, 4, 5	[34,38–40,46–49]
Cecum	-	1, 2, 3, 4, 5, 7, 8, 9, 10, 11, 12, 14, 15	-	-	[34]
Colon	-	1, 2, 3, 4, 5, 7, 8, 9, 10, 11, 12, 14, 15	1, 2, 3, 4, 5, 7, 8, 9, 12	1, 4	[34,39,40,46,52]
Ascending *	2, 3, 4, 7, 8, 12, 15, 18	-	-	-	[38]
Transverse *	2, 3, 4, 7, 8, 12, 15, 18	-	-	-	[38]
Descending *	2, 3, 4, 7, 8, 12, 15, 18	-	-	-	[38]
Sigmoid *	2, 3, 4, 7, 8, 12, 15, 18	-	-	-	[38]
rectum	1, 2, 3, 4, 7, 8, 12, 15, 18	-	3	-	[26,38,42]

4. Claudins in Gastrointestinal Cancer

As mentioned above, it is indisputable that tight junction proteins play a huge role in maintaining physiological homeostasis. Their abnormal expression very often correlates with many types of cancers, including the reproductive and digestive system, as well as many others (Figure 2) [2,53]. Recently, TJ proteins have gained more and more attention, due to their crucial function in the pathogenesis of various diseases and the high potential for both diagnosis and treatment. As shown in Table 2, modulation of claudin expression in different cancers may vary. Moreover, their expression may be affected by many substances, including mycotoxins, which we would like to present in the next step of our article.

Table 2. Summarized information about claudin proteins expression in GI cancer cell line. ↑—upregulated, ↓—downregulated.

Cancer	Claudin	Expression	Described Effects on Cells	References
Oral	1	↑	Invasiveness ↑ Proliferation ↑	[54–59]
	7	↓	Invasiveness ↑	[60,61]

Table 2. *Cont.*

Cancer	Claudin	Expression	Described Effects on Cells	References
Oesophageal	1	↑	Proliferation ↑ Metastasis ↑ Invasiveness ↑	[62]
	4	↓	Growth ↑ Colony formation ↑ Invasiveness ↑	[63]
	7	↓	Invasiveness ↑ Metastasis ↑ Tumour progression ↑	[64]
Liver	1	↓	Invasiveness ↑ Metastasis ↑	[65]
	3	↓	Invasiveness ↑ Metastasis ↑ Colony formation ↑	[66]
	10	↑	Angiogenesis ↑ Invasiveness ↑	[67]
Gastric	1	↑	Apoptosis ↑ Invasiveness ↑ Migration ↑ Colony formation ↑	[68–70]
	4	↑	Invasiveness ↑ Migration ↑	[71,72]
	4	↓	Migration ↑ Proliferation ↑ Invasiveness ↑	[73]
	6	↑	Migration ↑ Proliferation ↑ Invasiveness ↑ Colony formation ↑	[74,75]
	7	↑	Migration ↑ Proliferation ↑ Invasiveness ↑ Colony formation ↑ EMT ↑	[75,76]
	9	↑	Migration ↑ Proliferation ↑ Invasiveness ↑	[75]
	11	↓	Migration ↑ Invasiveness ↑	[77]
Colorectal	1	↑	Growth ↑ Colony formation ↑ Migration ↑ Invasiveness ↑	[78,79]
	2	↑	Colony formation ↑ Proliferation ↑	[80]
	3	↓	Proliferation ↑ Invasiveness ↑ EMT ↑	[81]
	7	↓	EMT ↑ Colony formation ↑ Growth ↑ Invasiveness ↑	[82,83]

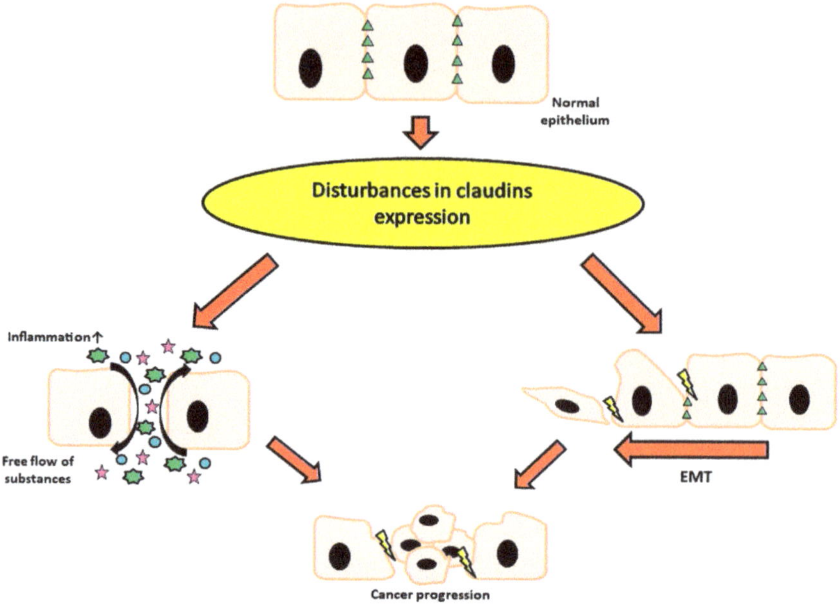

Figure 2. Diagram showing the consequences of abnormal claudin expression. EMT—epithelial to mesenchymal transition.

4.1. Oral Cancer

Oral cancer can attack any part of the oral cavity. It is widely believed that the most important risk factors of this type of cancer are smoking and high intake of alcohol. As oral cavity cancer has a very high mortality rate (~50%), understanding the role of claudins in this disease seems to be necessary to allow improvement of current therapeutic or diagnostic modalities [84]. Numerous studies reported that claudin-1 play a significant role in this type of cancer and its elevated expression is observed in most oral, carcinomas. Moreover, the higher the expression, the more advanced stage tumors are diagnosed, and what is worth emphasizing—it is also associated with a lower survival rate [54–57]. Patricia Pintor Dos Reis et al. showed that overexpression of *CLDN1* is associated with higher invasion of cells and aggressiveness observed in immunohistochemistry [58]. Naohisa Oku et al. explained that claudin-1 stimulates the invasiveness of oral cancer cells via activation of MT1-MMP and MMP-2 [85]. Silencing of *CLDN1* results in decreased invasive potential and proliferation of oral cancer cells [58,59]. Another very important claudin associated with the pathogenesis of oral cancer appears to be claudin-7, which has been reported to be downregulated in most cases of this type of cancer [60]. Furthermore, it was also postulated that downregulated expression of claudin-7 is associated with an increased risk of cancer recurrence and poor prognosis [61,86].

4.2. Esophageal Cancer

Esophageal cancer is not the most common cancer; however, the mortality rate is very high, and the prognosis is very poor, therefore this disease is considered a serious global health problem [87]. Some studies reported the involvement of TJ proteins in the character of this cancer. On the one hand, it was reported that the decreased expression of claudin-1 in tissue derived from patients was associated with poor prognosis and an increased risk of recurrence [88]. On the other hand, it was observed that increased expression of *CLDN1* results in increased proliferation and metastasis via stimulation of autophagy in human esophageal cancer cell lines. Moreover, the same team demonstrated in an in vivo model that claudin-1 is able to stimulate metastasis [62]. Elevated expression of claudin-2 was observed in the cancerous and pre-cancerous lesion [89]. Lower expression of *CLDN4* is postulated to be a risk factor and prognostic biomarker for cancer recurrence and survival rate [63,90]. Furthermore, in vitro studies have shown that claudin-4 expression regulates invasion and metastasis of cells [63]. It was also shown that Twist1 may modulate *CLDN4* expression which suggests that claudin-4 is directly associated with EMT in esophageal cancer [91]. Another claudin involved in esophageal cancer is claudin-7, which was reported to be downregulated and this disturbance may lead to cancer progression [64]. This hypothesis seems to be confirmed by later studies, which showed that claudin-7 influences the expression of E-cadherin, and thus may lead to loss of the "gate" function and stimulate EMT [42].

4.3. Liver Cancer

Among liver cancers, we can distinguish two major types: hepatocellular carcinoma (HCC) and bile duct cancer (cholangiocarcinoma, CC). The first one occurs more often and constitutes a serious global problem. The main risk factors are viruses, metabolic disorders, and, as was mentioned before, toxins, including aflatoxin. TJ proteins appear to be promising markers in this disease, however, we still do not know enough to use them in medical diagnosis. It was suggested that decreased expression of claudin-1 may be associated with poor survival rates and potential for metastasis and invasion [65]. At the same time, overexpression of *CLDN1* may be a positive prognostic marker after treatment [92]. However, it was also observed, that overexpression of this protein induces EMT in normal liver cells [93]. Liver metastases from primary colorectal cancer were characterized by an increased expression of claudin-1 [94]. Silencing of *CLDN1* sensitizes HepG2 to 5-fluorouracil [95]. Based on the abovementioned facts, it may be concluded that abnormal *CLDN1* expression has a very important impact on the liver. Similarly, as claudin-1, proper expression and localization of claudin-3 in cells are also very important to maintain homeostasis. Downregulation of this protein was observed in HCC; moreover, it was also found in most types of HCC cell lines. Lei Jiang et al. reported that silencing of claudin-3 in HepG2 and Huh7 cell lines stimulates changes in their morphology, increases the ability to migrate, induces foci formation in monolayer culture, and stimulates cells invasiveness [66]. All of these underline the importance of claudin-3, however, further studies are required. Claudin-7 expression is upregulated in HCC [96]. Yusuke Ono et al. postulated that claudin-4 and claudin-7 may be a useful immunohistochemical marker to distinguish HCC and CC, because the expression of these proteins is higher in these two cases than in control, and the expression of *CLDN4* and *CLDN7* is lower in HCC than in CC [97]. Earlier studies also provided evidence for this relationship [98,99]. Claudin-10 is overexpressed in most HCC patients and that expression is highly associated with poor prognosis after resection of the liver [67]. Studies conducted in in vitro models showed that abolition of *CLDN10* reduces the invasiveness of cancer cells what seems to be promising in the context of the possibility of using claudins as a target in the treatment of liver cancer [100].

4.4. Gastric Cancer

Gastric cancer (GC) also known as stomach cancer is one of the most common causes of human death. Each year 989,000 people are diagnosed with gastric cancer, out of which 738,000 patients die because of this disease [101]. Many factors contribute to its development, including age, tobacco smoking, diet, *Helicobacter pylori* infection, and many others [102]. Claudins are postulated to play an important role in this type of cancer as well. Claudin-1 was found to be elevated in GC. Moreover, it has been observed that increased expression of *CLDN1* gene is correlated with poor survival [68,103]. Huang J et al. showed that claudin-1 is regulated by β-catenin in gastric cancer samples [103]. Elevated expression of *CLDN1* in cancer cells results in increased proliferation, migration, invasion in gastric cancer cells, but also protects them against apoptosis [69,70]. Zhe Lin et al. reported that claudin-2 and claudin-6 are down-regulated, while claudin-11 was upregulated compared to normal tissue [104]. In contrast, Luoluo Yang et al. have not noticed any significant difference between claudin-2 expression in normal and cancerous tissue, but they found that claudin-5 expression was significantly higher and the levels of claudin-7 and claudin-8 were significantly lower than in normal tissue, both at the protein and mRNA levels [105]. Claudin-4 was observed to be overexpressed in GC, moreover, increased expression of *CLDN4* results in enhanced invasion and migration of gastric cancer cells [106]. Tsann Long Hwang et al. reported that this effect may be generated via activation of matrix metalloproteinase proteins (MMP) [71,106]. Satoshi Ohtani et al. reported that, on the one hand, decreased expression of claudin-4 was associated with lower tumor aggressiveness, but on the other hand, low expression of *CLDN4* was associated with poor prognosis and survival [72]. Studies conducted on the GC cell line demonstrated that silencing of *CLDN4* expression leads to increase resistance to chemotherapy [73]. Moreover, the abolition of claudin-4 stimulates migration, invasion, and proliferation [73]. Abnormal expression of *CLDN6* was also observed in tissue obtained from patients with gastric cancer. Furthermore, up-regulated expression of claudin-6 is connected with poor prognosis and survival [74,107]. Overexpression of claudin-6 both in gastric cancer cells and in vivo models results in increased migration, proliferation, and invasiveness [74,75]. Interestingly, claudin-6 may stimulate invasion and migration via claudin-1 and thus activate MMP proteins [108]. Site Yu et al. proposed that claudin-6 may stimulate EMT by affecting YAP1-SNAIL1 axis [74]. Overexpression of Claudin-7 correlates with poor prognosis and a high possibility of lymph nodes metastasis [18,76]. Studies conducted on cells and animal models showed that up-regulated claudin-7 expression has a great impact on cell proliferation, migration, invasion, and colony formation [75,76]. Claudin-11 was found to be down-regulated in GC tissue, furthermore, silencing of *CLDN11* in gastric cancer cell line was responsible for increased motility and invasiveness [77]. Kyong Hwa Jun et al. presented that negative regulation of claudin-11 may be useful as a marker of poor prognosis in patients with gastric cancer [18]. It seems that this thesis was confirmed by later research [109–111]. Claudin-10 and claudin-17 were down-regulated in GC tissues, claudin-14 was observed to be up-regulated [45].

4.5. Colon Cancer

Colorectal cancer (CRC) is one of the most common cancers worldwide. In the UK, 11% of new cancer cases are CRC [112]. Nevertheless, it is estimated that the incidence of colorectal cancer is still decreasing [113]. It may be associated with increased knowledge about nutrition. Claudins are undoubtedly very important in the formation, progression, and metastases of this cancer. Moreover, CRC appears to be the best-known neoplasm in terms of claudins. Already in 2005, Punita Ghawan et al. reported that claudin-1 has a significant role in colon cancer. They showed that the expression of *CLDN1* is elevated in commercially available cell lines (HT29, SW480, and SW620, but not in HCT116) and in samples obtained from patients; furthermore they demonstrated that localization of claudin-1 is different than in normal tissue. In vivo results reported by the same team, showed that abolishment of the expression of *CLDN1* stimulates liver metastasis and tumor size, at the same time providing that one of the mechanisms in the regulation of claudin-1 expression may be the regulation of E-cadherin and β-catenin/Tcf signaling pathway [78]. These results seem to be consistent with the other reports [114,115]. Amar B Singh et al. proposed another mechanism of action and showed that cldn1 may regulate E-cadherin via ZEB-1 (Zinc Finger E-box binding homeobox-box1) and thus increasing the invasiveness of cells [79]. In turn, Jillian L Pope with colleagues presented that claudin-1 is closely associated with the Notch-signaling pathway, and thus affects the behavior of cells [116]. Other authors showed that pyruvate kinase M2 (PKM2) increased the expression of cldn1 in Caco-2 and SW480 cell lines via the epidermal growth factor receptor (EGFR)- protein kinase C (PKC) pathway [117]. All these studies have clearly demonstrated how complex the mechanism of action of claudin-1 is. Recent results shed new light on the use of *CLDN1* in the diagnostics of CRC. It was indicated that *CLDN1* may be useful in the future as an imaging agent in fluorescence-guided surgery [118]. Interestingly, the use of claudin-1 during confocal endomicroscopy seems to be promising, as it can significantly increase the detection of the precancerous lesion within the intestine based on the expression of this protein—as it was shown in an animal model [119]. In contrary to previously described results, Murray B Resnick et al. reported that down-regulated expression of claudin-1 in tissue obtained from patients in II stage colon carcinoma was associated with poor prognosis and a high possibility of tumor recurrence [120]. All of this underlines the importance of proper expression of claudin-1 in our organism because both too high and too low expression may lead to homeostatic imbalance. Claudin-2 was also reported to have an important role in colon cancer progression. It was observed that its expression is higher in CRC than in normal, physiological tissue [80]. It was also suggested that *CLDN2* may act via EGFR transactivation and that forced claudin-2 expression leads to increased proliferation of cells and significantly greater tumor growth in vivo [80]. In turn, claudin-3 was observed to be decreased in CRC. Studies conducted on cells with silencing *CLDN3* expression showed that with the abolition of claudin-3 expression, an increase in cell invasion and migration occurs [81]. Similar to claudin-3, claudin-7 is also down-regulated in colon cancer tissues [82]. Bhat et al. demonstrated that claudin-7 may have anti-cancer properties because its forced expression in in vitro models led to reduced invasiveness, colonies formation, proliferation, and ability to grow, while abolition resulted in enhancement of these properties. In in vivo models decreased expression of claudin-7 led to the lower weight of the tumor [82]. Similar results, but on a different cell line have been shown recently, which seems to confirm previous studies [83]. Moreover, it was noticed that *CLDN7* expression correlates with the potential for metastasis in CRC-it has been proposed that decreased expression of claudin-7 may be a predictor of liver metastasis [121].

5. Environment Contamination by Mycotoxins and Their Occurrence in Food and Feed

Mycotoxins are chemical compounds considered as secondary metabolites of fungi, generally from *Fusarium*, *Aspergillus*, and *Penicillium* genus. Worldwide, mycotoxins have significant implications for human and animal health, as well as for the economy and international trade [122]. Mold that can produce mycotoxins grows on numerous foodstuffs such as cereals, dried fruits, nuts, and spices. Their growth might occur either before harvesting or after harvesting, during storage, on/in the food itself often under warm, damp, and humid conditions. It is worth emphasizing that most mycotoxins are chemically stable even during food processing, and that their neutralization is only available for feed [123]. The commonness of occurrence of fungi and their spreading is mainly due to the climate conditions that favor the growth and multiplication of fungi, food spoilage and then favor mycotoxin production. These conditions are very important in the harvesting, transport, and storage—the processes that depend on good practices and habits of people responsible for them [124]. Several hundred different mycotoxins are known, but the most commonly observed mycotoxins that present a concern to human health and livestock include aflatoxin B1 (AFB1), ochratoxin A (OTA), patulin (PAT), fumonisins (FBs), zearalenone (ZEA), nivalenol (NIV) and deoxynivalenol (DON) [125]. The presence of individual mycotoxins is also partially determined by the type of crops e.g., in the case of maize, widespread occurrence of fumonisins and deoxynivalenol is observed. This review aims to collect and analyze the information about the effects of individual mycotoxins on the gastrointestinal tract in the context of claudins as a core element of tight junctions.

6. Mycotoxins and Human Health Especially the Health of the Gut and the Entire Digestive Tract

It is generally known that mycotoxins possess harmful effects both on human and animal well-being. The influence on human health is mainly observed in the area of the reproductive system (hyper estrogenic syndrome, precocious puberty), hepatotoxicity (especially hepatocellular carcinoma) immunotoxicity, or genotoxicity (carcinogenic, mutagenic, teratogenic), and nephrotoxicity (nephropathies and urinary tract tumors) [126–130]. Interestingly, it was also reported that some toxins may stimulate neurodevelopmental toxicity. In vivo study showed that ochratoxin A affects differentiation of almost all neural cells [131]. Moreover, exposure to OTA was also reported to be connected with pathobiology of autism in autistic children what underlines the importance of controlling the presence of mycotoxins in our diet [132,133]. In zebrafish model, authors observed that aflatoxin B1 may influence neurobehavior and neurodevelopment [134,135]. Similar results were obtained in response to exposure to zearalenone [136].

The negative effect of mycotoxins is mainly observed in the case of long-time exposure to low doses of mycotoxins. It is also worth emphasizing that the gastrointestinal tract is the first element that is directly exposed to the toxic effects of mycotoxins. The limitation in exploring the impact of mycotoxins on the human body is primarily the quantification of lifetime individual exposure. Analysis of 74,821 samples of (feed and feed raw materials) from 100 different countries revealed the presence of at least one of the mycotoxins (64% of samples was co-contaminated with at least two mycotoxins) what seems to be important when we know that people may be exposed to these secondary metabolites after consuming food products derived from animals exposed to high levels of mycotoxins in their feed [124,137]. The analyzed data present the global scale of the problem and provoke reflection because of potential health risks. The agency responsible for the preparation of the guidelines is the European Food Safety Authority (EFSA). To protect people against adverse effects of mycotoxins, EFSA established a total tolerable daily intake for humans (Table 3).

Table 3. Total tolerable daily intake present for mycotoxins described in this review based on EFSA reports.

Mycotoxin	Total Tolerable Daily Intake	References
Aflatoxins	Not established	-
Fumonisin B1	1 μg/kg	[138]
Zearalenone	0.25 μg/kg	[139]
Deoxynivalenol	1 μg/kg	[140]
Patulin	Not established	-

The unavoidable presence and chronic human exposure to mycotoxins lead to the occurrence of many pathological conditions, ranging from disturbances in the composition of the gut microbiota to neoplasms, e.g., of the liver [141,142]. The role of mycotoxins in modulating the immune response is also not negligible. GALT, as the tissue responsible for regulating many immune processes, plays a crucial role. Its proper functioning allows maintaining the balance between many organs, which is why the intestine is called the second brain for a reason. Therefore, the next step in our review was to compile the information about the effects of the mycotoxins on a healthy gastrointestinal tract with emphasis on modulation of tight junction proteins if it was reported.

6.1. *Aflatoxins*

Aflatoxins (AF) are secondary metabolites of fungi produced by *Aspergillus* molds (*Aspergillus flavus*, *Aspergillus parasiticus*, *Aspergillus nomius*). In 1987, aflatoxins (including aflatoxin B1, B2, G1, G2, and M1) were classified as Group 1 on the basis of the International Agency for Research on Cancer (IARC) evaluation [143]. Group 1 classifies chemical compounds as clearly influencing the neoplastic process in humans. Aflatoxin B1 (AFB1) is considered one of the most harmful compounds produced by *Aspergillus*. Aflatoxin contamination is mainly reported in maize, peanuts (and generally ground nuts) and their products, oilseeds, and pulp [144]. Aflatoxin B1 is rapidly absorbed into the blood from the gastrointestinal tract (GIT). In the bloodstream, aflatoxin B1 is metabolized into its toxic metabolite AFB1-8,9-exo-epoxide (AFBO) [145]. AFs have carcinogenic, hepatotoxic, teratogenic, mutagenic, and immunosuppressive effects, with the liver as the organ responsible for detoxification is most affected by the negative effects of aflatoxins [146]. In vivo studies in pigs show a clear effect of AFB1 on the functioning of the digestive system, in particular the jejunum. Increased serum diamine oxidase (DAO) activity was observed in AFB1 treated group, indicating that AFB1 supplementation damages intestinal barrier integrity. Moreover, pigs fed with the AFB1 diet exhibited significantly decreased mRNA abundance of ZO-1 in jejunal mucosa, which supports the thesis made above, that aflatoxin B1 has a destructive effect on gut barrier integrity [147]. Reduced final body weight (BW) in pigs (AFB1 diet) may also be associated with a decrease in mRNA expression of sodium-glucose co-transporter 1 (SGLT1) and solute carrier family 7 member 1 (SLC7A1) in jejunal mucosa [147,148]. Both SGLT1 and SLC7A1 act as intestinal epithelium transporters of nutrients from the intestinal lumen (glucose and cationic amino acids respectively) [149–151]. In mice exposed to aflatoxin B1 and aflatoxin M1 separately and in combination, it was observed that villus height was reduced, while the crypth depth was deepened compared to non-treated mice. Moreover, the expression of claudin-1 and ZO-1 were significantly decreased [152]. Disturbed barrier function via affection claudin-1 expression was also observed in broiler chicks, but in this study, aflatoxin B1 stimulated the expression of *CLDN1* [153].

6.2. Fumonisins

Fumonisins are secondary metabolites produced by a number of *Fusarium* species, especially *Fusarium verticillioides* and *Fusarium proliferatum* [154]. Three main chemical compounds belong to the Fumonisis group; namely FB1, FB2, and FB3 [155]. Maize and its derivatives (corn-based food) are one of the main sources of fumonisins, which can potentially influence the occurrence of pathological conditions in both livestock and humans. Fumonisin B1 is the most prevalent member of the Fumosin family. According to the IARC opinion, fumonisin B1 was classified as possibly carcinogenic to humans. The last update of the data took place in 2002, and the decision was made based on the results of in vivo studies in mice and rats (at present, the extensive scope of publications clearly indicates the association of fumonisins and the cancer process in the human body). Currently, the vast majority of the data is focused on the association between fumonisin B1 and esophagus cancer (OC) [156–159]. According to available evidence, the molecular mechanism of action is linked to sphingolipid metabolism [160,161]. Due to the similarity to two critically important molecules (sphinganine and sphingosine), the ceramide synthesis pathway is blocked. The consequence of the inhibition process is the inhibition of the sphingolipid biosynthetic pathway (disrupted sphingolipid metabolism) [162,163]. In IPEC-J2 cells, FB1 leads to decreased viability, decreased expression of tight junction proteins (*CLDN1*, *OCLN*, and *ZO-1*), and altered expression of mucin genes (*MUC1*, *MUC2*). Moreover, the permeability of IPEC-J2 monolayer was also disturbed after exposition to fumosin B1 [164]. It was also observed that FB1 decreased transepithelial electrical resistance (TEER) value in IPEC-J2 cells [165]. Similar results were obtained on pig iliac endothelial cells (PIECs) [166]. Exposure to fumonisin was also postulated to be responsible for the decrease in the diversity of the bacterial flora [167].

6.3. Zearalenone

Zearalenone (ZEA) is a non-steroidal estrogenic mycotoxin. The species of fungi that are responsible for the biosynthesis and secretion of the ZEA mycotoxin are mainly *Fusarium graminearum* (*Gibberella zeae*), *Fusarium culmorum*, *Fusarium crook-wellense*, *Fusarium semitectum* and *Fusarium equiseti* [146]. The abovementioned fungi species are responsible for the contamination of most cereal crops worldwide. According to the IARC classification, ZEA is not classifiable as to its carcinogenicity to humans. It owes its estrogenic properties to the presence of a macrocyclic lactone ring with a spatial arrangement similar to that of steroid hormones (structurally similar to 17β-estradiol) [168–170]; hence the destructive effect of zearalenone on the reproductive system. Until now, 5 metabolites of the ZEA have been described, i.e., α-Zearalenol (α-ZEA), α-Zearalanol (α-ZAL), Zearalanone (ZON), β-Zearalenol (β-ZEA), β-Zearalanol (β-ZAL), with individual metabolites possessing different estrogenic properties [171]. Due to the presence of the lactone ring, ZEA is known to be heat stable up to 150 °C and does not degrade during food and feed processing; this presents some sort of challenge in terms of removing mycotoxins, particularly from food chain products [172]. The following events related to the presence of zearalenone in the human body are distinguished: altered progesterone level, hyper-estrogenic syndrome, precocious puberty, decreased sperm count, decreased serum testosterone level, and infertility [126,127,173–175]. Additionally, disruption of blood coagulation and changes in haematological parameters in rats were found [176,177]. Unfortunately, the context of the intestinal barrier is still little recognized. In the intestine of Juvenile Grass Carp (*Ctenopharyngodon idella*), ZEA affected intestinal integrity via affecting tight junctions [178]. In the intestine of piglets under the influence of ZEA, the expression of claudin-4 was reduced and the intestinal microbiota was disturbed [24]. Zearalenone was also postulated to decrease claudin-4, occludin, and connexin-43 in tissue derived from rats [179,180]. Similar to FB1, ZEA is able to affect mucin genes [180]. In addition, ZEA was reported to alter the morphological structure of the villi in the jejunum [179].

6.4. *Deoxynivalenol*

Deoxynivalenol, also known as vomitoxin, is the most common member of the trichothecenes group which include also nivalenol. *Fusarium, Myrothecium, Phomopsis, Stachybotrys, Trichoderma, Trichothecium*, and other species are responsible for their production [181,182]. DON frequently contaminates cereal grains such as maize, wheat, oats, barley, and rice. Among trichothecenes group, type A and type B are the most concerning due to their broad and highly toxic nature, with DON being classified as trichothecenes B. In the IARC assessment, DON was classified in Group 3 so it is not classifiable as to its carcinogenicity to humans. DON possesses the following effects: cytotoxicity, steroidogenesis disruption, affected mRNA expression of genes responsible for regulating the integrity and permeability of the intestinal barrier, increased reactive oxygen species (ROS) production, inhibition of cellular protein synthesis, and ribosomal stress syndrome [183–187]. Changes in the functioning of the intestinal barrier were observed both in in vitro and in vivo studies. It was observed that DON decreases TEER value in IPEC-2 cells [188,189]. The expression of occludin, claudin-3, and claudin-4 is also modulated by DON which was showed both in in vitro and in vivo models [188,190–192]. In human non-cancerous intestinal epithelial cell line (HIEC-6), decreased expression of claudin-1 and increased expression of inflammatory-related interleukins was observed [193]. Moreover, it was also found that DON may affect another very important compound associated with an intestinal barrier—trefoil factors (TFFs), which are involved in repairing and protecting intestinal mucus. Shuai Wang et al. showed that DON decreases expression of TFF2 and TFF3, which may suggest that it is another pathway altered by deoxynivalenol in the intestine [190]. Induction of immune response was also observed [24,191]. It was postulated that the pathway connected with an intestinal disturbance caused by DON may be MAPK p44/42 pathway [189,192,194]. In the context of changes in the intestinal biota, an increase in abundance of the *Lactobacillus* genus was observed (DON and ZEA treated group), suggesting that members of this genus could play a key role in the detoxification of dietary DON and ZEA in pigs [195]. Basolateral DON exposition caused inhibited intestinal stem cell activity through the Wnt/β-catenin pathway [196]. Based on this study, Hikaru Hanyu et al. decided to go one step further, demonstrating that basolateral exposure was more toxic than luminal DON exposure in terms of intestinal barrier functions and stem cells [197]. It is worth mentioning that DON toxicity observed in vitro and in vivo might be different, possibly due to biological barrier function. It was observed that in a zebrafish model, DON does not cause any toxic effect, as suggested by in vitro results. Shu Guan et al. demonstrated that DON is transformed via gut microbes to the depoxyated form of DON (DOM-1) which is less toxic than DON itself [198]. A similar effect was observed in zebrafish larvae exposed to DON microinjections and aqueous solutions, where no effect was observed, which in consequence suggests that DON is not capable to pass through biological barriers [199]. This fact underlines the importance of the intestinal barrier in response to exposition to mycotoxins.

6.5. Patulin

Patulin (4-hydroxy-4H-furo [3,2c]pyran-2(6H)-one, PAT), water-soluble polyketide lactone, is a secondary metabolite of filamentous fungi (toxigenic molds) such as *Penicillium*, *Aspergillus*, and *Byssochlamys* species. PAT was firstly isolated as a compound with antimicrobial activity in the 1940s from *Penicillium patulum*. Among all fungi listed above, *Penicillium expansum* is the main source of patulin contamination in apples, pears, and their derived products, and it also causes decay in fruits [200]. Over time, detrimental effects of patulin on animal organisms were discovered. Patulin has been classified by IARC in group-3 (not classifiable as to carcinogenicity to humans) [201,202]. The potential effect of patulin on the gastrointestinal tract results primarily from the reduction of TJs' mRNA (including ZO-1 and Occludin) and degeneration of intestinal villi [203,204]. In addition, in animal models, PAT exposure has been shown to lead to epithelial degeneration, hemorrhage, ulceration of gastric mucosa, reduction in the number of goblet cells in villi and crypts [205,206]. From the molecular biology point of view, change in the composition of tight junctions or degradation of epithelial cells leads to impaired permeability (TEER increase), which results in a loss of balance between the external environment and the organism [207,208]. Moreover, changes in the claudin distribution patterns are observed (for claudins 1, 3, and 4, the staining pattern became quite diffuse compared to the controls, large gaps were also observed to have appeared in the 'chicken wire' pattern) [209]. Nevertheless, the effect of patulin on non-cancerous junctional cells has not been evaluated yet.

Summarizing all the collected information, it can be concluded that the range of mycotoxins activity in the area of cells of the human body is very wide. A critical role in the penetration of mycotoxins into the body is played by the gastrointestinal barrier, which is the first protective element. Individual mycotoxins differently influence the integrity of the gastrointestinal barriers. Some of them may affect the expression of tight junction proteins and thus disturb homeostasis in the gut (Figure 3). Interestingly, a growing body of evidence shows that naturally co-occurring mycotoxins may have a greater impact on the intestinal barrier. For example, mixed doses of *Fusarium* toxins are more harmful and stimulate an immune response more than individual mycotoxins [210]. Many of the areas, such as the influence of mycotoxins on the nervous system of the digestive tract or the immune system, require better understanding. Furthermore, it seems necessary to investigate the effect of mixed doses of mycotoxins given that the presence of more than one toxin in food is very common. Expanding the scope of our knowledge about mycotoxins has the potential to provide a better understanding and the possibility to eliminate the side effects of mycotoxins both on human and animal health.

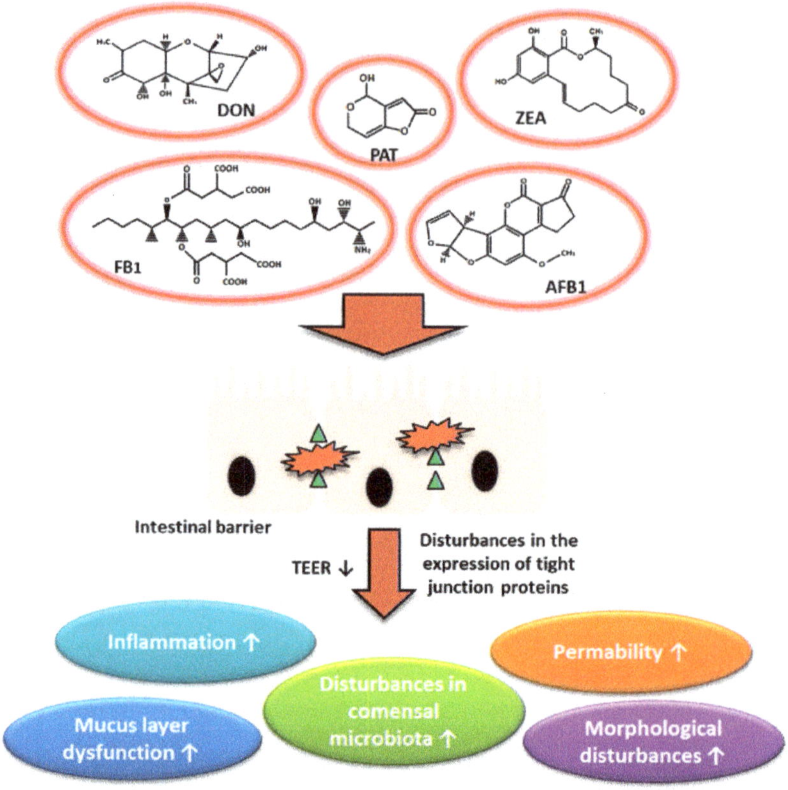

Figure 3. Mycotoxins and their influence on intestinal physiology. DON—deoxynivalenol, PAT—patulin, ZEA—Zearalenone, FB_1—Fumosin B1, AFB1—Aflatoxin B1.

7. Mycotoxins and Their Association with Claudins in Gastrointestinal Cancers

Over the past two decades, we can observe a growing body of evidence that highlights the importance of various mycotoxins in the everyday diet in different parts of the digestive system. For example, it was found that Aflatoxin B1 is one of the risk factors associated with hepatocellular carcinoma [142]. Nevertheless, there is still little known about the relationship between mycotoxins and claudins. Hereby, we described the effects of various mycotoxins on claudin expression in colorectal cancer cell lines as the most popular model in research focusing on cytotoxicity of mycotoxins and the barrier function. A lot of studies focused on molecular aspects and the effects of various mycotoxins on colon cancer cells in vitro. It is generally known that mycotoxins are able to reduce the viability of various cells, not only cancerous. Thus, it is important to understand their mechanism of action, but it is very difficult due to their variety. One of the mycotoxin mentioned in this review is deoxynivalenol. It was verified on HT-29 cell line that DON inhibits proliferation, stimulates DNA damage, increases expression of p53, leads to release of cytochrome c from the mitochondria, stimulates changes in *Bcl-2*, *Bax*, and *Bid* expression and then as a consequence induces caspase-dependent apoptosis. Moreover, deoxynivalenol in a study conducted on Caco-2 and T84 cell lines significantly decreased monolayer integrity in TEER assay [211]. This implies that DON leads to changes in tight junction, possibly by modulating the expression of various tight junction proteins. As an example, we can cite studies conducted by other researchers, where they showed that DON alters the expression of claudin-4 in Caco-2 cell line [188,212]. Other mycotoxins have also been investigated for modulation of tight junction protein expression. Alejandro Romero et al. reported that aflatoxin B1, fumonisin B1, ochratoxin A and T-2 toxin (T2) significantly reduced monolayer integrity and decreased the expression of claudin-3, claudin-4, and occludin in Caco-2 cell line [213]. Patulin leads to a reduction in transepithelial electrical resistance values, however, no changes in claudin expression were observed [204]. It is worth noticing that another study showed that although the claudin's expression was not disturbed by patulin, their localization has changed [209,214]. The combination of mycotoxins may either affect the intestinal barrier. ChenQing Wu et al. showed that the mixture of aflatoxin M1 (AFM1), ZEA, and OTA affects the morphology of TJ proteins and thus disturbs intestinal permeability [215]. It was also observed that a combination of AFM1 and OTA presents a synergistic effect together and that they may damage the intestinal barrier [216]. However, the expression of claudins in this study has not been investigated. Gao et al. presented that the mixture of AFM1 and OTA disturbs the expression of claudin-3 and claudin-4 in Caco-2 cells and that these toxins present a synergistic effect [217]. It has been postulated that the combination of AFM1 and AFB1 may also influence the intestinal barrier function [152]. Nevertheless, mycotoxins can act as a double-way axis. It was observed that zearalenone is able to stimulate the proliferation of colon cancer cells (HCT116), but what is worth emphasizing- it all depends on the concentration, because only low doses stimulate proliferation and migration, while at high doses ZEA is cytotoxic to these cells [218]. Similar results, including ZEA derivatives, were observed in MCF-7 cells and in prostate cancer cells [219,220]. It seems to be very worrying, not only for healthy people but especially for patients with cancer, who are exposed to low doses of this mycotoxin in their everyday diet. Nevertheless, it is hard to compare in vitro results with in vivo, because the absorption of toxins in cells culture may be different. Further studies would be necessary to discover the role of ZEA and its derivatives and other mycotoxins which are poorly understood in colon cancer cell lines including the role of tight junction and epithelial permeability. At the same time, it underlines the significance of the detection of toxins in our food and the need to understand the exact mechanisms of action of these toxins in order to consider them in the context of anticancer properties in general. All the information discussed above has been summarized in the Table 4.

Table 4. Summarized information about mycotoxins and their relationship to claudins in colon cancer cell lines. * only markedly affected, but distribution was disturbed, ↓ lower value/expression, AFM1—aflatoxin M1, ZEA—zearalenon, OTA—ochratoxin A, AFB1—aflatoxin B1.

Mycotoxin	Cell Line	TEER Values	Targeted Claudin	References
Aflatoxin B1	Caco-2	↓	*CLDN3* ↓	[213]
Ochratoxin A	Caco-2, HT-29-DR	↓	*CLDN3, CLDN4* ↓	[213,221,222]
Patulin	Caco-2	↓	*CLDN1, CLDN3, CLDN4)* ↓ *	[204,209,213]
T-2 toxin	Caco-2	↓	*CLDN3, CLDN4* ↓	[213]
Fumonisin B1	Caco-2	↓	*CLDN3, CLDN4* ↓	[213]
Deoxynivalenol	Caco-2, T84, HT-29-DR	↓	*CLDN4* ↓	[188,212]
AFM1 + ZEA + OTA (combination)	Caco-2	↓	*CLDN3, CLDN4*	[215]
AFM1 + OTA	Caco-2	↓	*CLDN3, CLDN4*	[217]
AFM1 + AFB1	Caco-2	↓	*CLDN1*	[153]

8. Conclusions

It is generally known that the intestinal barrier plays a crucial role in the functioning of the organism. Its disturbance may be associated with many pathological conditions. Some compounds, such as mycotoxin, may induce changes in their structure and thus lead to numerous disorders. As mycotoxins are known to be harmful to human and animal health, there is still little known about their effect on claudin expression in a healthy gut. It is also worth highlighting that active metabolites of main mycotoxins may be more toxic and cause even more detrimental biological effect in cells than a mycotoxin itself. For example, α-ZOL is reported to have even more estrogenic effect than ZEA itself. Thus, it should be also taken into account that not only mycotoxins but also their metabolites might participate in intestinal barrier dysfunction. In recent years, claudins gain more and more attention due to their diagnostic and therapeutic potential. Understanding the connection between mycotoxins and claudins may shed new light on both treatment options of gastrointestinal cancers and protection against adverse effects caused by mycotoxins present in our everyday diet.

Author Contributions: M.J.K., M.Z.; writing—original draft preparation, M.J.K., M.Z.; writing—review and editing, A.W.P.-C.; supervision. All authors have read and agreed to the published version of the manuscript.

Funding: This study was supported by Medical University of Lodz, Poland grants no. 503/0-078-03/503-01-001-19-00 and 503/0-078-03/503-99-001.

Institutional Review Board Statement: Not applicable.

Informed Consent Statement: Not applicable.

Data Availability Statement: Not applicable.

Conflicts of Interest: The authors declare no conflict of interest.

References

1. Vancamelbeke, M.; Vermeire, S. The intestinal barrier: A fundamental role in health and disease. *Expert Rev. Gastroenterol. Hepatol.* **2017**, *11*, 821–834. [CrossRef] [PubMed]
2. Kozieł, M.J.; Kowalska, K.; Piastowska-Ciesielska, A.W. Claudins: New players in human fertility and reproductive system cancers. *Cancers* **2020**, *12*, 711. [CrossRef]
3. Camara-Lemarroy, C.R.; Metz, L.; Meddings, J.B.; Sharkey, K.A.; Wee Yong, V. The intestinal barrier in multiple sclerosis: Implications for pathophysiology and therapeutics. *Brain* **2018**, *141*, 1900–1916. [CrossRef] [PubMed]
4. Lee, S.-E.; Massie, I.; Meran, L.; Li, V.S.W. Extracellular Matrix Remodeling in Intestinal Homeostasis and Disease. *Intest. Stem Cell Niche* **2018**, *2*, 99–140. [CrossRef]
5. Birchenough, G.M.H.; Johansson, M.E.V.; Gustafsson, J.K.; Bergström, J.H.; Hansson, G.C. New developments in goblet cell mucus secretion and function. *Mucosal Immunol.* **2015**, *8*, 712–719. [CrossRef]
6. Paone, P.; Cani, P.D. Mucus barrier, mucins and gut microbiota: The expected slimy partners? *Gut* **2020**, *69*, 2232–2243. [CrossRef] [PubMed]
7. Vetrano, S.; Danese, S. The role of JAM-A in inflammatory bowel disease: Unrevealing the ties that bind. *Ann. N. Y. Acad. Sci.* **2009**, *1165*, 308–313. [CrossRef] [PubMed]
8. Sharma, L.; Riva, A. Intestinal barrier function in health and disease—Any role of SARS-CoV-2? *Microorganisms* **2020**, *8*, 1744. [CrossRef]
9. Yang, E.; Shen, J. The roles and functions of Paneth cells in Crohn's disease: A critical review. *Cell Prolif.* **2021**, *54*, e12958. [CrossRef]
10. Nakajima, A.; Vogelzang, A.; Maruya, M.; Miyajima, M.; Murata, M.; Son, A.; Kuwahara, T.; Tsuruyama, T.; Yamada, S.; Matsuura, M.; et al. IgA regulates the composition and metabolic function of gut microbiota by promoting symbiosis between bacteria. *J. Exp. Med.* **2018**, *215*, 2019–2034. [CrossRef] [PubMed]
11. Liang, S.; Wu, X.; Hu, X.; Wang, T.; Jin, F. Recognizing depression from the microbiota-gut-brain axis. *Int. J. Mol. Sci.* **2018**, *19*, 1592. [CrossRef] [PubMed]
12. Zoledziewska, M. The gut microbiota perspective for interventions in MS. *Autoimmun. Rev.* **2019**, *18*, 814–824. [CrossRef]
13. Robinson, K.; Deng, Z.; Hou, Y.; Zhang, G. Regulation of the intestinal barrier function by host defense peptides. *Front. Vet. Sci.* **2015**, *2*, 57. [CrossRef] [PubMed]
14. Tsukita, S.; Furuse, M. Claudin-based barrier in simple and stratified cellular sheets. *Curr. Opin. Cell Biol.* **2002**, *14*, 531–536. [CrossRef]
15. Furuse, M. Molecular basis of the core structure of tight junctions. *Cold Spring Harb. Perspect. Biol.* **2010**, *2*, a002907. [CrossRef] [PubMed]
16. Anderson, J.M.; van Itallie, C.M. Physiology and function of the tight junction. *Cold Spring Harb. Perspect. Biol.* **2009**, *1*, a002584. [CrossRef] [PubMed]
17. Cunningham, S.C.; Kamangar, F.; Kim, M.P.; Hammoud, S.; Haque, R.; Iacobuzio-Donahue, C.A.; Maitra, A.; Ashfaq, R.; Hustinx, S.; Heitmiller, R.E.; et al. Claudin-4, mitogen-activated protein kinase kinase 4, and stratifin are markers of gastric adenocarcinoma precursor lesions. *Cancer Epidemiol. Biomark. Prev.* **2006**, *15*, 281–287. [CrossRef]

18. Jun, K.H.; Kim, J.H.; Jung, J.H.; Choi, H.J.; Chin, H.M. Expression of claudin-7 and loss of claudin-18 correlate with poor prognosis in gastric cancer. *Int. J. Surg.* **2014**, *12*, 156–162. [CrossRef]
19. Soini, Y.; Tommola, S.; Helin, H.; Martikainen, P. Claudins 1, 3, 4 and 5 in gastric carcinoma, loss of claudin expression associates with the diffuse subtype. *Virchows Arch.* **2006**, *448*, 52–58. [CrossRef] [PubMed]
20. Sheehan, G.M.; Kallakury, B.V.S.; Sheehan, C.E.; Fisher, H.A.G.; Kaufman, R.P.; Ross, J.S. Loss of claudins-1 and -7 and expression of claudins-3 and -4 correlate with prognostic variables in prostatic adenocarcinomas. *Hum. Pathol.* **2007**, *38*, 564–569. [CrossRef] [PubMed]
21. Sumida, H. Recent advances in roles of G-protein coupled receptors in intestinal intraepithelial lymphocytes. *Biosci. Microbiota Food Health* **2020**, *39*, 77–82. [CrossRef] [PubMed]
22. Wiarda, J.E.; Trachsel, J.M.; Bond, Z.F.; Byrne, K.A.; Gabler, N.K.; Loving, C.L. Intraepithelial T Cells Diverge by Intestinal Location as Pigs Age. *Front. Immunol.* **2020**, *11*, 1139. [CrossRef] [PubMed]
23. Tabariès, S.; Siegel, P.M. The role of claudins in cancer metastasis. *Oncogene* **2017**, *36*, 1176–1190. [CrossRef] [PubMed]
24. Jia, R.; Liu, W.; Zhao, L.; Cao, L.; Shen, Z. Low doses of individual and combined deoxynivalenol and zearalenone in naturally moldy diets impair intestinal functions via inducing inflammation and disrupting epithelial barrier in the intestine of piglets. *Toxicol. Lett.* **2020**, *333*, 159–169. [CrossRef] [PubMed]
25. Garcia-Hernandez, V.; Quiros, M.; Nusrat, A. Intestinal epithelial claudins: Expression and regulation in homeostasis and inflammation. *Ann. N. Y. Acad. Sci.* **2017**, *1397*, 66–79. [CrossRef]
26. Zeissig, S.; Bürgel, N.; Günzel, D.; Richter, J.; Mankertz, J.; Wahnschaffe, U.; Kroesen, A.J.; Zeitz, M.; Fromm, M.; Schulzke, J.D. Changes in expression and distribution of claudin 2, 5 and 8 lead to discontinuous tight junctions and barrier dysfunction in active Crohn's disease. *Gut* **2007**, *56*, 61–72. [CrossRef]
27. Weber, C.R.; Raleigh, D.R.; Su, L.; Shen, L.; Sullivan, E.A.; Wang, Y.; Turner, J.R. Epithelial myosin light chain kinase activation induces mucosal interleukin-13 expression to alter tight junction ion selectivity. *J. Biol. Chem.* **2010**, *285*, 12037–12046. [CrossRef] [PubMed]
28. Amasheh, S.; Meiri, N.; Gitter, A.H.; Schöneberg, T.; Mankertz, J.; Schulzke, J.D.; Fromm, M. Claudin-2 expression induces cation-selective channels in tight junctions of epithelial cells. *J. Cell Sci.* **2002**, *115*, 4969–4976. [CrossRef] [PubMed]
29. Rosenthal, R.; Milatz, S.; Krug, S.M.; Oelrich, B.; Schulzke, J.D.; Amasheh, S.; Günzel, D.; Fromm, M. Claudin-2, a component of the tight junction, forms a paracellular water channel. *J. Cell Sci.* **2010**, *123*, 1913–1921. [CrossRef] [PubMed]
30. Tsukita, S.; Furuse, M.; Itoh, M. Multifunctional strands in tight junctions. *Nat. Rev. Mol. Cell Biol.* **2001**, *2*, 285–293. [CrossRef] [PubMed]
31. Will, C.; Fromm, M.; Müller, D. Claudin tight junction proteins: Novel aspects in paracellular transport. *Perit. Dial. Int.* **2008**, *28*, 577–584. [CrossRef] [PubMed]
32. Furuse, M.; Hata, M.; Furuse, K.; Yoshida, Y.; Haratake, A.; Sugitani, Y.; Noda, T.; Kubo, A.; Tsukita, S. Claudin-based tight junctions are crucial for the mammalian epidermal barrier: A lesson from claudin-1-deficient mice. *J. Cell Biol.* **2002**, *156*, 1099–1111. [CrossRef]
33. Sasaki, H.; Matsui, C.; Furuse, K.; Mimori-Kiyosue, Y.; Furuse, M.; Tsukita, S. Dynamic behavior of paired claudin strands within apposing plasma membranes. *Proc. Natl. Acad. Sci. USA* **2003**, *100*, 3971–3976. [CrossRef] [PubMed]
34. Holmes, J.L.; van Itallie, C.M.; Rasmussen, J.E.; Anderson, J.M. Claudin profiling in the mouse during postnatal intestinal development and along the gastrointestinal tract reveals complex expression patterns. *Gene Expr. Patterns* **2006**, *6*, 581–588. [CrossRef] [PubMed]
35. Troy, T.C.; Arabzadeh, A.; Yerlikaya, S.; Turksen, K. Claudin immunolocalization in neonatal mouse epithelial tissues. *Cell Tissue Res.* **2007**, *330*, 381–388. [CrossRef]
36. Michlig, S.; Damak, S.; le Coutre, J. Claudin-based permeability barriers in taste buds. *J. Comp. Neurol.* **2007**, *502*, 1003–1011. [CrossRef] [PubMed]
37. Matsuda, Y.; Semba, S.; Ueda, J.; Fuku, T.; Hasuo, T.; Chiba, H.; Sawada, N.; Kuroda, Y.; Yokozaki, H. Gastric and intestinal claudin expression at the invasive front of gastric carcinoma. *Cancer Sci.* **2007**, *98*, 1014–1019. [CrossRef] [PubMed]
38. Lameris, A.L.; Huybers, S.; Kaukinen, K.; Mäkelä, T.H.; Bindels, R.J.; Hoenderop, J.G.; Nevalainen, P.I. Expression profiling of claudins in the human gastrointestinal tract in health and during inflammatory bowel disease. *Scand. J. Gastroenterol.* **2013**, *48*, 58–69. [CrossRef]
39. Markov, A.G.; Veshnyakova, A.; Fromm, M.; Amasheh, M.; Amasheh, S. Segmental expression of claudin proteins correlates with tight junction barrier properties in rat intestine. *J. Comp. Physiol. B Biochem. Syst. Environ. Physiol.* **2010**, *180*, 591–598. [CrossRef]
40. Fujita, H.; Chiba, H.; Yokozaki, H.; Sakai, N.; Sugimoto, K.; Wada, T.; Kojima, T.; Yamashita, T.; Sawada, N. Differential expression and subcellular localization of claudin-7, -8, -12, -13, and -15 along the mouse intestine. *J. Histochem. Cytochem.* **2006**, *54*, 933–944. [CrossRef] [PubMed]
41. Saitoh, M.; Kurashige, Y.; Nishimura, M.; Yamazaki, M.; Igarashi, S.; Kaku, T.; Abiko, Y. Expression of claudin-4 and -7 in porcine gingival junctional epithelium. *Med. Mol. Morphol.* **2009**, *42*, 212–215. [CrossRef] [PubMed]
42. Lioni, M.; Brafford, P.; Andl, C.; Rustgi, A.; El-Deiry, W.; Herlyn, M.; Smalley, K.S.M. Dysregulation of claudin-7 leads to loss of E-cadherin expression and the increased invasion of esophageal squamous cell carcinoma cells. *Am. J. Pathol.* **2007**, *170*, 709–721. [CrossRef]

43. Pinto-Sánchez, M.I.; Nachman, F.D.; Fuxman, C.; Iantorno, G.; Hwang, H.J.; Ditaranto, A.; Costa, F.; Longarini, G.; Wang, X.Y.; Huang, X.; et al. Altered Esophageal Mucosal Structure in Patients with Celiac Disease. *Can. J. Gastroenterol. Hepatol.* **2016**, *2016*, 1980686. [CrossRef]
44. Lu, Y.; Jing, J.; Sun, L.; Gong, Y.; Chen, M.; Wang, Z.; Sun, M.; Yuan, Y. Expression of claudin-11, -23 in different gastric tissues and its relationship with the risk and prognosis of gastric cancer. *PLoS ONE* **2017**, *12*, e0174476. [CrossRef] [PubMed]
45. Gao, M.; Li, W.; Wang, H.; Wang, G. The distinct expression patterns of claudin-10, -14, -17 and E-cadherin between adjacent non-neoplastic tissues and gastric cancer tissues. *Diagn. Pathol.* **2013**, *8*, 205. [CrossRef]
46. Lee, Y.J.; Hussain, Z.; Huh, C.W.; Lee, Y.J.; Park, H. Inflammation, impaired motility, and permeability in a Guinea Pig model of postoperative ileus. *J. Neurogastroenterol. Motil.* **2018**, *24*, 147–158. [CrossRef] [PubMed]
47. Wu, X.; Gao, L.M.; Liu, Y.L.; Xie, C.; Cai, L.; Xu, K.; Zhou, X.H. Maternal dietary uridine supplementation reduces diarrhea incidence in piglets by regulating the intestinal mucosal barrier and cytokine profiles. *J. Sci. Food Agric.* **2020**, *100*, 3709–3718. [CrossRef] [PubMed]
48. Deluco, B.; Fourie, K.R.; Simko, O.M.; Wilson, H.L. Localization of Claudin-3 and Claudin-4 within the Small Intestine of newborn piglets. *Physiol. Rep.* **2021**, *9*, e14717. [CrossRef]
49. Zong, Q.F.; Huang, Y.J.; Wu, L.S.; Wu, Z.C.; Wu, S.L.; Bao, W.B. Effects of porcine epidemic diarrhea virus infection on tight junction protein gene expression and morphology of the intestinal mucosa in pigs. *Pol. J. Vet. Sci.* **2019**, *22*, 345–353. [CrossRef]
50. Martínez, C.; Rodinõ-Janeiro, B.K.; Lobo, B.; Stanifer, M.L.; Klaus, B.; Granzow, M.; González-Castro, A.M.; Salvo-Romero, E.; Alonso-Cotoner, C.; Pigrau, M.; et al. MiR-16 and miR-125b are involved in barrier function dysregulation through the modulation of claudin-2 and cingulin expression in the jejunum in IBS with diarrhoea. *Gut* **2017**, *66*, 1597–1610. [CrossRef] [PubMed]
51. He, Y.; Sang, Z.; Zhuo, Y.; Wang, X.; Guo, Z.; He, L.; Zeng, C.; Dai, H. Transport stress induces pig jejunum tissue oxidative damage and results in autophagy/mitophagy activation. *J. Anim. Physiol. Anim. Nutr.* **2019**, *103*, 1521–1529. [CrossRef]
52. Yong, Y.; Li, J.; Gong, D.; Yu, T.; Wu, L.; Hu, C.; Liu, X.; Yu, Z.; Ma, X.; Gooneratne, R.; et al. ERK1/2 mitogen-activated protein kinase mediates downregulation of intestinal tight junction proteins in heat stress-induced IBD model in pig. *J. Therm. Biol.* **2021**, *101*, 103103. [CrossRef]
53. Zeisel, M.B.; Dhawan, P.; Baumert, T.F. Tight junction proteins in gastrointestinal and liver disease. *Gut* **2019**, *68*, 547–561. [CrossRef] [PubMed]
54. Warner, G.C.; Reis, P.P.; Jurisica, I.; Sultan, M.; Arora, S.; Macmillan, C.; Makitie, A.A.; Grénman, R.; Reid, N.; Sukhai, M.; et al. Molecular classification of oral cancer by cDNA microarrays identifies overexpressed genes correlated with nodal metastasis. *Int. J. Cancer* **2004**, *110*, 857–868. [CrossRef]
55. Sappayatosok, K.; Phattarataratip, E. Overexpression of Claudin-1 is Associated with Advanced Clinical Stage and Invasive Pathologic Characteristics of Oral Squamous Cell Carcinoma. *Head Neck Pathol.* **2015**, *9*, 173–180. [CrossRef] [PubMed]
56. Ouban, A.; Ahmed, A. Analysis of the Distribution and Expression of Claudin-1 Tight Junction Protein in the Oral Cavity. *Appl. Immunohistochem. Mol. Morphol.* **2015**, *23*, 444–448. [CrossRef]
57. Upadhaya, P.; Barhoi, D.; Giri, A.; Bhattacharjee, A.; Giri, S. Joint detection of claudin-1 and junctional adhesion molecule-A as a therapeutic target in oral epithelial dysplasia and oral squamous cell carcinoma. *J. Cell. Biochem.* **2019**, *120*, 18117–18127. [CrossRef] [PubMed]
58. Dos Reis, P.P.; Bharadwaj, R.R.; Machado, J.; MacMillan, C.; Pintilie, M.; Sukhai, M.A.; Perez-Ordonez, B.; Gullane, P.; Irish, J.; Kamel-Reid, S.; et al. Claudin 1 overexpression increases invasion and is associated with aggressive histological features in oral squamous cell carcinoma. *Cancer* **2008**, *113*, 3169–3180. [CrossRef]
59. Babkair, H.; Yamazaki, M.; Uddin, M.S.; Maruyama, S.; Abé, T.; Essa, A.; Sumita, Y.; Ahsan, M.S.; Swelam, W.; Cheng, J.; et al. Aberrant expression of the tight junction molecules claudin-1 and zonula occludens-1 mediates cell growth and invasion in oral squamous cell carcinoma. *Hum. Pathol.* **2016**, *57*, 51–60. [CrossRef]
60. Lourenço, S.V.; Coutinho-Camillo, C.M.; Buim, M.E.C.; de Carvalho, A.C.; Lessa, R.C.; Pereira, C.M.; Vettore, A.L.; Carvalho, A.L.; Fregnani, J.H.; Kowalski, L.P.; et al. Claudin-7 down-regulation is an important feature in oral squamous cell carcinoma. *Histopathology* **2010**, *57*, 689–698. [CrossRef] [PubMed]
61. Yoshizawa, K.; Nozaki, S.; Kato, A.; Hirai, M.; Yanase, M.; Yoshimoto, T.; Kimura, I.; Sugiura, S.; Okamune, A.; Kitahara, H.; et al. Loss of claudin-7 is a negative prognostic factor for invasion and metastasis in oral squamous cell carcinoma. *Oncol. Rep.* **2013**, *29*, 445–450. [CrossRef]
62. Wu, J.; Gao, F.X.; Xu, T.; Li, J.; Hu, Z.; Wang, C.; Long, Y.; He, X.M.; Deng, X.; Ren, D.L.; et al. CLDN1 induces autophagy to promote proliferation and metastasis of esophageal squamous carcinoma through AMPK/STAT1/ULK1 signaling. *J. Cell. Physiol.* **2020**, *235*, 2245–2259. [CrossRef] [PubMed]
63. Shi, M.; Wang, Z.; Song, L.; Wang, D.; Sun, Z. Low expression of claudin-4: An indicator of recurrence in esophageal squamous cell carcinoma after Ivor Lewis esophagectomy? *Med. Oncol.* **2014**, *31*, 951. [CrossRef] [PubMed]
64. Usami, Y.; Chiba, H.; Nakayama, F.; Ueda, J.; Matsuda, Y.; Sawada, N.; Komori, T.; Ito, A.; Yokozaki, H. Reduced expression of claudin-7 correlates with invasion and metastasis in squamous cell carcinoma of the esophagus. *Hum. Pathol.* **2006**, *37*, 569–577. [CrossRef] [PubMed]
65. Higashi, Y.; Suzuki, S.; Sakaguchi, T.; Nakamura, T.; Baba, S.; Reinecker, H.C.; Nakamura, S.; Konno, H. Loss of Claudin-1 Expression Correlates with Malignancy of Hepatocellular Carcinoma. *J. Surg. Res.* **2007**, *139*, 68–76. [CrossRef] [PubMed]

66. Jiang, L.; Yang, Y.D.; Fu, L.; Xu, W.; Liu, D.; Liang, Q.; Zhang, X.; Xu, L.; Guan, X.Y.; Wu, B.; et al. CLDN3 inhibits cancer aggressiveness via Wnt-EMT signaling and is a potential prognostic biomarker for hepatocellular carcinoma. *Oncotarget* **2014**, *5*, 7663–7676. [CrossRef] [PubMed]
67. Huang, G.W.; Ding, X.; Chen, S.L.; Zeng, L. Expression of claudin 10 protein in hepatocellular carcinoma: Impact on survival. *J. Cancer Res. Clin. Oncol.* **2011**, *137*, 1213–1218. [CrossRef] [PubMed]
68. Eftang, L.L.; Esbensen, Y.; Tannæs, T.M.; Blom, G.P.; Bukholm, I.R.K.; Bukholm, G. Up-regulation of CLDN1 in gastric cancer is correlated with reduced survival. *BMC Cancer* **2013**, *13*, 586. [CrossRef]
69. Shiozaki, A.; Shimizu, H.; Ichikawa, D.; Konishi, H.; Komatsu, S.; Kubota, T.; Fujiwara, H.; Okamoto, K.; Iitaka, D.; Nakashima, S.; et al. Claudin 1 mediates tumor necrosis factor alpha-induced cell migration in human gastric cancer cells. *World J. Gastroenterol.* **2014**, *20*, 17863–17876. [CrossRef] [PubMed]
70. Huang, J.; Zhang, L.; He, C.; Qu, Y.; Li, J.; Zhang, J.; Du, T.; Chen, X.; Yu, Y.; Liu, B.; et al. Claudin-1 enhances tumor proliferation and metastasis by regulating cell anoikis in gastric cancer. *Oncotarget* **2015**, *6*, 1652–1665. [CrossRef]
71. Hwang, T.L.; Lee, L.Y.; Wang, C.C.; Liang, Y.; Huang, S.F.; Wu, C.M. Claudin-4 expression is associated with tumor invasion, MMP-2 and MMP-9 expression in gastric cancer. *Exp. Ther. Med.* **2010**, *1*, 789–797. [CrossRef]
72. Ohtani, S.; Terashima, M.; Satoh, J.; Soeta, N.; Saze, Z.; Kashimura, S.; Ohsuka, F.; Hoshino, Y.; Kogure, M.; Gotoh, M. Expression of tight-junction-associated proteins in human gastric cancer: Downregulation of claudin-4 correlates with tumor aggressiveness and survival. *Gastric Cancer* **2009**, *12*, 43–51. [CrossRef] [PubMed]
73. Luo, J.; Wang, H.; Chen, H.; Gan, G.; Zheng, Y. CLDN4 silencing promotes proliferation and reduces chemotherapy sensitivity of gastric cancer cells through activation of the PI3K/Akt signalling pathway. *Exp. Physiol.* **2020**, *105*, 979–988. [CrossRef]
74. Yu, S.; Zhang, Y.; Li, Q.; Zhang, Z.; Zhao, G.; Xu, J. CLDN6 promotes tumor progression through the YAP1-snail1 axis in gastric cancer. *Cell Death Dis.* **2019**, *10*, 949. [CrossRef]
75. Zavala-Zendejas, V.E.; Torres-Martinez, A.C.; Salas-Morales, B.; Fortoul, T.I.; Montaño, L.F.; Rendon-Huerta, E.P. Claudin-6, 7, or 9 overexpression in the human gastric adenocarcinoma cell line AGS increases its invasiveness, migration, and proliferation rate. *Cancer Investig.* **2011**, *29*, 1–11. [CrossRef] [PubMed]
76. Wu, Z.; Shi, J.; Song, Y.; Zhao, J.; Sun, J.; Chen, X.; Gao, P.; Wang, Z. Claudin-7 (CLDN7) is overexpressed in gastric cancer and promotes gastric cancer cell proliferation, invasion and maintains mesenchymal state. *Neoplasma* **2018**, *65*, 349–359. [CrossRef] [PubMed]
77. Agarwal, R.; Mori, Y.; Cheng, Y.; Jin, Z.; Olaru, A.V.; Hamilton, J.P.; David, S.; Selaru, F.M.; Yang, J.; Abraham, J.M.; et al. Silencing of claudin-11 is associated with increased invasiveness of gastric cancer cells. *PLoS ONE* **2009**, *4*, e8002. [CrossRef] [PubMed]
78. Dhawan, P.; Singh, A.B.; Deane, N.G.; No, Y.R.; Shiou, S.R.; Schmidt, C.; Neff, J.; Washington, M.K.; Beauchamp, R.D. Claudin-1 regulates cellular transformation and metastatic behavior in colon cancer. *J. Clin. Investig.* **2005**, *115*, 1765–1776. [CrossRef] [PubMed]
79. Singh, A.B.; Sharma, A.; Smith, J.J.; Krishnan, M.; Chen, X.; Eschrich, S.; Washington, M.K.; Yeatman, T.J.; Beauchamp, R.D.; Dhawan, P. Claudin-1 up-regulates the repressor ZEB-1 to inhibit E-cadherin expression in colon cancer cells. *Gastroenterology* **2011**, *141*, 2140–2153. [CrossRef] [PubMed]
80. Dhawan, P.; Ahmad, R.; Chaturvedi, R.; Smith, J.J.; Midha, R.; Mittal, M.K.; Krishnan, M.; Chen, X.; Eschrich, S.; Yeatman, T.J.; et al. Claudin-2 expression increases tumorigenicity of colon cancer cells: Role of epidermal growth factor receptor activation. *Oncogene* **2011**, *30*, 3234–3247. [CrossRef]
81. Ahmad, R.; Kumar, B.; Chen, Z.; Chen, X.; Müller, D.; Lele, S.M.; Washington, M.K.; Batra, S.K.; Dhawan, P.; Singh, A.B. Loss of claudin-3 expression induces IL6/gp130/Stat3 signaling to promote colon cancer malignancy by hyperactivating Wnt/β-catenin signaling. *Oncogene* **2017**, *36*, 6592–6604. [CrossRef] [PubMed]
82. Bhat, A.A.; Pope, J.L.; Smith, J.J.; Ahmad, R.; Chen, X.; Washington, M.K.; Beauchamp, R.D.; Singh, A.B.; Dhawan, P. Claudin-7 expression induces mesenchymal to epithelial transformation (MET) to inhibit colon tumorigenesis. *Oncogene* **2015**, *34*, 4570–4580. [CrossRef] [PubMed]
83. Wang, K.; Li, T.; Xu, C.; Ding, Y.; Li, W.; Ding, L. Claudin-7 downregulation induces metastasis and invasion in colorectal cancer via the promotion of epithelial-mesenchymal transition. *Biochem. Biophys. Res. Commun.* **2019**, *508*, 797–804. [CrossRef] [PubMed]
84. Wong, T.S.C.; Wiesenfeld, D. Oral Cancer. *Aust. Dent. J.* **2018**, *63*, S91–S99. [CrossRef]
85. Oku, N.; Sasabe, E.; Ueta, E.; Yamamoto, T.; Osaki, T. Tight junction protein claudin-1 enhances the invasive activity of oral squamous cell carcinoma cells by promoting cleavage of laminin-5 $\gamma 2$ chain via matrix metalloproteinase (MMP)-2 and membrane-type MMP-1. *Cancer Res.* **2006**, *66*, 5251–5257. [CrossRef] [PubMed]
86. Melchers, L.J.; Bruine De Bruin, L.; Schnell, U.; Slagter-Menkema, L.; Mastik, M.F.; de Bock, G.H.; van Dijk, B.A.C.; Giepmans, B.N.G.; van der Laan, B.F.A.M.; van der Wal, J.E.; et al. Lack of claudin-7 is a strong predictor of regional recurrence in oral and oropharyngeal squamous cell carcinoma. *Oral Oncol.* **2013**, *49*, 998–1005. [CrossRef]
87. Smyth, E.C.; Lagergren, J.; Fitzgerald, R.C.; Lordick, F.; Shah, M.A.; Lagergren, P.; Cunningham, D. Oesophageal cancer. *Nat. Rev. Dis. Prim.* **2017**, *3*, 17048. [CrossRef] [PubMed]
88. Miyamoto, K.; Kusumi, T.; Sato, F.; Kawasaki, H.; Shibata, S.; Ohashi, M.; Hakamada, K.; Sasaki, M.; Kijima, H. Decreased expression of claudin-1 is correlated with recurrence status in Esophageal squamous cell carcinoma. *Biomed. Res.* **2008**, *29*, 71–76. [CrossRef] [PubMed]

89. Abu-Farsakh, S.; Wu, T.; Lalonde, A.; Sun, J.; Zhou, Z. High expression of Claudin-2 in esophageal carcinoma and precancerous lesions is significantly associated with the bile salt receptors VDR and TGR5. *BMC Gastroenterol.* **2017**, *17*, 33. [CrossRef]
90. Sung, C.O.; Han, S.Y.; Kim, S.H. Low expression of claudin-4 is associated with poor prognosis in esophageal squamous cell carcinoma. *Ann. Surg. Oncol.* **2011**, *18*, 273–281. [CrossRef]
91. Lee, K.W.; Lee, N.K.; Kim, J.H.; Kang, M.S.; Yoo, H.Y.; Kim, H.H.; Um, S.H.; Kim, S.H. Twist1 causes the transcriptional repression of claudin-4 with prognostic significance in esophageal cancer. *Biochem. Biophys. Res. Commun.* **2012**, *423*, 454–460. [CrossRef] [PubMed]
92. Bouchagier, K.A.; Assimakopoulos, S.F.; Karavias, D.D.; Maroulis, I.; Tzelepi, V.; Kalofonos, H.; Karavias, D.D.; Kardamakis, D.; Scopa, C.D.; Tsamandas, A.C. Expression of claudins-1, -4, -5, -7 and occludin in hepatocellular carcinoma and their relation with classic clinicopathological features and patients' survival. *In Vivo* **2014**, *28*, 315–326. [PubMed]
93. Suh, Y.; Yoon, C.H.; Kim, R.K.; Lim, E.J.; Oh, Y.S.; Hwang, S.G.; An, S.; Yoon, G.; Gye, M.C.; Yi, J.M.; et al. Claudin-1 induces epithelial-mesenchymal transition through activation of the c-Abl-ERK signaling pathway in human liver cells. *Oncogene* **2013**, *32*, 4873–4882. [CrossRef] [PubMed]
94. Kinugasa, T.; Akagi, Y.; Ochi, T.; Tanaka, N.; Kawahara, A.; Ishibashi, Y.; Gotanda, Y.; Yamaguchi, K.; Shiratuchi, I.; Oka, Y.; et al. Increased claudin-1 protein expression in hepatic metastatic lesions of colorectal cancer. *Anticancer Res.* **2012**, *32*, 2309–2314. [PubMed]
95. Tong, H.; Li, T.; Qiu, W.; Zhu, Z. Claudin-1 silencing increases sensitivity of liver cancer HepG2 cells to 5-fluorouracil by inhibiting autophagy. *Oncol. Lett.* **2019**, *18*, 5709–5716. [CrossRef] [PubMed]
96. Brokalaki, E.I.; Weber, F.; Sotiropoulos, G.C.; Daoudaki, M.; Cicinnati, V.R.; Beckebaum, S. Claudin-7 expression in hepatocellular carcinoma. In *Transplantation Proceedings*; Elsevier: Amsterdam, The Netherlands, 2012; Volume 44, pp. 2737–2740.
97. Ono, Y.; Hiratsuka, Y.; Murata, M.; Takasawa, A.; Fukuda, R.; Nojima, M.; Tanaka, S.; Osanai, M.; Hirata, K.; Sawada, N. Claudins-4 and -7 might be valuable markers to distinguish hepatocellular carcinoma from cholangiocarcinoma. *Virchows Arch.* **2016**, *469*, 417–426. [CrossRef] [PubMed]
98. Jakab, C.; Kiss, A.; Schaff, Z.; Szabó, Z.; Rusvai, M.; Gálfi, P.; Szabára, A.; Sterczer, Á.; Kulka, J. Claudin-7 protein differentiates canine cholangiocarcinoma from hepatocellular carcinoma. *Histol. Histopathol.* **2010**, *25*, 857–864. [CrossRef]
99. Lódi, C.; Szabó, E.; Holczbauer, A.; Batmunkh, E.; Szíjártó, A.; Kupcsulik, P.; Kovalszky, I.; Paku, S.; Illyés, G.; Kiss, A.; et al. Claudin-4 differentiates biliary tract cancers from hepatocellular carcinomas. *Mod. Pathol.* **2006**, *19*, 460–469. [CrossRef] [PubMed]
100. Ying, C.I.; Siu, T.C.; Yuk, T.L.; Ho, J.C.; Sheung, T.F. Inhibition of hepatocellular carcinoma invasion by suppression of claudin-10 in HLE cells. *Mol. Cancer Ther.* **2007**, *6*, 2858–2867. [CrossRef]
101. Ferlay, J.; Shin, H.R.; Bray, F.; Forman, D.; Mathers, C.; Parkin, D.M. Estimates of worldwide burden of cancer in 2008: GLOBOCAN 2008. *Int. J. Cancer* **2010**, *127*, 2893–2917. [CrossRef] [PubMed]
102. Karimi, P.; Islami, F.; Anandasabapathy, S.; Freedman, N.D.; Kamangar, F. Gastric cancer: Descriptive epidemiology, risk factors, screening, and prevention. *Cancer Epidemiol. Biomark. Prev.* **2014**, *23*, 700–713. [CrossRef] [PubMed]
103. Huang, J.; Li, J.; Qu, Y.; Zhang, J.; Zhang, L.; Chen, X.; Liu, B.; Zhu, Z. The expression of Claudin 1 correlates with β-catenin and is a prognostic factor of poor outcome in gastric cancer. *Int. J. Oncol.* **2014**, *44*, 1293–1301. [CrossRef]
104. Lin, Z.; Zhang, X.; Liu, Z.; Liu, Q.; Wang, L.; Lu, Y.; Liu, Y.; Wang, M.; Yang, M.; Jin, X.; et al. The distinct expression patterns of claudin-2, -6, and -11 between human gastric neoplasms and adjacent non-neoplastic tissues. *Diagn. Pathol.* **2013**, *8*, 133. [CrossRef]
105. Yang, L.; Sun, X.; Meng, X. Differences in the expression profiles of claudin proteins in human gastric carcinoma compared with non-neoplastic mucosa. *Mol. Med. Rep.* **2018**, *18*, 1271–1278. [CrossRef] [PubMed]
106. Hwang, T.L.; Changchien, T.T.; Wang, C.C.; Wu, C.M. Claudin-4 expression in gastric cancer cells enhances the invasion and is associated with the increased level of matrix metalloproteinase-2 and -9 expression. *Oncol. Lett.* **2014**, *8*, 1367–1371. [CrossRef] [PubMed]
107. Kohmoto, T.; Masuda, K.; Shoda, K.; Takahashi, R.; Ujiro, S.; Tange, S.; Ichikawa, D.; Otsuji, E.; Imoto, I. Claudin-6 is a single prognostic marker and functions as a tumor-promoting gene in a subgroup of intestinal type gastric cancer. *Gastric Cancer* **2020**, *23*, 403–417. [CrossRef] [PubMed]
108. Torres-Martínez, A.C.; Gallardo-Vera, J.F.; Lara-Holguin, A.N.; Montaño, L.F.; Rendón-Huerta, E.P. Claudin-6 enhances cell invasiveness through claudin-1 in AGS human adenocarcinoma gastric cancer cells. *Exp. Cell Res.* **2017**, *350*, 226–235. [CrossRef]
109. Matsuda, M.; Sentani, K.; Noguchi, T.; Hinoi, T.; Okajima, M.; Matsusaki, K.; Sakamoto, N.; Anami, K.; Naito, Y.; Oue, N.; et al. Immunohistochemical analysis of colorectal cancer with gastric phenotype: Claudin-18 is associated with poor prognosis. *Pathol. Int.* **2010**, *60*, 673–680. [CrossRef]
110. Lu, Y.; Wu, T.; Sheng, Y.; Dai, Y.; Xia, B.; Xue, Y. Correlation between Claudin-18 expression and clinicopathological features and prognosis in patients with gastric cancer. *J. Gastrointest. Oncol.* **2020**, *11*, 1253–1260. [CrossRef]
111. Li, J.; Zhang, Y.; Hu, D.; Gong, T.; Xu, R.; Gao, J. Analysis of the expression and genetic alteration of CLDN18 in gastric cancer. *Aging* **2020**, *12*, 14271–14284. [CrossRef] [PubMed]
112. Office for National Statistics. *Public Health England—National Cancer Registration and Analysis Service*; Office for National Statistics: London, UK, 2017.
113. Siegel, R.L.; Miller, K.D.; Jemal, A. Cancer statistics, 2019. *CA Cancer J. Clin.* **2019**, *69*, 7–34. [CrossRef] [PubMed]

114. Miwa, N.; Furuse, M.; Tsukita, S.; Niikawa, N.; Nakamura, Y.; Furukawa, Y. Involvement of claudin-1 in the β-catenin/Tcf signaling pathway and its frequent upregulation in human colorectal cancers. *Oncol. Res.* **2000**, *12*, 469–476. [CrossRef] [PubMed]
115. Qun, H.; Kinugasa, T.; Lin, W.; Huang, J.; Jun, Z.; Shibaguchi, H.; Kuroki, M.; Tanaka, T.; Yamashita, Y.; Nabeshima, K.; et al. Claudin-1 protein is a major factor involved in the tumorigenesis of colorectal cancer. *Anticancer Res.* **2009**, *29*, 851–858.
116. Pope, J.L.; Bhat, A.A.; Sharma, A.; Ahmad, R.; Krishnan, M.; Washington, M.K.; Beauchamp, R.D.; Singh, A.B.; Dhawan, P. Claudin-1 regulates intestinal epithelial homeostasis through the modulation of Notch-signalling. *Gut* **2014**, *63*, 622–634. [CrossRef] [PubMed]
117. Kim, H.; Kim, S.H.; Hwang, D.; An, J.; Chung, H.S.; Yang, E.G.; Kim, S.Y. Extracellular pyruvate kinase M2 facilitates cell migration by upregulating claudin-1 expression in colon cancer cells. *Biochem. Cell Biol.* **2020**, *98*, 219–226. [CrossRef] [PubMed]
118. Hollandsworth, H.M.; Lwin, T.M.; Amirfakhri, S.; Filemoni, F.; Batra, S.K.; Hoffman, R.M.; Dhawan, P.; Bouvet, M. Anti-Claudin-1 Conjugated to a Near-Infrared Fluorophore Targets Colon Cancer in PDOX Mouse Models. *J. Surg. Res.* **2019**, *242*, 145–150. [CrossRef] [PubMed]
119. Wang, F.; Duan, X.; Chen, J.; Gao, Z.; Zhou, J.; Wu, X.; Chang, T.S.; Lee, M.; Li, G.; Nusrat, A.; et al. Integrated Imaging Methodology Detects Claudin-1 Expression in Premalignant Nonpolypoid and Polypoid Colonic Epithelium in Mice. *Clin. Transl. Gastroenterol.* **2020**, *11*, e00089. [CrossRef]
120. Resnick, M.B.; Konkin, T.; Routhier, J.; Sabo, E.; Pricolo, V.E. Claudin-1 is a strong prognostic indicator in stage II colonic cancer: A tissue microarray study. *Mod. Pathol.* **2005**, *18*, 511–518. [CrossRef] [PubMed]
121. Oshima, T.; Kunisaki, C.; Yoshihara, K.; Yamada, R.; Yamamoto, N.; Sato, T.; Makino, H.; Yamagishi, S.; Nagano, Y.; Fujii, S.; et al. Reduced expression of the claudin-7 gene correlates with venous invasion and liver metastasis in colorectal cancer. *Oncol. Rep.* **2008**, *19*, 953–959. [CrossRef] [PubMed]
122. Bryden, W.L. Mycotoxin contamination of the feed supply chain: Implications for animal productivity and feed security. *Anim. Feed Sci. Technol.* **2012**, *173*, 134–158. [CrossRef]
123. Rogowska, A.; Pomastowski, P.; Sagandykova, G.; Buszewski, B. Zearalenone and its metabolites: Effect on human health, metabolism and neutralisation methods. *Toxicon* **2019**, *162*, 46–56. [CrossRef] [PubMed]
124. Gruber-Dorninger, C.; Jenkins, T.; Schatzmayr, G. Global Mycotoxin Occurrence in Feed. *Toxins* **2019**, *11*, 375. [CrossRef]
125. World Health Organization. Mycotoxins. Available online: https://www.who.int/news-room/fact-sheets/detail/mycotoxins (accessed on 13 July 2021).
126. Massart, F.; Saggese, G. Oestrogenic mycotoxin exposures and precocious pubertal development. *Int. J. Androl.* **2010**, *33*, 369–376. [CrossRef] [PubMed]
127. Massart, F.; Meucci, V.; Saggese, G.; Soldani, G. High Growth Rate of Girls with Precocious Puberty Exposed to Estrogenic Mycotoxins. *J. Pediatr.* **2008**, *152*, 690–695. [CrossRef] [PubMed]
128. Liu, Y.; Wu, F. Global burden of Aflatoxin-induced hepatocellular carcinoma: A risk assessment. Environ. *Health Perspect.* **2010**, *118*, 818–824. [CrossRef]
129. Turner, P.C.; Moore, S.E.; Hall, A.J.; Prentice, A.M.; Wild, C.P. Modification of immune function through exposure to dietary aflatoxin in Gambian children. *Environ. Health Perspect.* **2003**, *111*, 217–220. [CrossRef] [PubMed]
130. Radovanovic, Z.; Jankovic, S.; Jevremovic, I. Incidence of tumors of urinary organs in a focus of Balkan endemic nephropathy. *Kidney Int.* **1991**, *40*, S75–S76.
131. Gill, S.; Kumara, V.M.R. Detecting neurodevelopmental toxicity of domoic acid and ochratoxin a using rat fetal neural stem cells. *Mar. Drugs* **2019**, *17*, 566. [CrossRef] [PubMed]
132. De Santis, B.; Brera, C.; Mezzelani, A.; Soricelli, S.; Ciceri, F.; Moretti, G.; Debegnach, F.; Bonaglia, M.C.; Villa, L.; Molteni, M.; et al. Role of mycotoxins in the pathobiology of autism: A first evidence. *Nutr. Neurosci.* **2019**, *22*, 132–144. [CrossRef]
133. De Santis, B.; Raggi, M.E.; Moretti, G.; Facchiano, F.; Mezzelani, A.; Villa, L.; Bonfanti, A.; Campioni, A.; Rossi, S.; Camposeo, S.; et al. Study on the association among mycotoxins and other variables in children with autism. *Toxins* **2017**, *9*, 203. [CrossRef] [PubMed]
134. Zuberi, Z.; Eeza, M.N.H.; Matysik, J.; Berry, J.P.; Alia, A. NMR-based metabolic profiles of intact zebrafish embryos exposed to aflatoxin b1 recapitulates hepatotoxicity and supports possible neurotoxicity. *Toxins* **2019**, *11*, 258. [CrossRef] [PubMed]
135. Wu, T.S.; Cheng, Y.C.; Chen, P.J.; Huang, Y.T.; Yu, F.Y.; Liu, B.H. Exposure to aflatoxin B1 interferes with locomotion and neural development in zebrafish embryos and larvae. *Chemosphere* **2019**, *217*, 905–913. [CrossRef] [PubMed]
136. Muthulakshmi, S.; Maharajan, K.; Habibi, H.R.; Kadirvelu, K.; Venkataramana, M. Zearalenone induced embryo and neurotoxicity in zebrafish model (Danio rerio): Role of oxidative stress revealed by a multi biomarker study. *Chemosphere* **2018**, *198*, 111–121. [CrossRef]
137. Kozieł, M.J.; Kowalska, K.; Piastowska-Ciesielska, A.W. Nrf2: A main responsive element in cells to mycotoxin-induced toxicity. *Arch. Toxicol.* **2021**, *95*, 1521–1533. [CrossRef] [PubMed]
138. Knutsen, H.K.; Barregård, L.; Bignami, M.; Brüschweiler, B.; Ceccatelli, S.; Cottrill, B.; Dinovi, M.; Edler, L.; Grasl-Kraupp, B.; Hogstrand, C.; et al. Appropriateness to set a group health-based guidance value for fumonisins and their modified forms. *EFSA J.* **2018**, *16*, e05172. [CrossRef] [PubMed]
139. Scientific opinion on the risks for public health related to the presence of zearalenone in food. *EFSA J.* **2011**, *9*, 2197. [CrossRef]
140. Food, E.; Authority, S. Deoxynivalenol in food and feed: Occurrence and exposure. *EFSA J.* **2013**, *11*, 3379. [CrossRef]

141. Yang, X.; Liu, L.; Chen, J.; Xiao, A. Response of intestinal bacterial flora to the long-term feeding of aflatoxin B1 (AFB1) in mice. *Toxins* **2017**, *9*, 317. [CrossRef] [PubMed]
142. Kew, M.C. Aflatoxins as a cause of hepatocellular carcinoma. *J. Gastrointest. Liver Dis.* **2013**, *22*, 305–310.
143. WHO. *IARC Monographs on the Evaluation of Carcinogenic Risks to Humans: 1,3-Butadiene, Ethylene Oxide and Vinyl Halides (Vinyl Fluoride, Vinyl Chloride and Vinyl Bromide)*; IARC Publications: Lyon, France, 2008; Volume 97, pp. 3–471.
144. Kolenda, M.; Mroczkowski, S. Fusarium mycotoxins and methods of assessing the mycotoxicity: A review. *J. Cent. Eur. Agric.* **2013**, *14*, 169–180. [CrossRef]
145. Eaton, D.L.; Gallagher, E.P. Mechanisms of aflatoxin carcinogenesis. *Annu. Rev. Pharmacol. Toxicol.* **1994**, *34*, 135–172. [CrossRef]
146. Bennett, J.W.; Klich, M. Mycotoxins. *Clin. Microbiol. Rev.* **2003**, *16*, 497–516. [CrossRef] [PubMed]
147. Pu, J.; Yuan, Q.; Yan, H.; Tian, G.; Chen, D.; He, J.; Zheng, P.; Yu, J.; Mao, X.; Huang, Z.; et al. Effects of chronic exposure to low levels of dietary aflatoxin b1 on growth performance, apparent total tract digestibility and intestinal health in pigs. *Animals* **2021**, *11*, 336. [CrossRef]
148. Breves, G.; Kock, J.; Schröder, B. Transport of nutrients and electrolytes across the intestinal wall in pigs. *Livest. Sci.* **2007**, *109*, 4–13. [CrossRef]
149. Röder, P.V.; Geillinger, K.E.; Zietek, T.S.; Thorens, B.; Koepsell, H.; Daniel, H. The role of SGLT1 and GLUT2 in intestinal glucose transport and sensing. *PLoS ONE* **2014**, *9*, e89977. [CrossRef] [PubMed]
150. Hatzoglou, M.; Fernandez, J.; Yaman, I.; Closs, E. Regulation of cationic amino acid transport: The story of the CAT-1 transporter. *Annu. Rev. Nutr.* **2004**, *24*, 377–399. [CrossRef]
151. Yang, Z.; Venardos, K.; Jones, E.; Morris, B.J.; Chin-Dusting, J.; Kaye, D.M. Identification of a novel polymorphism in the 3′UTR of the L-arginine transporter gene SLC7A1: Contribution to hypertension and endothelial dysfunction. *Circulation* **2007**, *115*, 1269–1274. [CrossRef] [PubMed]
152. Gao, Y.; Bao, X.; Meng, L.; Liu, H.; Wang, J.; Zheng, N. Aflatoxin B1 and Aflatoxin M1 Induce Compromised Intestinal Integrity through Clathrin-Mediated Endocytosis. *Toxins* **2021**, *13*, 184. [CrossRef]
153. Chen, X.; Naehrer, K.; Applegate, T.J. Interactive effects of dietary protein concentration and aflatoxin B1 on performance, nutrient digestibility, and gut health in broiler chicks. *Poult. Sci.* **2016**, *95*, 1312–1325. [CrossRef]
154. Rheeder, J.P.; Marasas, W.F.O.; Vismer, H.F. Production of fumonisin analogs by Fusarium species. *Appl. Environ. Microbiol.* **2002**, *68*, 2101–2105. [CrossRef] [PubMed]
155. Henry, M.H.; Wyatt, R.D. Environment and Health: The toxicity of fumonisin B1, B2, and B3, individually and in combination, in chicken embryos. *Poult. Sci.* **2001**, *80*, 401–407. [CrossRef] [PubMed]
156. Sydenham, E.W.; Shephard, G.S.; Thiel, P.G.; Marasas, W.F.O.; Stockenström, S. Fumonisin Contamination of Commercial Corn-Based Human Foodstuffs. *J. Agric. Food Chem.* **1991**, *39*, 2014–2018. [CrossRef]
157. Khan, R.B.; Phulukdaree, A.; Chuturgoon, A.A. Concentration-dependent effect of fumonisin B1 on apoptosis in oesophageal cancer cells. *Hum. Exp. Toxicol.* **2018**, *37*, 762–771. [CrossRef] [PubMed]
158. Khan, R.B.; Phulukdaree, A.; Chuturgoon, A.A. Fumonisin B1 induces oxidative stress in oesophageal (SNO) cancer cells. *Toxicon* **2018**, *141*, 104–111. [CrossRef] [PubMed]
159. Peraica, M.; Radić, B.; Lucić, A.; Pavlović, M. Toxic effects of mycotoxins in humans. *Bull. World Health Organ.* **1999**, *77*, 754–766.
160. Richard, J.L.; Meerdink, G.; Maragos, C.M.; Tumbleson, M.; Bordson, G.; Rice, L.G.; Ross, P.F. Absence of detectable fumonisins in the milk of cows fed Fusarium proliferatum (Matsushima) Nirenberg culture material. *Mycopathologia* **1996**, *133*, 123–126. [CrossRef] [PubMed]
161. Richard, E.; Heutte, N.; Bouchart, V.; Garon, D. Evaluation of fungal contamination and mycotoxin production in maize silage. *Anim. Feed Sci. Technol.* **2009**, *148*, 309–320. [CrossRef]
162. Marasas, W.F.O. Fumonisins: Their implications for human and animal health. *Nat. Toxins* **1995**, *3*, 193–198. [CrossRef] [PubMed]
163. Merrill, A.H.; Sullards, M.C.; Wang, E.; Voss, K.A.; Riley, R.T. Sphingolipid metabolism: Roles in signal transduction and disruption by fumonisins. Environ. *Health Perspect.* **2001**, *109*, 283–289. [CrossRef]
164. Chen, Z.; Chen, H.; Li, X.; Yuan, Q.; Su, J.; Yang, L.; Ning, L.; Lei, H. Fumonisin B1 damages the barrier functions of porcine intestinal epithelial cells in vitro. *J. Biochem. Mol. Toxicol.* **2019**, *33*, e22397. [CrossRef] [PubMed]
165. Bouhet, S.; Hourcade, E.; Loiseau, N.; Fikry, A.; Martinez, S.; Roselli, M.; Galtier, P.; Mengheri, E.; Oswald, I.P. The mycotoxin fumonisin B1 alters the proliferation and the barrier function of porcine intestinal epithelial cells. *Toxicol. Sci.* **2004**, *77*, 165–171. [CrossRef]
166. Yuan, Q.; Jiang, Y.; Fan, Y.; Ma, Y.; Lei, H.; Su, J. Fumonisin B1 induces oxidative stress and breaks barrier functions in pig iliac endothelium cells. *Toxins* **2019**, *11*, 387. [CrossRef]
167. Mateos, I.; Combes, S.; Pascal, G.; Cauquil, L.; Barilly, C.; Cossalter, A.M.; Laffitte, J.; Botti, S.; Pinton, P.; Oswald, I.P. Fumonisin-exposure impairs age-related ecological succession of bacterial species in weaned pig gut microbiota. *Toxins* **2018**, *10*, 230. [CrossRef]
168. Adibnia, E.; Razi, M.; Malekinejad, H. Zearalenone and 17 β-estradiol induced damages in male rats reproduction potential; evidence for ERα and ERβ receptors expression and steroidogenesis. *Toxicon* **2016**, *120*, 133–146. [CrossRef] [PubMed]
169. Briones-Reyes, D.; Gómez-Martinez, L.; Cueva-Rolón, R. Zearalenone contamination in corn for human consumption in the state of Tlaxcala, Mexico. *Food Chem.* **2007**, *100*, 693–698. [CrossRef]

170. Zinedine, A.; Soriano, J.M.; Moltó, J.C.; Mañes, J. Review on the toxicity, occurrence, metabolism, detoxification, regulations and intake of zearalenone: An oestrogenic mycotoxin. *Food Chem. Toxicol.* **2007**, *45*, 1–18. [CrossRef]
171. Rai, A.; Das, M.; Tripathi, A. Occurrence and toxicity of a fusarium mycotoxin, zearalenone. *Crit. Rev. Food Sci. Nutr.* **2020**, *60*, 2710–2729. [CrossRef] [PubMed]
172. Yumbe-Guevara, B.E.; Imoto, T.; Yoshizawa, T. Effects of heating procedures on deoxynivalenol, nivalenol and zearalenone levels in naturally contaminated barley and wheat. *Food Addit. Contam.* **2003**, *20*, 1132–1140. [CrossRef]
173. Wang, Y.C.; Deng, J.L.; Xu, S.W.; Peng, X.; Zuo, Z.C.; Cui, H.M.; Wang, Y.; Ren, Z.H. Effects of zearalenone on IL-2, IL-6, and IFN-γ mRNA levels in the splenic lymphocytes of chickens. *Sci. World J.* **2012**, *2012*, 567327. [CrossRef] [PubMed]
174. Shier, W.T.; Shier, A.C.; Xie, W.; Mirocha, C.J. Structure-activity relationships for human estrogenic activity in zearalenone mycotoxins. *Toxicon* **2001**, *39*, 1435–1438. [CrossRef]
175. Fink-Gremmels, J.; Malekinejad, H. Clinical effects and biochemical mechanisms associated with exposure to the mycoestrogen zearalenone. *Anim. Feed Sci. Technol.* **2007**, *137*, 326–341. [CrossRef]
176. Maaroufi, K.; Chekir, L.; Creppy, E.E.; Ellouz, F.; Bacha, H. Zearalenone induces modifications of haematological and biochemical parameters in rats. *Toxicon* **1996**, *34*, 535–540. [CrossRef]
177. Yang, L.; Yang, W.; Feng, Q.; Huang, L.; Zhang, G.; Liu, F.; Jiang, S.; Yang, Z. Effects of purified zearalenone on selected immunological measurements of blood in post-weaning gilts. *Anim. Nutr.* **2016**, *2*, 142–148. [CrossRef]
178. Wang, Y.L.; Zhou, X.Q.; Jiang, W.D.; Wu, P.; Liu, Y.; Jiang, J.; Wang, S.W.; Kuang, S.Y.; Tang, L.; Feng, L. Effects of dietary zearalenone on oxidative stress, cell apoptosis, and tight junction in the intestine of juvenile grass carp (Ctenopharyngodon idella). *Toxins* **2019**, *11*, 333. [CrossRef] [PubMed]
179. Liu, M.; Gao, R.; Meng, Q.; Zhang, Y.; Bi, C.; Shan, A. Toxic effects of maternal zearalenone exposure on intestinal oxidative stress, barrier function, immunological and morphological changes in rats. *PLoS ONE* **2014**, *9*, e106412. [CrossRef]
180. Zhang, W.; Zhang, S.; Wang, J.; Shan, A.; Xu, L. Changes in intestinal barrier functions and gut microbiota in rats exposed to zearalenone. *Ecotoxicol. Environ. Saf.* **2020**, *204*, 111072. [CrossRef] [PubMed]
181. Cole, R.J.; Cox, R.H. *Handbook of Toxic Fungal Metabolites*; Academic Press: Cambridge, MA, USA, 1981. [CrossRef]
182. Wu, Q.; Dohnal, V.; Kuca, K.; Yuan, Z. Trichothecenes: Structure-Toxic Activity Relationships. *Curr. Drug Metab.* **2013**, *14*, 641–660. [CrossRef]
183. Urbanek, K.A.; Habrowska-Górczyńska, D.E.; Kowalska, K.; Stańczyk, A.; Domińska, K.; Piastowska-Ciesielska, A.W. Deoxynivalenol as potential modulator of human steroidogenesis. *J. Appl. Toxicol.* **2018**, *38*, 1450–1459. [CrossRef]
184. Akbari, P.; Braber, S.; Gremmels, H.; Koelink, P.J.; Verheijden, K.A.T.; Garssen, J.; Fink-Gremmels, J. Deoxynivalenol: A trigger for intestinal integrity breakdown. *FASEB J.* **2014**, *28*, 2414–2429. [CrossRef] [PubMed]
185. Pestka, J.J. Deoxynivalenol-induced proinflammatory gene expression: Mechanisms and pathological sequelae. *Toxins* **2010**, *2*, 1300–1317. [CrossRef]
186. Ranzenigo, G.; Caloni, F.; Cremonesi, F.; Aad, P.Y.; Spicer, L.J. Effects of Fusarium mycotoxins on steroid production by porcine granulosa cells. *Anim. Reprod. Sci.* **2008**, *107*, 115–130. [CrossRef]
187. Habrowska-Górczyńska, D.E.; Kowalska, K.; Urbanek, K.A.; Domińska, K.; Sakowicz, A.; Piastowska-Ciesielska, A.W. Deoxynivalenol modulates the viability, ROS production and apoptosis in prostate cancer cells. *Toxins* **2019**, *11*, 265. [CrossRef]
188. Pinton, P.; Nougayrède, J.P.; del Rio, J.C.; Moreno, C.; Marin, D.E.; Ferrier, L.; Bracarense, A.P.; Kolf-Clauw, M.; Oswald, I.P. The food contaminant deoxynivalenol, decreases intestinal barrier permeability and reduces claudin expression. *Toxicol. Appl. Pharmacol.* **2009**, *237*, 41–48. [CrossRef] [PubMed]
189. Springler, A.; Hessenberger, S.; Schatzmayr, G.; Mayer, E. Early activation of MAPK p44/42 is partially involved in DON-induced disruption of the intestinal barrier function and tight junction network. *Toxins* **2016**, *8*, 264. [CrossRef]
190. Wang, S.; Zhang, C.; Wang, X.; Yang, J.; Wu, K.; Zhang, J.; Zhang, B.; Yang, A.; Qi, D. Deoxynivalenol inhibits porcine intestinal trefoil factors expression in weanling piglets and IPEC-J2 cells. *Toxins* **2019**, *11*, 670. [CrossRef] [PubMed]
191. Lessard, M.; Savard, C.; Deschene, K.; Lauzon, K.; Pinilla, V.A.; Gagnon, C.A.; Lapointe, J.; Guay, F.; Chorfi, Y. Impact of deoxynivalenol (DON) contaminated feed on intestinal integrity and immune response in swine. *Food Chem. Toxicol.* **2015**, *80*, 7–16. [CrossRef]
192. Pinton, P.; Braicu, C.; Nougayrede, J.P.; Laffitte, J.; Taranu, I.; Oswald, I.P. Deoxynivalenol impairs porcine intestinal barrier function and decreases the protein expression of claudin-4 through a mitogen-activated protein kinase-dependent mechanism. *J. Nutr.* **2010**, *140*, 1956–1962. [CrossRef]
193. Pomothy, J.M.; Szabó, O.; Czimmermann, Á.E.; Babiczky, Á.; Jerzsele, Á.; Pászti-Gere, E. Investigation of the inflammatory and oxidative stress-inducing effects of deoxynivalenol and T-2 toxin exposure in non-tumorigenic human intestinal cell model. *Toxicon* **2021**, *200*, 78–86. [CrossRef] [PubMed]
194. Yu, Y.H.; Lai, Y.H.; Hsiao, F.S.H.; Cheng, Y.H. Effects of deoxynivalenol and mycotoxin adsorbent agents on mitogen-activated protein kinase signaling pathways and inflammation-associated gene expression in porcine intestinal epithelial cells. *Toxins* **2021**, *13*, 301. [CrossRef] [PubMed]
195. Reddy, K.E.; Jeong, J.Y.; Song, J.; Lee, Y.; Lee, H.J.; Kim, D.W.; Jung, H.J.; Kim, K.H.; Kim, M.; Oh, Y.K.; et al. Colon microbiome of pigs fed diet contaminated with commercial purified deoxynivalenol and zearalenone. *Toxins* **2018**, *10*, 347. [CrossRef]

196. Li, X.G.; Zhu, M.; Chen, M.X.; Fan, H.B.; Fu, H.L.; Zhou, J.Y.; Zhai, Z.Y.; Gao, C.Q.; Yan, H.C.; Wang, X.Q. Acute exposure to deoxynivalenol inhibits porcine enteroid activity via suppression of the Wnt/β-catenin pathway. *Toxicol. Lett.* **2019**, *305*, 19–31. [CrossRef]
197. Hanyu, H.; Yokoi, Y.; Nakamura, K.; Ayabe, T.; Tanaka, K.; Uno, K.; Miyajima, K.; Saito, Y.; Iwatsuki, K.; Shimizu, M.; et al. Mycotoxin deoxynivalenol has different impacts on intestinal barrier and stem cells by its route of exposure. *Toxins* **2020**, *12*, 610. [CrossRef] [PubMed]
198. Guan, S.; He, J.; Young, J.C.; Zhu, H.; Li, X.Z.; Ji, C.; Zhou, T. Transformation of trichothecene mycotoxins by microorganisms from fish digesta. *Aquaculture* **2009**, *290*, 290–295. [CrossRef]
199. Khezri, A.; Herranz-Jusdado, J.G.; Ropstad, E.; Fraser, T.W. Mycotoxins induce developmental toxicity and behavioural aberrations in zebrafish larvae. *Environ. Pollut.* **2018**, *242*, 500–506. [CrossRef] [PubMed]
200. Saleh, I.; Goktepe, I. The characteristics, occurrence, and toxicological effects of patulin. *Food Chem. Toxicol.* **2019**, *129*, 301–311. [CrossRef] [PubMed]
201. Zhang, X.; Guo, Y.; Ma, Y.; Chai, Y.; Li, Y. Biodegradation of patulin by a Byssochlamys nivea strain. *Food Control.* **2016**, *64*, 142–150. [CrossRef]
202. Joshi, V.K.; Lakhanpal, P.; Kumar, V. Occurrence of Patulin its Dietary Intake through Consumption of Apple and Apple Products and Methods of its Removal. *Int. J. Food Ferment. Technol.* **2013**, *3*, 15. [CrossRef]
203. Zhai, Q.; Gong, X.; Wang, C.; Zhao, J.; Zhang, H.; Tian, F.; Chen, W. Food-borne patulin toxicity is related to gut barrier disruption and can be prevented by docosahexaenoic acid and probiotic supplementation. *Food Funct.* **2019**, *10*, 1330–1339. [CrossRef] [PubMed]
204. Assunção, R.; Alvito, P.; Kleiveland, C.R.; Lea, T.E. Characterization of in vitro effects of patulin on intestinal epithelial and immune cells. *Toxicol. Lett.* **2016**, *250–251*, 47–56. [CrossRef] [PubMed]
205. Singh, N.; Bansal, M.; Pal, S.; Alam, S.; Jagdale, P.; Ayanur, A.; Ansari, K.M. COX-2/EP2-EP4/β-catenin signaling regulates patulin-induced intestinal cell proliferation and inflammation. *Toxicol. Appl. Pharmacol.* **2018**, *356*, 224–234. [CrossRef]
206. Maidana, L.; Gerez, J.R.; El Khoury, R.; Pinho, F.; Puel, O.; Oswald, I.P.; Bracarense, A.P.F.R.L. Effects of patulin and ascladiol on porcine intestinal mucosa: An ex vivo approach. *Food Chem. Toxicol.* **2016**, *98*, 189–194. [CrossRef]
207. McKinley, E.R.; Carlton, W.W. Patulin mycotoxicosis in Swiss ICR mice. *Food Cosmet. Toxicol.* **1980**, *18*, 181–187. [CrossRef]
208. McKinley, E.R.; Carlton, W.W.; Boon, G.D. Patulin mycotoxicosis in the rat: Toxicology, pathology and clinical pathology. *Food Chem. Toxicol.* **1982**, *20*, 289–300. [CrossRef]
209. McLaughlin, J.; Lambert, D.; Padfield, P.J.; Burt, J.P.H.; O'Neill, C.A. The mycotoxin patulin, modulates tight junctions in caco-2 cells. *Toxicol. Vitr.* **2009**, *23*, 83–89. [CrossRef]
210. Wan, L.Y.M.; Woo, C.S.J.; Turner, P.C.; Wan, J.M.F.; El-Nezami, H. Individual and combined effects of Fusarium toxins on the mRNA expression of pro-inflammatory cytokines in swine jejunal epithelial cells. *Toxicol. Lett.* **2013**, *220*, 238–246. [CrossRef] [PubMed]
211. Kasuga, F.; Kara-Kudo, Y.; Saito, N.; Kumagai, S.; Sugita-Konishi, Y. In vitro effect of deoxynivalenol on the differentiation of human colonic cell lines Caco-2 and T84. *Mycopathologia* **1998**, *142*, 161–167. [CrossRef]
212. Van de Walle, J.; Sergent, T.; Piront, N.; Toussaint, O.; Schneider, Y.J.; Larondelle, Y. Deoxynivalenol affects in vitro intestinal epithelial cell barrier integrity through inhibition of protein synthesis. *Toxicol. Appl. Pharmacol.* **2010**, *245*, 291–298. [CrossRef]
213. Romero, A.; Ares, I.; Ramos, E.; Castellano, V.; Martínez, M.; Martínez-Larrañaga, M.R.; Anadón, A.; Martínez, M.A. Mycotoxins modify the barrier function of Caco-2 cells through differential gene expression of specific claudin isoforms: Protective effect of illite mineral clay. *Toxicology* **2016**, *353–354*, 21–33. [CrossRef] [PubMed]
214. Kawauchiya, T.; Takumi, R.; Kudo, Y.; Takamori, A.; Sasagawa, T.; Takahashi, K.; Kikuchi, H. Correlation between the destruction of tight junction by patulin treatment and increase of phosphorylation of ZO-1 in Caco-2 human colon cancer cells. *Toxicol. Lett.* **2011**, *205*, 196–202. [CrossRef] [PubMed]
215. Wu, C.Q.; Gao, Y.N.; Li, S.L.; Huang, X.; Bao, X.Y.; Wang, J.Q.; Zheng, N. Modulation of intestinal epithelial permeability and mucin mRNA (MUC2, MUC5AC, and MUC5B) expression and protein secretion in Caco-2/HT29-MTX co-cultures exposed to aflatoxin M1, ochratoxin A, and zearalenone individually or collectively. *Toxicol. Lett.* **2019**, *309*, 1–9. [CrossRef]
216. Huang, X.; Gao, Y.; Li, S.; Wu, C.; Wang, J.; Zheng, N. Modulation of Mucin (MUC2, MUC5AC and MUC5B) mRNA Expression and Protein Production and Secretion in Caco-2/HT29-MTX Co-Cultures Following Exposure to Individual and Combined Aflatoxin M1 and Ochratoxin A. *Toxins* **2019**, *11*, 132. [CrossRef]
217. Gao, Y.; Li, S.; Wang, J.; Luo, C.; Zhao, S.; Zheng, N. Modulation of intestinal epithelial permeability in differentiated caco-2 cells exposed to aflatoxin M1 and ochratoxin a individually or collectively. *Toxins* **2018**, *10*, 13. [CrossRef] [PubMed]
218. Abassi, H.; Ayed-Boussema, I.; Shirley, S.; Abid, S.; Bacha, H.; Micheau, O. The mycotoxin zearalenone enhances cell proliferation, colony formation and promotes cell migration in the human colon carcinoma cell line HCT116. *Toxicol. Lett.* **2016**, *254*, 1–7. [CrossRef]
219. Yip, K.Y.; Wan, M.L.Y.; Wong, A.S.T.; Korach, K.S.; El-Nezami, H. Combined low-dose zearalenone and aflatoxin B1 on cell growth and cell-cycle progression in breast cancer MCF-7 cells. *Toxicol. Lett.* **2017**, *281*, 139–151. [CrossRef] [PubMed]
220. Kowalska, K.; Habrowska-Górczyńska, D.E.; Domińska, K.; Piastowska-Ciesielska, A.W. The dose-dependent effect of zearalenone on mitochondrial metabolism, plasma membrane permeabilization and cell cycle in human prostate cancer cell lines. *Chemosphere* **2017**, *180*, 455–466. [CrossRef]

221. McLaughlin, J.; Padfield, P.J.; Burt, J.P.H.; O'Neill, C.A. Ochratoxin A increases permeability through tight junctions by removal of specific claudin isoforms. *Am. J. Physiol. Cell Physiol.* **2004**, *287*, C1412–C1417. [CrossRef]
222. Maresca, M.; Mahfoud, R.; Pfohl-Leszkowicz, A.; Fantini, J. The mycotoxin ochratoxin A alters intestinal barrier and absorption functions but has no effect on chloride secretion. *Toxicol. Appl. Pharmacol.* **2001**, *176*, 54–63. [CrossRef]

toxins

MDPI

Article

Cytoprotective, Antiproliferative, and Anti-Oxidant Potential of the Hydroethanolic Extract of *Fridericia chica* Leaves on Human Cancer Cell Lines Exposed to α- and β-Zearalenol

Neda Alvarez-Ortega [1,2], Karina Caballero-Gallardo [1,2], Cristina Juan [3], Ana Juan-Garcia [3,*] and Jesus Olivero-Verbel [1,*]

1 Environmental and Computational Chemistry Group, School of Pharmaceutical Sciences, Zaragocilla Campus, University of Cartagena, Cartagena 130014, Colombia
2 Functional Toxicology Group, School of Pharmaceutical Sciences, Zaragocilla Campus, University of Cartagena, Cartagena 130014, Colombia
3 Laboratory of Food Chemistry and Toxicology, Faculty of Pharmacy, University of Valencia (Spain)—Avda, Vicent Andrés Estellés, s/n. Burjassot, 46100 València, Spain
* Correspondence: ana.juan@uv.es (A.J.-G.); joliverov@unicartagena.edu.co (J.O.-V.)

Abstract: *Fridericia chica* (Bignoniaceae) is a Colombian Caribbean plant with numerous health benefits, including properties such as wound healing, immune system stimulation, and antioxidant capacity, among others. Mycotoxins alpha-zearalenol (α-ZEL) and beta-zearalenol (β-ZEL) are phase I metabolites of zearalenone, a natural product involved in endocrine disruption and cell proliferation processes. This study aimed to investigate the cytotoxic potential of the hydroethanolic extract of *F. chica* leaves (HEFc) and determine their protective effects against proliferation induced by α-ZEL and β-ZEL on human hepatoma HepG2, lung cancer Calu-1, and primary normal human epidermal keratinocytes, neonatal (HEKn). The cytotoxicity of HEFc was measured in a range from 4 to 1000 μg/mL and from 0.4 to 100 μM for both α-ZEL and β-ZEL. Cell production of intracellular ROS was monitored using the H_2-DCFDA probe. The cells exposed to HEFc presented IC_{50} of 128, 249, and 602 μg/mL for the HepG2, Calu-1, and HEKn cells, respectively. A greater selectivity was seen in HepG2 cells [selectivity index (SI) = 3.5] than in Calu-1 cells (SI = 2.4). Cells treated with mycotoxins remained viable during the first day, and cell proliferation increased at low tested concentrations (0.4-6.3 μM) in all three cell lines. However, after 48 h treatment, cells exposed to 50 and 100 μM of α-ZEL and β-ZEL displayed decreased viability. HEFc at 16 μg/mL was able to give some protection against cytotoxicity induced by high concentrations of β-ZEL in HepG2, reducing also cell proliferation elicited at low levels of α-ZEL and β-ZEL. ROS production was not observed in cells treated with this HEFc concentration; however, it prevented ROS formation induced by treatment with 50 μM α-ZEL or β-ZEL. In summary, HEFc isolated from plants grown in northern Colombia displayed promising results against cell proliferation and oxidative stress caused by mycotoxins.

Keywords: mycotoxins; ROS; cytotoxicity; plant extracts

Key Contribution: HEFc diminishes the proliferation and cytotoxicity of Zearalenone metabolites on human liver and lung cancer cells.

Citation: Alvarez-Ortega, N.; Caballero-Gallardo, K.; Juan, C.; Juan-Garcia, A.; Olivero-Verbel, J. Cytoprotective, Antiproliferative, and Anti-Oxidant Potential of the Hydroethanolic Extract of *Fridericia chica* Leaves on Human Cancer Cell Lines Exposed to α- and β-Zearalenol. *Toxins* **2023**, 15, 36. https://doi.org/10.3390/toxins15010036

Received: 24 October 2022
Revised: 16 November 2022
Accepted: 25 November 2022
Published: 3 January 2023

Copyright: © 2023 by the authors. Licensee MDPI, Basel, Switzerland. This article is an open access article distributed under the terms and conditions of the Creative Commons Attribution (CC BY) license (https://creativecommons.org/licenses/by/4.0/).

1. Introduction

Medicinal plants have become a topic of global interest, as they are been used to treat various health problems in many countries [1]. In developing countries, many communities utilize phytomedicines as the first treatment when any sign or symptom appears. However, the study of the pharmacological potential of plants continues to be limited, but it is a relevant field to develop due to the possible production of new substances that improve human health. Only about 10% of medicinal plants have been analyzed to treat various diseases.

Fridericia chica, a shrub of the Bignoniaceae family, has several benefits for human health. This plant is commonly found in the humid tropical forests of Central America, Mexico, and the Amazon. Fermentation of its leaves produces a red dye that is frequently used by the natives to paint their bodies and utensils. Moreover, the infusion of this plant turns into a red liquid used in their traditional medicine as a pain killer [2], astringent, and as a treatment for diarrhea, anemia, and leukemia [1]. Recent studies have displayed the chemical complexity of the *F. chica* extract. Its activity has been reported as antimicrobial, antioxidant [3], anti-inflammatory [4], healing [5], leishmanicidal [6], and antigenotoxic [7].

Mycotoxins are small molecules produced by fungi, mainly by the genera *Penicillium* sp., *Aspergillus* sp., and *Fusarium* sp. [8]. Mycotoxins are one of the biotic factors that may appear in cereals before harvest and during storage. According to the FAO, the production of these foods is increasing, with an estimated production of 2799 million tons [9]. The European Food Safety Authority (EFSA) has estimated that approximately up to 80% of food products are contaminated with these toxins each year [10]. One of the most worrying mycotoxins is Zearalenone (ZEN). This mycotoxin is produced by *Fusarium* sp., and through different biotransformation pathways two different metabolites are generated, α-Zearalenol (α-ZEL) and β-Zearalenol (β-ZEL) [11]. The biotransformation of ZEN occurs in the liver, lung, kidney, and prostate, among other tissues of various animal species [12].

Mycotoxins are of special concern for human health [13]. Several studies have shown their occurrence and effects on human and animal health after exposure to ZEN metabolites [14–17], aflatoxin B1 [18], deoxynivalenol [19], and Ochratoxin A [19]. Hydroxyl derivatives, such as α-ZEL and β-ZEL, have endocrine disruption potential [17], and capacity to induce oxidative stress [20] and DNA damage [21], although according to the IARC, ZEN is a Group 3 carcinogen [22]. There are different in vitro studies that have shown its role in cell proliferation processes, including human hepatoma [23–26], endometrial adenocarcinoma [27], and breast cancer cells [28].

The present study investigated the antitumor potential of the hydroethanolic extract of *F. chica* leaves (HEFc) in lung (Calu-1) and liver cancer cells (HepG2) versus non-cancer skin cells (HEKn). In addition, cell viability was evaluated against mycotoxins α-ZEL and β-ZEL in these same cells, followed by the antiproliferative and anti ROS effect of the extract.

2. Results

2.1. Viability of HEFc on Cells Lines

The results on the cytotoxicity of cancerous (HepG2 and Calu-1) and non-cancerous cells exposed to HEFc are shown in Figure 1. At the highest tested concentrations (500 and 1000 μg/mL) the extract exhibited cytotoxicity compared to the control ($p < 0.05$) during the first 24 h, but the IC_{50} value could not be determined for any of the three cell lines. However, the cytotoxic effect at 48 h showed a concentration-dependent trend, mainly in cancer cells. No increases in cell proliferation were observed in any case. In concordance with these results, the concentration of 16 μg/mL was selected for the cytoprotection assay. The selectivity index (SI) calculated for HEFc on cancer cell lines is presented in Table 1. The SIs for HEFc was greater than 1 for the two tumor cell lines.

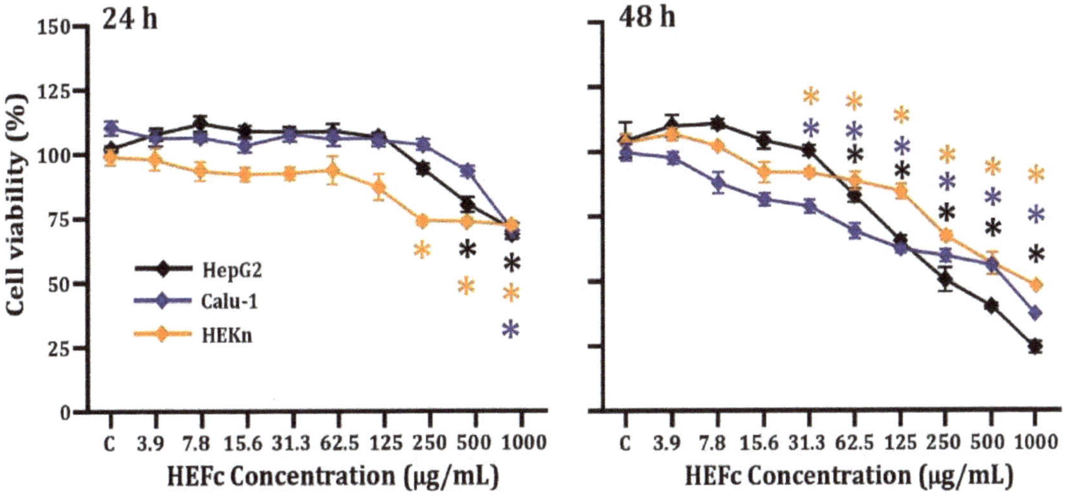

Figure 1. Cytotoxic activity of HEFc on cancerous (HepG2 and Calu-1) and non-cancerous (HEKn) cells after 24 h and 48 h exposure. Each result represents the mean ± standard error of the mean. * Significant difference ($p < 0.05$) when compared to the control group (C).

Table 1. Selectivity index (SI) of HEFc on HEKn relative to cancer cell lines (HepG2 and Calu-1).

	IC_{50} (48 h)	SI
HepG2	168	3.5
Calu-1	249	2.4
HEKn	602.9	-

2.2. Viability of Cells Exposed to Mycotoxins

The viability of the cell lines exposed to mycotoxins after 24 h and 48 h of treatment are presented in Figure 2. A reduction in viability was observed in HepG2 cells exposed to 100 μM α-ZEL by approximately 25% and 50% after incubation for one and two days, respectively. Nonetheless, the IC_{50} could not be calculated. In contrast, the lowest tested concentrations (0.4, 0.8, and 1.5 μM) showed a significant increase in cell growth ($p < 0.05$).

For β-ZEL, after 24 h and 48 h of exposure, cell viability was significantly reduced in the 50 μM and 100 μM treatments ($p < 0.05$) compared to the control. In contrast, after 48 h at 0.8 and 1.5 μM, cell proliferation significantly increased in HepG2 cells ($p < 0.05$). The IC_{50} was found in the HepG2 cell line after 48 h of exposure. Photographs of cell morphology changes in HepG2 cells exposed to 50 μM α-ZEL and β-ZEL after 24 h exposure are shown in Figure S1.

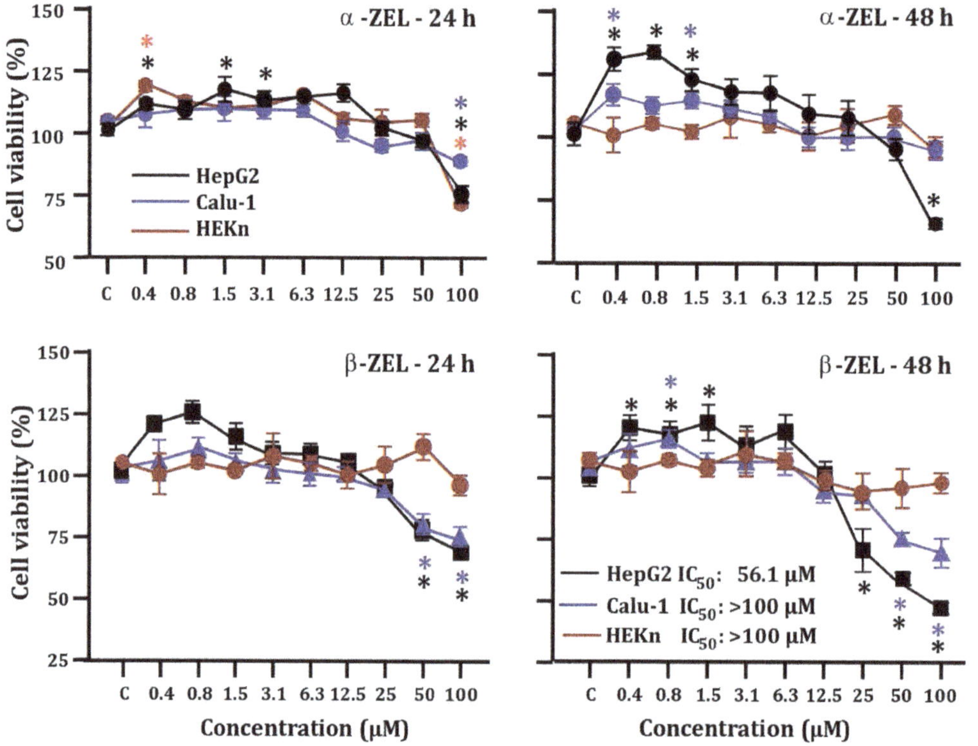

Figure 2. Cytotoxicity effects of mycotoxins (α-ZEL and β-ZEL) on HepG2, Calu-1, and HEKn cells after 24 h and 48 h exposure. Each result represents the mean ± standard error of the mean. * Significant difference ($p < 0.05$) when compared to the control group (C).

2.3. Antiproliferative Effects of HEFc against α-ZEL and β-ZEL

To develop the cytoprotection assay, HepG2 cells were selected since it was the only cell line with IC_{50} detection after 48 h of exposure (Figure 2). The cell proliferation induced by the mycotoxins α-ZEL and β-ZEL was markedly reduced in the presence of 16 μg/mL HEFc, after treatment for 24 or 48 h. In addition, the IC_{50} of the combination showed a considerable difference from that of β-ZEL alone, with a value of 63.6 μM (Figure 3).

2.4. ROS Levels of HepG2 Cells

The results of the ROS generation by HEFc (8, 16, and 32 μg/mL) on HepG2 cells are presented in Figure 4. DCF-DA fluorescence levels in cells treated with HEFc (8, 16, and 32 μg/mL) did not show significant changes when compared to the negative control. Afterwards, the anti-ROS effect of the extract (16 μg/mL) was evaluated. Incubation of HepG2 cells with 16 μg/mL HEFc for 6 h and 24 h reduced ROS generation induced by α-ZEL and β-ZEL (Figure 5).

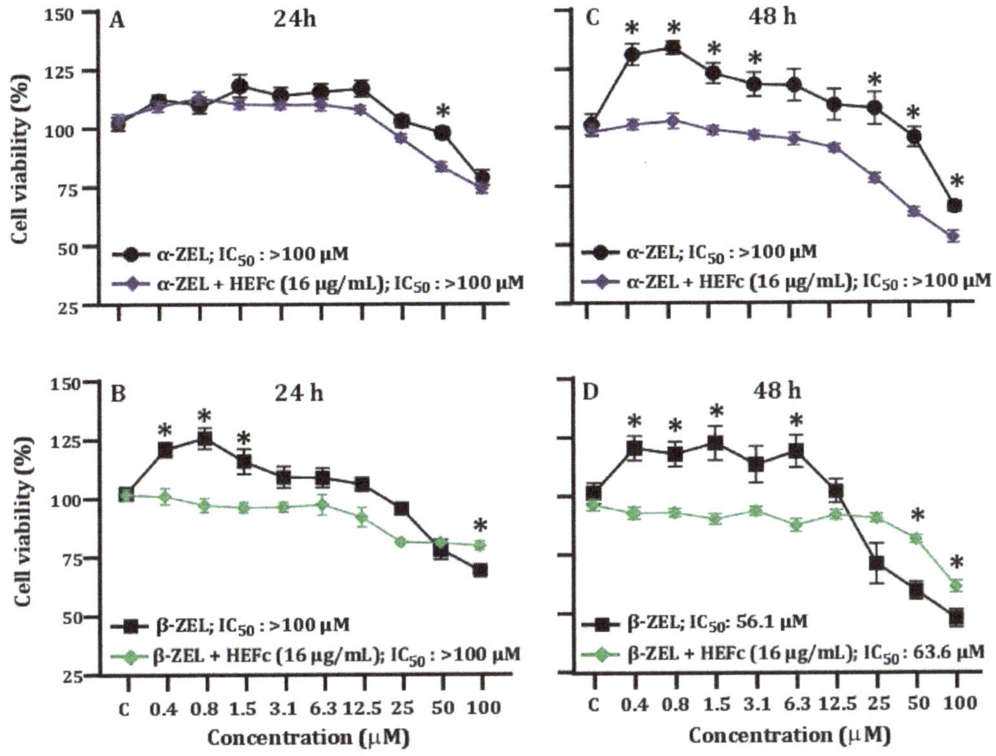

Figure 3. Anti-proliferative effects of HEFc on HepG2 cells exposed to α-ZEL or β-ZEL after treatment for 24 h (**A**,**B**) and 48 h (**C**,**D**). Each result represents the mean ± standard error of the mean. * Significant difference compared to the corresponding control group (C), $p < 0.05$.

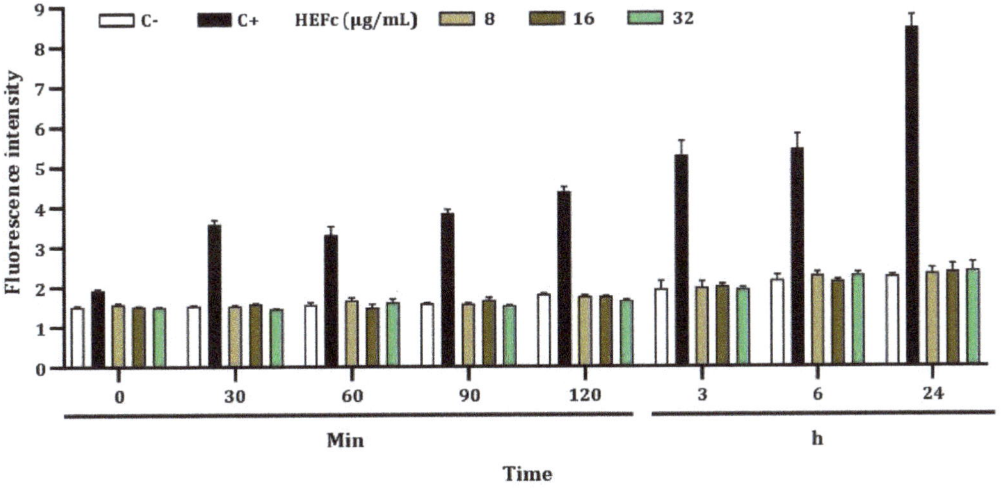

Figure 4. ROS levels of HepG2 cells exposed to HEFc (8, 16, and 32 µg/mL) after different exposure times. Each result represents the mean ± standard error of the mean. C- represents the negative control (DMSO 1%) and C+ the positive control (200 µM hydrogen peroxide).

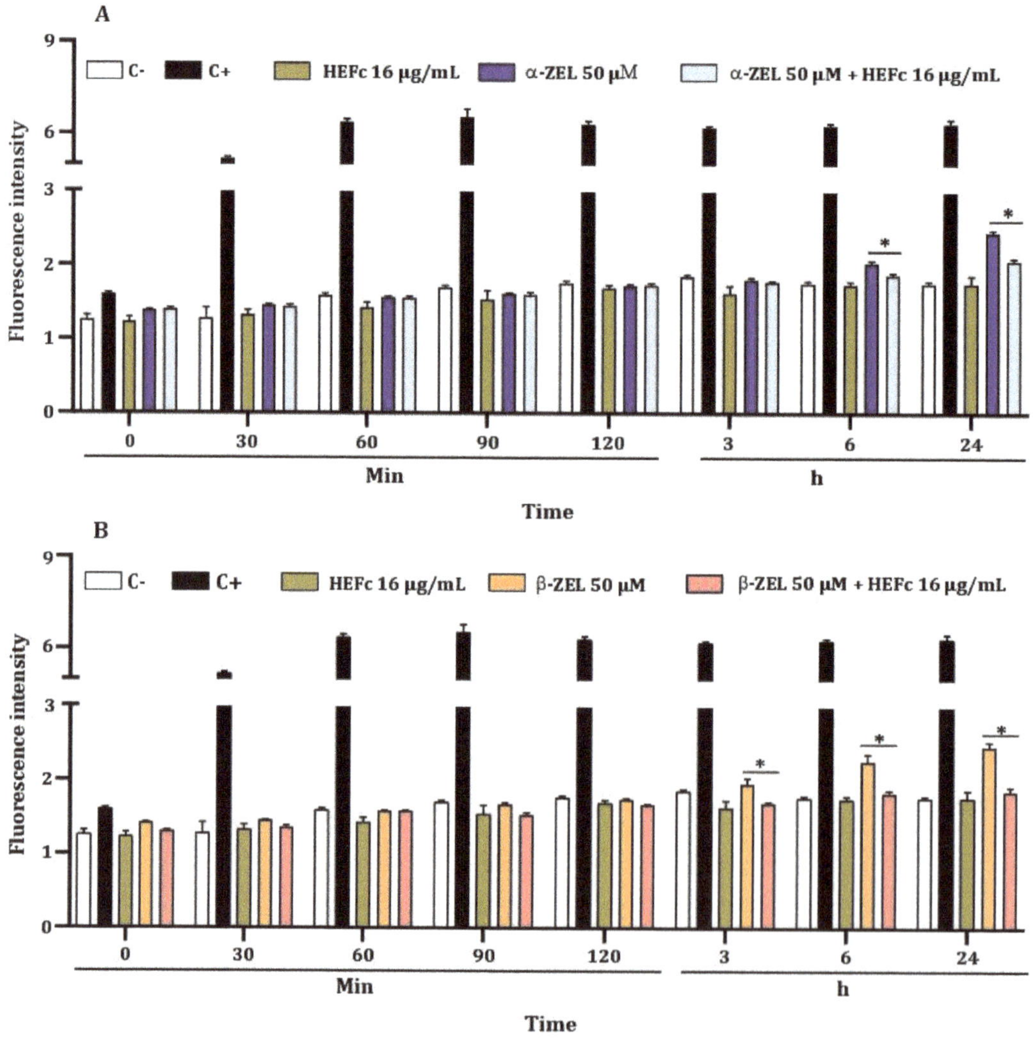

Figure 5. ROS levels of HepG2 cells. (**A**) ROS levels of HepG2 cells treated with HEFc (16 μg/mL), α-ZEL, or HEFc (16 μg/mL) + α -ZEL (50 μM). (**B**) ROS levels of HepG2 cells treated with HEFc (16 μg/mL), β-ZEL, or HEFc (16 μg/mL) + β-ZEL (50 μM). Negative control (C-), culture medium; positive control (C+), hydrogen peroxide (200 μM). * Significant differences ($p < 0.05$) compared to α-ZEL or β-ZEL alone.

3. Discussion

The increasing use of medicinal plants makes evaluating their toxicity a necessary step for developing new synthetic or natural pharmaceutical products. Toxicity tests are recommended to ensure the safe use of extracts obtained from medicinal plants. Due to the wide use of *F. chica* as a medicine for numerous health conditions, in this study, we developed tests to detect the SI of HEFc on hepatocarcinoma cells (HepG2) and lung cancer cells (Calu-1). The most potent effect on cell viability, with the lowest IC_{50} values, occurred in HepG2 cells. Based on these findings, we selected a concentration of 16 μg/mL of HEFc to perform bioassays on HepG2 with the combination mycotoxin-extract. Moreover, the

HepG2 cell line was used due to its endogenous capacity for bioactivation, and their role on the metabolism of xenobiotics [29].

Cancer remains a main global health problem affecting both sexes [30]. This complex disease has diverse etiologies and multiple stages [31,32]. Its treatment is focused on cells with rapid division, although it affects normal cells [33]. Therefore, it is essential to search for bioactive molecules from various sources that can target specific molecular signaling pathways in tumor cells without damaging normal cells.

Extracts of *F. chica* have been used in traditional medicine as an anticancer agent against leukemia [34]. Furthermore, ethanolic and aqueous extracts of *F. chica* have immunomodulatory and antitumor activities on solid Ehrlich tumors, probably attributed to the presence of flavonoids. In the present study, the ethanolic extract of *F. chica* leaves exerted a more significant cytotoxic effect on cancer cells than on non-cancer cells. IC_{50} values were much lower in liver cancer cells than in the tested Calu-1. Previous studies with *F. chica* extracts, or its components, in cancer cells, such as Jurkat-1, HL60 [35] and SH5YSY cell lines [36] showed results consistent with our results.

Zearalenone is an estrogenic mycotoxin whose role in carcinogenesis remains questionable. In this study, we evaluated the effect of exposure to ZEN metabolites on the proliferation of HepG2, Calu-1, and HEKn cells. Our results showed that the mycotoxins studied decreased cell viability at high concentrations, but stimulated cell proliferation at lower concentrations in HepG2 and Calu-1 cells. Similarly, this behavior has been reported for SH-SY5Y [37], MCF-7 [38,39], PALMA [40], MDA-MB-231, and T47D cells [13]. The cell proliferation exerted by ZEN and its metabolites may play a role in promoting hormone-dependent tumors because these compounds bind to and activate the ER (estrogen receptor) [37,41] due to their similar structure with the body's natural hormone, 17-β-estradiol (E2). The finding in terms of proliferation reported here may be due to the expression of two ER (α-ER and β-ER), which has also been reported for other xenoestrogens acting on liver cancer cells [42]; however, signaling pathways are not yet fully established. It is plausible that micro-RNAs are also involved in stimulating cell proliferation by silencing various cell cycle inhibitors; accordingly, it has been stated that ZEN induces oxidative stress, which may be a regulator of miRNAs, interfering with micro-RNA expression [43].

Several pharmacological properties have been reported for HEFc, one of them has been the reduction of α-ZEL and β-ZEL-induced cell proliferation in SH5YSY [36]. The results presented here make possible to suggest that the antiproliferative activity of HEFc may be link to its flavonoid content, for instance, 6-hydroxyluteolin, which has shown good antiproliferative activity in cell lines such as MCF7, HepG2, and Caco2, among others [44]. The extract of *F. chica* is also a good source of Vicenin-2, a flavonoid with antioxidant, hepatoprotective, anti-inflammatory, and anticancer properties. Its action, as a potent inhibitor of cell proliferation, has been shown in colon cancer cells [45] and hepatocellular carcinoma [46]. Different components of *F. chica* have been tested for their antimicrobial action [47]. In addition to the properties mentioned, it should be noted that other authors have demonstrated the safety of the extract in an *in vivo* model with rats, where a dose greater than 200 mg/kg did not induce changes in biochemical and hematological parameters [47,48]. Similarly, Takenaka et al. [49] did not observe signs of toxicity after the treatment with 300 mg/Kg of hydroethanolic extract of *F. chica* in BALB/c mice.

In the present study, treatment with HEFc significantly protected cells from oxidative damage by inhibiting the generation of ROS induced by ZEN metabolites. This can be explained due to the leaves' reported antioxidant properties given their chemical composition [50].

4. Conclusions

HEFc offers protection against the cytotoxic effects provoked by α-ZEL and β-ZEL on human liver and lung cancer cells. It also reduces ROS production on liver cancer cells. Further investigations are required to understand the mechanisms behind the hepatoprotective effect of *F. chica*.

5. Materials and Methods

5.1. Reagents

The reagents used were Eagle's Minimum Essential Medium (EMEN) (Quality Biological, Gaithersburg, MD, USA); Fetal Bovine Serum (FBS) (Biowest, South America Origin); Keratinocyte Medium (KM) (ScienCell Research Laboratories, Carlsbad, CA, USA); dimethyl sulfoxide (DMSO), hydrogen peroxide 30% *v*/*v*, and phosphate-buffered saline (PBS) were supplied by Panreac (Applichem, Barcelona, Spain); α-ZEL and β-ZEL (MW: 320.38 g/mol) standards, penicillin, streptomycin, Trypsin–EDTA, and 2.7-dichlorofluorescein diacetate (DCFH-DA) were purchased from Sigma-Aldrich (St. Louis, MO, USA).

5.2. Extract Preparation and Characterization

Fallen leaves of *F. chica* samples were collected in Sincelejo at the University of Sucre (Sincelejo, Colombia). The species was identified by Pedro Alvarez Perez in the Herbarium at the University of Sucre (Sincelejo, Colombia). A voucher specimen was deposited in the Herbarium at the University of Sucre (Sincelejo, Colombia) under the number 004537. The hydroethanolic extract (ethanol:water; 7:3) preparation from leaves, as well as its characterization, were carried out as reported elsewhere [36]. The extract was reconstituted in DMSO and kept at −20 °C until use.

5.3. Maintaining Cell Culture

The HepG2 (ATCC-HB-8065) and Calu-1 cells were routinely cultured using EMEN, and supplemented with 10% FBS and 1% penicillin/streptomycin (Sigma-Aldrich, St. Louis, MI, USA) as previously described [45,51]. Additionally, 5×10^5 primary HEKn cells were thawed in KM culture medium (ScienCell Research Laboratories, Carlsbad, CA, USA) supplemented with KGS 100× (ScienCell Research Laboratories, Carlsbad, CA, USA) and 1% penicillin/streptomycin solution (ScienCell Research Laboratories, Carlsbad, CA, USA). After cells reached ~80–90% confluency, they were washed with DPBS and trypsinized to detach from the culture flask. Trypsin-EDTA 0.025% (2 mL) was added to the monolayer and incubated for 3–5 min, and once detached, 2 mL of TNS was added to the cell suspension and centrifuged at 1200 rpm for 3 min. The supernatant was discarded, and the pellet was resuspended in growth medium and replaced in a 1:4 ratio. Passages 2–4 were used for the experiments. All cell lines were grown at 37 °C in 5% CO_2.

5.4. Exposure of HepG2, Calu-1, and HEKn Cells to F. chica Extract and Mycotoxins

The cell lines were cultured in 96-well plates at a density of 2×10^4 cells/well for 24 h. Then, the cells were treated at 24 h and 48 h with 100 µL of fresh medium containing different concentrations of (a) HEFc (from 1000 to 3.9 µg/mL, 1:2 dilutions), (b) α-ZEL (from 100 to 0.4 µM, 1:2 dilutions), or (c) β-ZEL (from 100 to 0.4 µM, 1:2 dilutions). These concentrations have been reported previously. The 1% DMSO vehicle was used as a control. After exposure to mycotoxins or extracts, the viability assay was performed using MTT in the same way as reported in our previous study [46,52]. Optical density was measured on a spectrophotometer Varioskan™ LUX (Thermo Fisher Scientific, Inc., Waltham, MA, USA) at 620 nm. The experiments were developed three times with four replicates each. Viability was defined as the ratio (expressed as a percentage) of the absorbance of treated cells to control cells.

5.5. Cytoprotective Effects of HEFc against ZEN Metabolites

To determine the cytoprotective effect of HEFc on mycotoxin-induced toxicity, the working concentration of the extract was selected considering the largest concentration at which no significant cytotoxicity was observed (16 μg/mL). Two independent treatment combinations were carried out. They consisted of a mixture of HEFc at 16 μg/mL (fixed concentration) with eight concentrations of α-ZEL or β-ZEL (from 0.4 to 100 μM, 1:2 dilutions). The plates were incubated for 24 h and 48 h at 37 °C in a 5% CO_2 atmosphere, and viability was examined using the MTT assay. Three experiments were carried out with four replicates each. The selectivity index (SI) was obtained using the ratio of the IC_{50} of non-cancer cells versus the IC_{50} of cancer cells.

5.6. ROS Assay

Intracellular ROS levels were monitored in HepG2 cells using the 2,7-dichloro-fluorescein diacetate probe (DCFH-DA; Sigma–Aldrich, St. Louis, MO, USA). Briefly, 2×10^6 cells/plate were seeded in a 96-well black culture microplate. After incubation for 24 h, the culture medium was removed and supplemented, and EMEN containing 20 μM of DCFH-DA probe was added for 20 min under the same incubation conditions [47]. Later, the cells were washed with PBS (Panreac Applichem®, Barcelona, Spain) and exposed to HEFc (16 μg/mL). A solution of 200 μM H_2O_2 (Panreac Applichem®, Barcelona, Spain) was used as a positive control. The increase in fluorescence was measured for 120 min after 3, 6, and 24 hours of exposure using a Varioskan TM LUX Multimode Microplate Reader (Thermo Fisher Scientific, Inc., Waltham, MA, USA). The fluorescence intensity was determined by employing 485 nm for excitation and 530 nm for emission. The results were expressed as the fluorescence values normalized over the control (untreated cells). Three independent experiments were performed with three replicates each.

5.7. Statistical Analysis

Cell viability data are presented as mean±SEM, and normality was assessed using Shapiro-Wilk test. Statistical comparisons between means were carried out using one-way analysis of variance (ANOVA), followed by Dunnett's test. IC_{50} values for each treatment were obtained utilizing nonlinear sigmoid curve fitting. All results were processed employing GraphPad 5.0 Instat Software (San Diego, CA, USA). The results were considered significant at $p < 0.05$.

Supplementary Materials: The following supporting information can be downloaded at: https://www.mdpi.com/article/10.3390/toxins15010036/s1, Figure S1: Morphology of HepG2 cells after 24 h incubation. Photographs taken under phase contrast. Control (1% DMSO). Scale bar: 200 μm.

Author Contributions: Data curation, N.A.-O.; formal analysis, N.A.-O., K.C.-G., and J.O.-V.; funding acquisition, J.O.-V., C.J., and A.J.-G.; investigation, J.O.-V., K.C.-G., and A.J.-G.; methodology, N.A.-O.; supervision, J.O.-V., K.C.-G., C.J., and A.J.-G.; writing—original draft, N.A.-O., K.C.-G., and J.O.-V.; writing—review & editing, J.O.-V., K.C.-G., C.J., and A.J.-G. All authors have read and agreed to the published version of the manuscript.

Funding: The authors thank the Ministry of Science, Technology and Innovation (Minciencias), the Ministry of Education, the Ministry of Industry, Commerce and Tourism, and ICETEX, Programme Ecosistema Científico-Colombia Científica, from the Francisco José de Caldas Fund (Grant RC-FP44842-212-2018); Spanish Ministry of Science and Innovation (PID2020-115871RB-I00); Minciencias (Grant: 466-2021); Minciencias, Sistema General de Regalías de Colombia (BPIN 2020000100093-2020, Gobernación de Bolívar).

Institutional Review Board Statement: Not applicable.

Informed Consent Statement: Not applicable.

Data Availability Statement: The datasets used and analyzed during the current study are available from the corresponding author on reasonable request.

Acknowledgments: The authors thank the Program for Doctoral Studies in Colombia (Minciencias, Grant 727-2015).

Conflicts of Interest: The authors declare no conflict of interest.

References

1. de Souza, A.S.; Pagadigorria, C.L.S.; Ishii-Iwamoto, E.L.; Bracht, A.; Cortez, D.A.G.; Yamamoto, N.S. Effects of the *Arrabidaea chica* extract on energy metabolism in the rat liver. *Pharm. Biol.* **2009**, *47*, 154–161. [CrossRef]
2. Vasconcelos, C.C.; Lopes, A.J.O.; Sousa, E.L.F.; Camelo, D.S.; Lima, F.C.V.M.; Rocha, C.Q.D.; Silva, G.E.B.; Garcia, J.B.S.; Cartágenes, M.D.S.D.S. Effects of extract of *Arrabidaea chica* Verlot on an experimental model of osteoarthritis. *Int. J. Mol. Sci.* **2019**, *20*, 4717. [CrossRef] [PubMed]
3. Torres, C.A.; Pérez Zamora, C.M.; Nuñez, M.B.; Gonzalez, A.M. In vitro antioxidant, antilipoxygenase and antimicrobial activities of extracts from seven climbing plants belonging to the Bignoniaceae. *J. Integr. Med.* **2018**, *16*, 255–262. [CrossRef] [PubMed]
4. Lima, J.C.S.; de Oliveira, R.G.; Silva, V.C.; de Sousa, P.T., Jr.; Violante, I.M.P.; Macho, A.; Martins, D.T.O. Anti-inflammatory activity of 4′,6,7-trihydroxy-5-methoxyflavone from *Fridericia chica* (Bonpl.) L.G.Lohmann. *Nat. Prod. Res.* **2020**, *34*, 726–730. [CrossRef] [PubMed]
5. Aro, A.A.; Freitas, K.M.; Foglio, M.A.; Carvalho, J.E.; Dolder, H.; Gomes, L.; Vidal, B.C.; Pimentel, E.R. Effect of the *Arrabidaea chica* extract on collagen fiber organization during healing of partially transected tendon. *Life Sci.* **2013**, *92*, 799–807. [CrossRef]
6. Moragas-Tellis, C.J.; Almeida-Souza, F.; Chagas, M.; Souza, P.V.R.; Silva-Silva, J.V.; Ramos, Y.J.; Moreira, D.L.; Calabrese, K.D.S.; Behrens, M.D. The Influence of anthocyanidin profile on antileishmanial activity of *Arrabidaea chica* Morphotypes. *Molecules* **2020**, *25*, 3547. [CrossRef]
7. dos Santos, V.C.; Longo, T.B.; Garcia, A.L.; Richter, M.F.; Guecheva, T.N.; Henriques, J.A.; Ferraz Ade, B.; Picada, J.N. Evaluation of the mutagenicity and genotoxicity of *Arrabidaea chica* Verlot (Bignoneaceae), an Amazon plant with medicinal properties. *J. Toxicol. Environ. Health Part A* **2013**, *76*, 381–390. [CrossRef]
8. Juan-García, A.J.T. Introduction to the toxins special issue on toxicological effects of mycotoxin on target cells. *Toxins* **2020**, *12*, 446. [CrossRef]
9. FAO. *FAO Cereal Supply and Demand Brief*; FAO: Rome, Italy, 2022.
10. Eskola, M.; Kos, G.; Elliott, C.T.; Hajšlová, J.; Mayar, S.; Krska, R. Worldwide contamination of food-crops with mycotoxins: Validity of the widely cited 'FAO estimate'of 25%. *Crit. Rev. Food Sci. Nutr.* **2020**, *60*, 2773–2789. [CrossRef]
11. Rogowska, A.; Pomastowski, P.; Sagandykova, G.; Buszewski, B. Zearalenone and its metabolites: Effect on human health, metabolism and neutralisation methods. *Toxicon* **2019**, *162*, 46–56. [CrossRef]
12. Dong, M.; Tulayakul, P.; Li, J.-Y.; Dong, K.-S.; Manabe, N.; Kumagai, S. Metabolic Conversion of Zearalenone to α-Zearalenol by Goat Tissues. *J. Vet. Med. Sci.* **2010**, *72*, 307–312. [CrossRef]
13. Khosrokhavar, R.; Rahimifard, N.; Shoeibi, S.; Hamedani, M.P.; Hosseini, M.-J. Effects of zearalenone and α-Zearalenol in comparison with Raloxifene on T47D cells. *Toxicol. Mech. Methods* **2009**, *19*, 246–250. [CrossRef] [PubMed]
14. Kuciel-Lisieska, G.; Obremski, K.; Stelmachów, J.; Gajecka, M.; Zielonka, Ł.; Jakimiuk, E.; Gajecki, M. Presence of zearalenone in blood plasma in women with neoplastic lesions in the mammary gland. *Bull. Vet. Inst. Pulawy* **2008**, *52*, 671–674.
15. Ali, N.; Degen, G.H. Biomonitoring of zearalenone and its main metabolites in urines of Bangladeshi adults. Food and chemical toxicology: An international journal published for the British Industrial Biological Research Association. *Food Chem. Toxicol.* **2019**, *130*, 276–283. [CrossRef]
16. Fan, K.; Xu, J.; Jiang, K.; Liu, X.; Meng, J.; Di Mavungu, J.D.; Guo, W.; Zhang, Z.; Jing, J.; Li, H.; et al. Determination of multiple mycotoxins in paired plasma and urine samples to assess human exposure in Nanjing, China. *Environ. Pollut.* **2019**, *248*, 865–873. [CrossRef] [PubMed]
17. Tiemann, U.; Viergutz, T.; Jonas, L.; Schneider, F. Influence of the mycotoxins α- and β-zearalenol and deoxynivalenol on the cell cycle of cultured porcine endometrial cells. *Reprod. Toxicol.* **2003**, *17*, 209–218. [CrossRef] [PubMed]
18. Jin, S.J.; Yang, H.; Jiao, Y.H.; Pang, Q.; Wang, Y.J.; Wang, M.; Shan, A.S.; Feng, X.J. Dietary curcumin alleviated acute ileum damage of ducks (*Anas platyrhynchos*) induced by AFB1 through regulating Nrf2-ARE and NF-κB signaling pathways. *Foods* **2021**, *10*, 1370. [CrossRef]
19. Carballo, D.; Pallarés, N.; Ferrer, E.; Barba, F.J.; Berrada, H. Assessment of human exposure to deoxynivalenol, ochratoxin A, zearalenone and their metabolites biomarker in urine samples using LC-ESI-qTOF. *Toxins* **2021**, *13*, 530. [CrossRef]
20. Agahi, F.; Álvarez-Ortega, N.; Font, G.; Juan-García, A.; Juan, C. Oxidative stress, glutathione, and gene expression as key indicators in SH-SY5Y cells exposed to zearalenone metabolites and beauvericin. *Toxicol. Lett.* **2020**, *334*, 44–52. [CrossRef]
21. Fleck, S.C.; Hildebrand, A.A.; Müller, E.; Pfeiffer, E.; Metzler, M. Genotoxicity and inactivation of catechol metabolites of the mycotoxin zearalenone. *Mycotoxin Res.* **2012**, *28*, 267–273. [CrossRef]
22. IARC. IARC Monographs on the Evaluation of Car-Cinogenic Risks to Humans, 56, 599. 1993. Available online: https://monographs.iarc.fr/iarc-monographs-on-the-evaluation-of-carcinogenic-risks-to-humans-65/ (accessed on 14 November 2022).
23. Juan-García, A.; Agahi, F.; Drakonaki, M.; Tedeschi, P.; Font, G.; Juan, C. Cytoprotection assessment against mycotoxins on HepG2 cells by extracts from *Allium sativum* L. *Food Chem. Toxicol.* **2021**, *151*, 112129. [CrossRef] [PubMed]

24. Hu, X.-M.; Wang, Y.-M.; Zhao, Y.-Q.; Chi, C.-F.; Wang, B. Antioxidant peptides from the protein hydrolysate of monkfish (*Lophius litulon*) Muscle: Purification, identification, and cytoprotective function on HepG2 cells damage by H_2O_2. *Mar. Drugs* **2020**, *18*, 153. [CrossRef] [PubMed]
25. Li, Z.; Lan, Y.; Miao, J.; Chen, X.; Chen, B.; Liu, G.; Wu, X.; Zhu, X.; Cao, Y. Phytochemicals, antioxidant capacity and cytoprotective effects of jackfruit (*Artocarpus heterophyllus* Lam.) axis extracts on HepG2 cells. *Food Biosci.* **2021**, *41*, 100933. [CrossRef]
26. Pires, F.C.S.; Oliveira, J.C.D.; Menezes, E.G.O.; Ferreira, M.C.R.; Siqueira, L.M.M.; Almada-Vilhena, A.O.; Pieczarka, J.C.; Nagamachi, C.Y.; Carvalho Junior, R.N.D. Bioactive compounds and evaluation of antioxidant, cytotoxic and cytoprotective effects of murici pulp extracts (*Byrsonima crassifolia*) obtained by supercritical extraction in HepG2 cells treated with H_2O_2. *Foods* **2021**, *10*, 737. [CrossRef] [PubMed]
27. Nishida, M.; Kasahara, K.; Kaneko, M.; Iwasaki, H.; Hayashi, K. Establishment of a new human endometrial adenocarcinoma cell line, Ishikawa cells, containing estrogen and progesterone receptors. *Nihon Sanka Fujinka Gakkai Zasshi* **1985**, *37*, 1103–1111.
28. Malekinejad, H.; Maas-Bakker, R.F.; Fink-Gremmels, J. Bioactivation of zearalenone by porcine hepatic biotransformation. *Vet. Res.* **2005**, *36*, 799–810. [CrossRef]
29. Scolastici, C.; de Lima, R.A.; Barbisan, L.F.; Ferreira, A.L.D.A.; Ribeiro, D.A.; Salvadori, D.M.F. Antigenotoxicity and antimutagenicity of lycopene in HepG2 cell line evaluated by the comet assay and micronucleus test. *Toxicol. Vitr.* **2008**, *22*, 510–514. [CrossRef]
30. Samec, M.; Liskova, A.; Koklesova, L.; Samuel, S.M.; Murin, R.; Zubor, P.; Bujnak, J.; Kwon, T.K.; Büsselberg, D.; Prosecky, R.; et al. The role of plant-derived natural substances as immunomodulatory agents in carcinogenesis. *J. Cancer Res. Clin. Oncol.* **2020**, *146*, 3137–3154. [CrossRef]
31. Al-Khedhairy, A.A.; Wahab, R. Silver Nanoparticles: An instantaneous solution for anticancer activity against human liver (HepG2) and breast (MCF-7) cancer cells. *Metals* **2022**, *12*, 148. [CrossRef]
32. Moolgavkar, S.H.; Luebeck, E.G.J.G. Multistage carcinogenesis and the incidence of human cancer. *Genes Chromosom. Cancer* **2003**, *38*, 302–306. [CrossRef]
33. Abbas, Z.; Rehman, S. An overview of cancer treatment modalities. *Neoplasm* **2018**, *1*, 139–157.
34. Ribeiro, A.F.C.; Telles, T.C.; Ferraz, V.P.; Souza-Fagundes, E.M.; Cassali, G.D.; Carvalho, A.T.; Melo, M.M. Effect of *Arrabidaea chica* extracts on the Ehrlich solid tumor development. *Rev. Bras. Farmacogn.* **2012**, *22*, 364–373. [CrossRef]
35. Michel, A.F.R.M.; Melo, M.M.; Campos, P.P.; Oliveira, M.S.; Oliveira, F.A.S.; Cassali, G.D.; Ferraz, V.P.; Cota, B.B.; Andrade, S.P.; Souza-Fagundes, E.M. Evaluation of anti-inflammatory, antiangiogenic and antiproliferative activities of *Arrabidaea chica* crude extracts. *J. Ethnopharmacol.* **2015**, *165*, 29–38. [CrossRef] [PubMed]
36. Alvarez-Ortega, N.; Caballero-Gallardo, K.; Taboada-Alquerque, M.; Franco, J.; Stashenko, E.E.; Juan, C.; Juan-García, A.; Olivero-Verbel, J.J.T. Protective effects of the hydroethanolic extract of *Fridericia chica* on undifferentiated human neuroblastoma cells exposed to α-Zearalenol (α-ZEL) and β-Zearalenol (β-ZEL). *Toxins* **2021**, *13*, 748. [CrossRef] [PubMed]
37. Agahi, F.; Penalva-Olcina, R.; Font, G.; Juan-García, A.; Juan, C. Effects of *Voghiera garlic* extracts in neuronal human cell line against zearalenone's derivates and beauvericin. *Food Chem. Toxicol.* **2022**, *162*, 112905. [CrossRef] [PubMed]
38. Tatay, E.; Espín, S.; García-Fernández, A.J.; Ruiz, M.J. Estrogenic activity of zearalenone, α-zearalenol and β-zearalenol assessed using the E-screen assay in MCF-7 cells. *Toxicol. Mech. Methods* **2018**, *28*, 239–242. [CrossRef]
39. Yu, Z.; Hu, D.; Li, Y. Effects of zearalenone on mRNA expression and activity of cytochrome P450 1A1 and 1B1 in MCF-7 cells. *Ecotoxicol. Environ. Saf.* **2004**, *58*, 187–193. [CrossRef]
40. Molina-Molina, J.-M.; Real, M.; Jimenez-Diaz, I.; Belhassen, H.; Hedhili, A.; Torné, P.; Fernández, M.F.; Olea, N. Assessment of estrogenic and anti-androgenic activities of the mycotoxin zearalenone and its metabolites using in vitro receptor-specific bioassays. *Food Chem. Toxicol.* **2014**, *74*, 233–239. [CrossRef]
41. Flasch, M.; Bueschl, C.; Del Favero, G.; Adam, G.; Schuhmacher, R.; Marko, D.; Warth, B. Elucidation of xenoestrogen metabolism by non-targeted, stable isotope-assisted mass spectrometry in breast cancer cells. *Environ. Int.* **2022**, *158*, 106940. [CrossRef]
42. Shen, M.; Shi, H. Estradiol and estrogen receptor agonists oppose oncogenic actions of leptin in HepG2 cells. *PLoS ONE* **2016**, *11*, e0151455. [CrossRef]
43. Zheng, W.; Fan, W.; Feng, N.; Lu, N.; Zou, H.; Gu, J.; Yuan, Y.; Liu, X.; Bai, J.; Bian, J.; et al. The Role of miRNAs in Zearalenone-Promotion of TM3 Cell Proliferation. *Int. J. Environ. Res. Public Health* **2019**, *16*, 1517. [CrossRef] [PubMed]
44. Androutsopoulos, V.P.; Spandidos, D.A. The flavonoids diosmetin and luteolin exert synergistic cytostatic effects in human hepatoma HepG2 cells via CYP1A-catalyzed metabolism, activation of JNK and ERK and P53/P21 up-regulation. *J. Nutr. Biochem.* **2013**, *24*, 496–504. [CrossRef] [PubMed]
45. Yang, D.; Zhang, X.; Zhang, W.; Rengarajan, T. Vicenin-2 inhibits Wnt/β-catenin signaling and induces apoptosis in HT-29 human colon cancer cell line. *Drug Des. Dev. Ther.* **2018**, *12*, 1303–1310. [CrossRef] [PubMed]
46. Huang, G.; Li, S.; Zhang, Y.; Zhou, X.; Chen, W. Vicenin-2 is a novel inhibitor of STAT3 signaling pathway in human hepatocellular carcinoma. *J. Funct. Foods* **2020**, *69*, 103921. [CrossRef]
47. Mafioleti, L.; da Silva Junior, I.F.; Colodel, E.M.; Flach, A.; de Oliveira Martins, D.T. Evaluation of the toxicity and antimicrobial activity of hydroethanolic extract of *Arrabidaea chica* (Humb. & Bonpl.) B. Verl. *J. Ethnopharmacol.* **2013**, *150*, 576–582.
48. de Medeiros, B.J.L.; dos Santos Costa, K.; Ribeiro, J.F.A.; Silva Jr, J.O.C.; Barbosa, W.L.R.; Carvalho, J.C.T. Liver protective activity of a hydroethanolic extract of *Arrabidaea chica* (Humb. and Bonpl.) B. Verl.(pariri). *Pharmacogn. Res.* **2011**, *3*, 79–84.

49. Takenaka, I.K.T.M.; Amorim, J.M.; de Barros, P.A.V.; Brandão, G.C.; Contarini, S.M.L.; de Sales Souza e Melo, É.; Fernandes, S.O.A. Chemical characterization and anti-inflammatory assessment of the hydroethanolic extract of Fridericia chica. *Rev. Bras. Farmacogn.* **2020**, *30*, 559–567. [CrossRef]
50. Siraichi, J.T.G.; Felipe, D.F.; Brambilla, L.Z.S.; Gatto, M.J.; Terra, V.N.A.; Cecchini, A.L.; Cortez, L.E.R.; Rodrigues-Filho, E.; Cortez, D.A. Antioxidant capacity of the leaf extract obtained from *Arrabidaea chica* cultivated in Southern Brazil. *PLoS ONE* **2013**, *8*, e72733. [CrossRef]
51. Tirado-Ballestas, I.; Alvarez-Ortega, N.; Maldonado-Rojas, W.; Olivero-Verbel, J.; Caballero-Gallardo, K. Oxidative stress and alterations in the expression of genes related to inflammation, DNA damage, and metal exposure in lung cells exposed to a hydroethanolic coal dust extract. *Mol. Biol. Rep.* **2022**, *49*, 4861–4871. [CrossRef]
52. Agahi, F.; Font, G.; Juan, C.; Juan-García, A.J.T. Individual and combined effect of zearalenone derivates and beauvericin mycotoxins on SH-SY5Y cells. *Toxins* **2020**, *12*, 212. [CrossRef]

Disclaimer/Publisher's Note: The statements, opinions and data contained in all publications are solely those of the individual author(s) and contributor(s) and not of MDPI and/or the editor(s). MDPI and/or the editor(s) disclaim responsibility for any injury to people or property resulting from any ideas, methods, instructions or products referred to in the content.

toxins

MDPI

Article

Evaluation of Zearalenones and Their Metabolites in Chicken, Pig and Lamb Liver Samples

Paula Llorens Castelló [1], Matteo Antonio Sacco [2], Isabella Aquila [2], Juan Carlos Moltó Cortés [1] and Cristina Juan García [1,*]

[1] Laboratory of Food Chemistry and Toxicology, Faculty of Pharmacy, University of Valencia, 46100 Burjassot, Spain

[2] Institute of Legal Medicine, Department of Medical and Surgical Sciences, "Magna Graecia", Università degli Studi "Magna Graecia" di Catanzaro, 88100 Catanzaro, Italy

* Correspondence: cristina.juan@uv.es

Abstract: Zearalenone (ZON), zearalanone (ZAN) and their phase I metabolites: α-zearalenol (α-ZOL), β-zearalenol (β-ZOL), α-zearalalanol (α-ZAL) and β-zearalalanol (β-ZAL) are compounds with estrogenic activity that are metabolized and distributed by the circulatory system in animals and can access the food chain through meat products from livestock. Furthermore, biomonitoring of zearalenones in biological matrices can provide useful information to directly assess mycotoxin exposure; therefore, their metabolites may be suitable biomarkers. The aim of this study was to determine the presence of ZON, ZAN and their metabolites in alternative biological matrices, such as liver, from three different animals: chicken, pig and lamb, in order to evaluate their exposure. A solid–liquid extraction procedure coupled to a GC-MS/MS analysis was performed. The results showed that 69% of the samples were contaminated with at least one mycotoxin or metabolite at varying levels. The highest value (max. 152.62 ng/g of β-ZOL) observed, and the most contaminated livers (42%), were the chicken liver samples. However, pig liver samples presented a high incidence of ZAN (33%) and lamb liver samples presented a high incidence of α-ZOL (40%). The values indicate that there is exposure to these mycotoxins and, although the values are low (ranged to 0.11–152.6 ng/g for α-ZOL and β-ZOL, respectively), analysis and continuous monitoring are necessary to avoid exceeding the regulatory limits and to control the presence of these mycotoxins in order to protect animal and human health.

Keywords: liver; zearalenone; zearalanone; chicken; lamb; pig

Key Contribution: Zearalenone (ZON), zearalanone (ZAN) and their phase I metabolites: α-zearalenol (α-ZOL), β-zearalenol (β-ZOL), α-zearalalanol (α-ZAL) and β-zearalalanol (β-ZAL) were analyzed in chicken, pig, and lamb liver in order to evaluate their exposure. The results indicated that there was exposure to these mycotoxins and the values were low.

Citation: Llorens Castelló, P.; Sacco, M.A.; Aquila, I.; Moltó Cortés, J.C.; Juan García, C. Evaluation of Zearalenones and Their Metabolites in Chicken, Pig and Lamb Liver Samples. *Toxins* **2022**, *14*, 782. https://doi.org/10.3390/toxins14110782

Received: 11 October 2022
Accepted: 9 November 2022
Published: 11 November 2022

Publisher's Note: MDPI stays neutral with regard to jurisdictional claims in published maps and institutional affiliations.

Copyright: © 2022 by the authors. Licensee MDPI, Basel, Switzerland. This article is an open access article distributed under the terms and conditions of the Creative Commons Attribution (CC BY) license (https://creativecommons.org/licenses/by/4.0/).

1. Introduction

Zearalenone (ZON) is a common non-steroidal estrogen mycotoxin that was isolated for the first time from maize contaminated by *Fusarium* genera [1]. Different fungi species such as *F. culmorum*, *F. graminearum*, *F. crookwellense*, *F. equiseti* can produce this mycotoxin [2,3]. ZON is a common and potent contaminant of cereals and grain-derived products. It has been detected worldwide in various products including corn, peas, maize, eggs, fish feed, and fibrous feed [4,5]. Corn is the main contaminated cereal and the literature reports a mean value of 14 ng/g of ZON in samples from Morocco [6], 48 ng/g in samples from Germany [7], and 9.45 ng/g in samples from Pakistan [8].

ZON can directly affect foodstuffs intended for mammals. Maximum standards levels have been assessed by Commission Regulation (EC) No. 1881/2006 and Commission

Recommendation No. 2006/576/EC, i.e., 0.1 ng/g in feed for piglets and 0.25 ng/g in feed for porkers [9]. Various adverse effects of ZON on mammals are reported, including endocrine and reproductive disorders but also immunotoxicity, hematotoxicity, genotoxicity, hepatotoxicity [10]. ZON-estrogenic and anabolic properties are due to the binding to estrogen receptors with high affinity. In animals that are more sensitive to this mycotoxin, ZON causes estrogenic disorders such as infertility, uterine hypertrophy, feminization, testicular atrophy, miscarriage, vaginal prolapse or breast enlargement [11,12]. In addition, ZON can cause a decrease in sperm count, and disorders of progesterone and testosterone levels. Direct hepatotoxic effects have also been assessed. Dolenšek et al. have reported increases in hepatocellular necrosis, apoptosis and inflammation of hepatic lobules in pigs fed with contaminated food [13]. For piglets or pigs, WHO has reported the lowest observed adverse effect level (LOAEL) in the range from 17.6 μg/kg bw/day to 200 μg/kg bw/day, while the no effect level (NOEL) ranges from 10.4 μg/kg bw/day to 40 μg/kg [1,14,15].

After oral exposure, ZON is rapidly absorbed in the gastrointestinal tract and then distributed in various organs with slow body elimination. The liver is the most important organ for ZON metabolism, followed by the kidneys, bowel, and reproductive organs. The metabolism of ZON occurs in the intestinal cells, involving the production of two metabolites, i.e., α-zearalenol (α-ZOL) and β-zearalenol (β-ZOL), that are produced through a reduction catalyzed by 3α- and 3β-hydroxy-5-steroid dehydrogenases (HSDs), and then α-zearalanol (α-ZAL) and β-zearalanol (β-ZAL) via double reduction. In addition, zearalanone (ZAN) is produced through a reversible reduction. The metabolites, particularly α-ZAL and α-ZOL, show the main estrogenic activity, while β-ZOL, has shown a lower estrogenic activity [1,2].

Along with mycotoxin analysis in food, biomonitoring of ZON through biological matrices can provide useful information thus directly assessing mycotoxin exposure and metabolites levels [16]. In this way, mycotoxin biomarkers, represented by ZON phase I or phase II metabolites, may be suitable candidates to acquire exposure information for biomonitoring [17,18]. Nevertheless, as there is a lack of standards for quantification of conjugates, cleavage of ZON products by using β-glucuronidase may be a proper alternative to quantify the conjugated mycotoxins. Methods for analyzing ZEN and metabolite presence in biological samples have been developed to study its exposure in animals for clinical or research purposes [19]. Kwaśniewska et al. (2015) reviewed the methodology for determination of ZEN and metabolites in tissue samples [20]. Many methods require a similar procedure of extraction, and clean-up steps prior to instrumental analysis. The most common instrumental methods capable of separating ZEN and metabolites from other compounds are liquid (LC) and gas chromatography (GC) combined with mass spectrometry (MS) to identify them by ion mass. Until now, LC methods are the most used [20,21]; however, recent studies have indicated that GC may be capable of providing similar sensitivity [22]. Methods proposed for ZON quantification in tissue samples are gas chromatography (GC), which can provide good sensitivity for ZON, and metabolites evaluation. However, this does not avoid the necessity of having optimized homogenization and extraction methods prior to its analysis [23,24]. Among biological matrices, body fluids have been widely used to measure ZON intake and risk assessment, including urine, serum, and breast milk. Studies carried out in pigs' urine, the incidence was 100% in Croatian samples (mean ± SD 238 ± 30 μg/L) and 92% in Swedish samples (mean ± SD 2.44 ± 4.39 ng/mL) [25,26]; in pig serum, an incidence of 50% was found in Bulgarian samples (mean ± SD 0.24 ± 0.12 μg/L) and an incidence of 17.3% was found in Romanian samples (mean 0.8 ng/mL) [27,28].

Despite numerous studies focused on body fluids, few papers have measured the levels of ZON and metabolites in animal organs. The liver and kidneys represent important targets of ZON, with the possible accumulation of mycotoxins contributing to slow elimination from the body. Also, these two organs from animal origins are edible products that may be used in numerous preparations and are easily found in markets or supermarkets. Furthermore, illegal commerce of these products in a rural context is a concrete possibility.

For these reasons, determination of mycotoxins on animal organs, such as the liver and kidneys, is mandatory, and the development of adequate procedures for extraction and analytical techniques are highly desirable.

The aim of this paper was to develop and validate a useful method for determining the presence of ZON and metabolites in alternative biological matrices, represented by animal liver and kidney samples (chicken, pig, and lamb) to allow the exposure assessment of ZON, α-ZOL, β-ZOL, ZAN, α-ZAL and β-ZAL. In this work, a solid-liquid extraction procedure coupled with GC-MS/MS analysis was developed for direct determination of ZON and its main metabolites in liver and kidney samples easily obtained from Spanish markets. Critical factors that could affect the extraction efficiency were studied. The procedure was validated and used to quantify the concentrations of free and conjugated mycotoxins in different liver samples by using β-glucuronidase.

2. Results and Discussion

2.1. Validation of the Mycotoxin Determination in Animal Liver

A GC-MS/MS method was used for analysis of ZON, α-ZOL, β-ZOL, ZAN, α-ZAL and β-ZAL in pig, chicken, and lamb liver samples. Previous studies have used LC-MS for quantification of ZON and its metabolites in different tissues; however, GC-MS has been used for analysis of ZON in grains [29]. The concentrations of ZON and its metabolites in liver samples usually occur in units of μg/L; hence it is important to optimize the GC-MS/MS method to determine these possible levels. The suitability of the quantitation method for liver mycotoxin levels was evaluated by a validation study. The GC-MS/MS method was performed for linearity, matrix effect, accuracy, repeatability (intraday and interday precision) and sensitivity, following the EU Commission Decision 2002/657/EC [30].

The limit of detection (LD) and limit of quantification (LQ) were estimated for a signal-to-noise ratio (S/N) ≥ 3 and ≥ 10, respectively, from chromatograms of samples spiked at the lowest level validated. LD and LQ values were established as a mean of the LD and LQ for a mix with all studied matrices, taking into account the possible heterogeneity of the samples (Table 1).

Table 1. Quantification and confirmation transitions, retention time (Rt) and limits of quantification (LQ) and detection (LD) of the analyzed mycotoxins.

Mycotoxin *	Rt (min)	Quantification Transition (m/z)	Confirmation Transition (m/z)	LD (ng/g)	LQ (ng/g)
ZON	16.77	462 > 151	462 > 151	0.061	1.25
α-ZOL	16.74	305 > 289	305 > 289	0.031	0.31
β-ZOL	15.83	536 > 446	536 > 446	0.118	0.63
ZAN	15.84	449 > 335	449 > 335	0.017	0.31
α-ZAL	15.84	433 > 309	433 > 309	0.239	2.50
β-ZAL	15.89	307 > 292	307 > 292	0.361	1.25

* Zearalenone (ZON), zearalanone (ZAN) and their phase I metabolites: α-zearalenol (α-ZOL), β-zearalenol (β-ZOL), α-zearalalanol (α-ZAL) and β-zearalalanol (β-ZAL).

Matrix-matched calibration curves were constructed at concentration levels between the LQ to 1 μg/mL. Correlation between the response and the amount of analytes was verified by plotting signal intensity against analyte concentrations. Good linearity was achieved in all cases with regression coefficients higher than 0.9997. Calibration curves were checked at the end of the analysis to assess the response drift of the method. The matrix effect (ME) was assessed for each analyte by comparing the slope of the standard calibration curve (standard) with that of the matrix-matched calibration curve (matrix), for the same concentration levels (Figure 1). The ME values were suppression signals and ranged between (11 ± 4)% and (27 ± 4)% for α-ZOL in pork liver and ZAN and α-ZAL in chicken liver, respectively. The accuracy of the studied mycotoxins extraction from liver samples was determined by a liver sample fortification procedure (Figure 1). The values of recovery ranged between (104 ± 7)% and (76 ± 9)% for β-ZAL in chicken liver and β-ZAL in pork liver, respectively. The blank was initially prepared and tested negative, and was fortified before the extraction procedure with three different mycotoxin levels at LQ, 2 LQ and 10 LQ (n = 6). Method precision was estimated by calculating the relative standard deviation (RSD_R) using the results obtained during intra-day and inter-day replicates analysis (n = 9). The RSD_R values were bellow to 11% and proved good intra-day and inter-day precision.

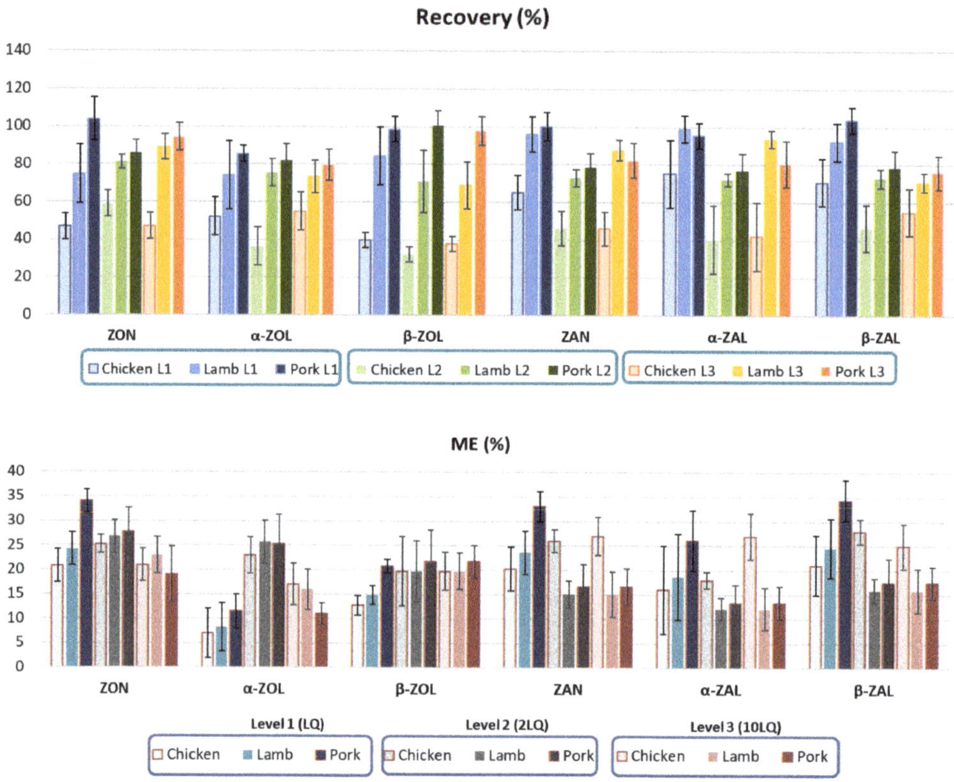

Figure 1. Accuracy (recovery), precision (bar borders RSD_R) and matrix effect (ME) data study at three levels (L1: LQ; L2: 2LQ; L3: 10LQ) in studied livers (Zearalenone (ZON), zearalanone (ZAN) and their phase I metabolites: α-zearalenol (α-ZOL), β-zearalenol (β-ZOL), α-zearalalanol (α-ZAL) and β-zearalalanol (β-ZAL)).

2.2. *Presence of Zearalenone (ZON), Zearalanone (ZAN) and Their Metabolites in Liver Samples*

The natural occurrence of six different mycotoxins (ZON, α-ZOL, β-ZOL, ZAN, α-ZAL and β-ZAL) was investigated in livers from chicken (n = 31), pig (n = 30) and lamb (n = 30). All samples were bought in different supermarkets from the Valencian Community in Spain during the period comprised between 2021 and 2022.

Mycotoxins and metabolites detected in the analyzed liver samples are presented in Table 2. Results show that 63 out of 91 samples (69%) were contaminated with at least one mycotoxin or metabolite at variable levels. The most present mycotoxin by investigated animal samples was β-ZOL (42%) in chicken liver samples, followed by α-ZOL in lamb liver samples (39%), and ZAN for pig liver samples (33%) (Figure 2). The number of positive samples with one mycotoxin was 69% and α-ZAL values were below the sensitivity of the method. A total of 60%, 54% and 6% of the samples of pig, chicken, and lamb, respectively, were not positive for any mycotoxins.

Table 2. Incidence (I), mean (M ± SD), mean of positive samples (Mp ± SD) and range results of detected mycotoxins in analyzed liver samples.

	Chicken Liver (n = 31)				Pig Liver (n = 30)				Lamb Liver (n = 30)			
Analyte *	I	M ± SD (ng/g)	Mp ± SD (ng/g)	Range (ng/g)	I	M ± SD (ng/g)	Mp ± SD (ng/g)	Range (ng/g)	I	M ± SD (ng/g)	Mp ± SD (ng/g)	Range (ng/g)
ZON	9	1.95 ± 5.96	8.94 ± 13.15	0.09–30.79	6	0.02 ± 0.04	0.13 ± 0.02	0.09–0.13	8	0.82 ±1.51	4.08 ± 1.70	1.94–5.91
α-ZOL	8	1.44 ± 4.21	7.46 ± 9.37	0.11–21.50	4	0.05 ± 0.16	0.48 ± 0.43	0.11–0.83	12	6.19 ± 9.16	20.64 ± 10.68	6.62–23.81
β-ZOL	13	7.15 ± 27.46	22.73 ± 55.01	0.25–152.62	2	0.02 ± 0.10	0.49 ± 0.11	0.31–0.43	9	0.58 ± 1.47	2.60 ± 2.94	0.19–4.97
ZAN	12	2.85 ± 7.92	9.83 ± 15.48	0.78–43.33	10	0.29 ± 0.43	1.17 ± 0.26	0.75–1.36	11	1.03 ± 1.49	3.75 ± 1.31	1.50–4.94
β-ZAL	3	1.49 ± 6.21	20.54 ± 21.38	6.01–33.92	2	0.04 ± 0.19	0.87 ± 0.66	0.30–1.00	9	5.61 ± 8.96	24.92 ± 5.33	15.33–24.71

* Zearalenone (ZON), zearalanone (ZAN) and their phase I metabolites: α-zearalenol (α-ZOL), β-zearalenol (β-ZOL), α-zearalalanol (α-ZAL) and β-zearalalanol (β-ZAL); n: number of samples analysed.

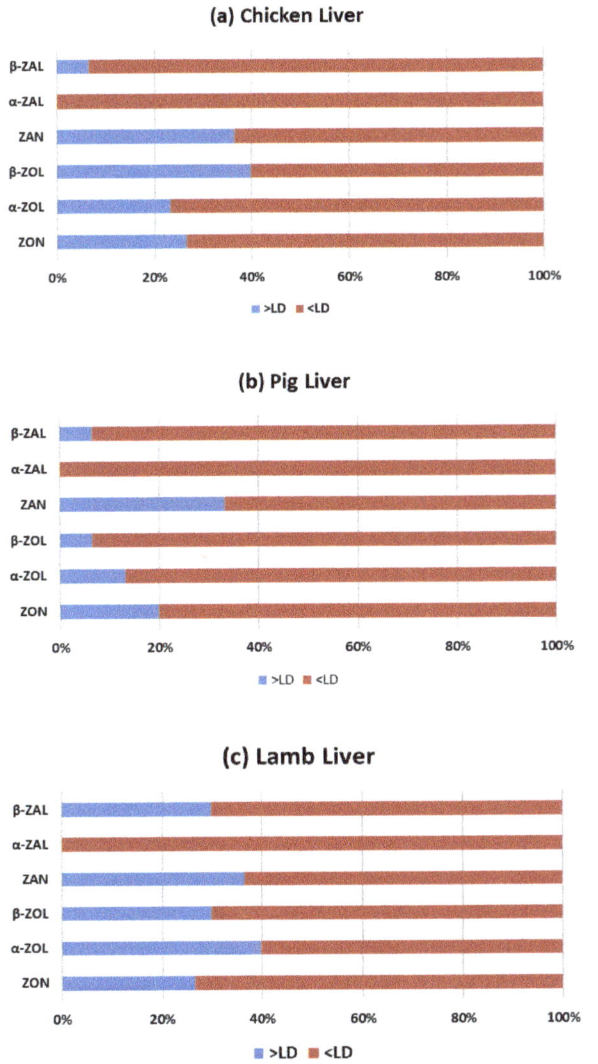

Figure 2. Frequency (%) of the studied mycotoxins and its metabolites in chicken (**a**), pig (**b**) and lamb (**c**) livers samples. (LD: limit of detection; Zearalenone (ZON), zearalanone (ZAN) and their phase I metabolites: α-zearalenol (α-ZOL), β-zearalenol (β-ZOL), α-zearalalanol (α-ZAL) and β-zearalalanol (β-ZAL).

Regarding the animal tissue used for the analysis, the highest incidence of mycotoxins was associated with lamb liver samples (93%), followed by chicken (46%) and pig (40%). Concerning the ranges, the highest ranges were found in chicken liver samples (LD-152.62 ng/g), followed by lamb liver samples (LD-24.71 ng/g) and pig liver samples (LD-1.36 ng/g), with the highest levels of β-ZOL, β-ZAL and ZAN, respectively (Table 2).

α- and β-ZON are important and accurate markers for exposure to these mycotoxins and usually analyzed in urine, due to, their values change rapidly in urine [31]. The detected mycotoxins in animal livers indicate a chronic exposure to these mycotoxins and are a possible risk to the animal's health due to the adverse effects that they can produce [32], specifically in pigs, the most sensitive animal species, in terms of the oestrogenic activity of zearalenone and its metabolites. Furthermore, Gajęcka et al., 2016 [33] reported that α-ZOL has a higher binding affinity to estrogen receptors than ZON [32].

2.2.1. Zearalenone (ZON) and Its Metabolites Occurrence

The analytical data showed that the ZON metabolite with the highest incidence was β-ZOL for chicken liver samples, with 42% (13 out of 31 samples) at maximum levels of 152.62 ng/g, ZON for pig liver samples with 20% (6 out of 30 samples) at maximum levels of 0.13 ng/g and α-ZOL for lamb liver samples with 39% (12 out of 30 samples) at maximum levels of 23.81 ng/g. Levels detected for ZON, α-ZOL and β-ZOL were below 30.79 ng/g, 23.81 ng/g and 152.61 ng/g, respectively (Table 2). In cases where the samples were being cooked and eaten by consumers, the exposure would be small because if it is calculated as the daily intake, the value will be below the TDI reported by EFSA of 0.25 μg/kg bw for ZON [34].

In previous studies, very limited reports have been documented for the presence of ZON contaminating chicken, pig or lamb liver. However, Iqbal et al. (2014) [35] reported low levels of ZON at 2.97 μg/kg, 4.91 μg/kg and 5.10 μg/kg in domestic chicken liver, boiler breed chicken liver, and layer breed chicken liver, respectively. Other studies reported the presence of ZON in the bile of breeding cows with a contamination rate of 96.2% [36].

Liver and enterocytes play an important role in ZON metabolism; in fact, it varies depending on the animal. This variation may be related to hepatic biotransformation. The literature reveals that in guinea pigs, both α-ZOL and β-ZOL were formed in equal amounts [37]; in pigs, α-ZOL is formed in a higher amount compared to β-ZOL; whereas, in chicken, β-ZOL is produced in high quantities by the hepatic microsomes [37,38]. According to these studies, our results in chicken livers showed higher amounts of β-ZOL (22.73 ± 55.0 ng/g) than α-ZOL (7.46 ± 9.37 ng/g). It is important to note that β-ZOL presented higher potential estrogenic amounts than ZON and α-ZOL [2].

2.2.2. Zearalanone (ZAN) and Its Metabolites Occurrence

The highest incidence was for ZAN for all liver samples: 12 out of 31 chicken liver samples (39%) (max. 7.92 ng/g), 11 out 30 lamb liver samples (37%) (max. 4.94 ng/g)) and 10 out of 30 pig liver samples (33%) (max. 1.36 ng/g) (Table 2).

α-ZAL, a resorcyl lactone, was not detected in any of the samples analyzed, and these results are in agreement with the literature [31]. It is important to remar, that α-ZAL was used as a growth promoter in the United States many years ago but nowadays it is banned [39].

Other authors have detected smaller quantities for α-ZAL than the rest of analyzed zearalenones. Döll et al. (2003) reveal that in piglets' liver, ZAN, α-ZAL and β-ZAL were below 100, 50 and 200 ng/g, respectively [40]. Moreover, it has been shown that in the enzymatic reduction of ZAN, smaller amounts of zearalenols are produced. Malekinejad et al. (2006) [37] reported differences between mammalian species in the hepatic transformation of ZAN to its reduced and glucuronide metabolites. All these mammalian species converted large percentages of ZAN and metabolites to the corresponding glucuronides [37]. On the other hand, the comparison between species suggests that pigs, which preferentially produce α-ZAL over the β analogue by five-fold, are predicted to be more sensitive to the oestrogenic effects of ZAN than other animal species [37].

In others studies, few levels of α-ZAL were observed in pig; indeed, a significant fraction of ZAN was found in the form of α-ZAL and its respective glucuronide conjugates [41], while cows converted ZAN predominantly to β-ZAL [42]. Smaller amounts of further reduced metabolites (i.e., α- and β-ZAL) were observed in other ruminant species [31]. However, ovine metabolism of ZON produces at least five compounds, including α- and β-ZOL, α- and β-ZAL, and ZAN [43].

2.2.3. Simultaneous Presence of Analyzed Mycotoxins and Metabolites

The natural copresence of analyzed mycotoxins was evaluated in all animal livers bought in the Valencian Community (Spain) in order to have an approximation of the oral exposure of these animals. Previous studies indicate that ZON is usually found to co-occurrence with its metabolites [19].

All in all, from a descriptive standpoint, a co-occurence of different metabolites was found in 30% of analyzed samples. The results show a combination of five mycotoxins in 2.2% different samples (ZON + α-ZOL + β-ZOL + ZAN + β-ZAL) (Figure 3). While, four and three associations were also observed in 8.8% (ZON + α-ZOL + β-ZOL + ZAN and ZON + α-ZOL + ZAN + β-ZAL) and 5.5% samples (β-ZOL + ZAN + β-ZAL, ZON + β-ZOL + β-ZAL and ZON + β-ZOL + ZAN), respectively.

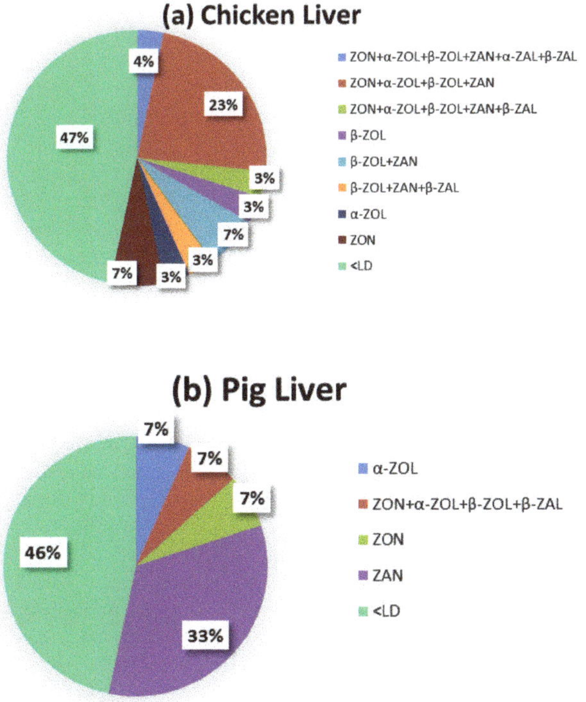

Figure 3. *Cont.*

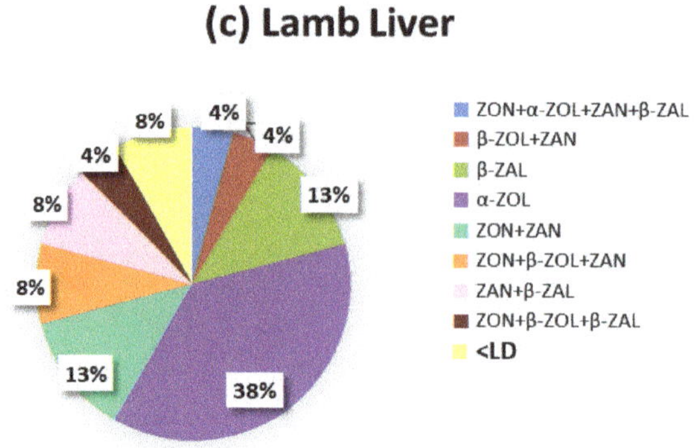

Figure 3. Co-occurrence of mycotoxins and metabolites in the three of analyzed animal liver samples (**a**) chicken, (**b**) pig and (**c**) lamb. (Zearalenone (ZON), zearalanone (ZAN) and their phase I metabolites: α-zearalenol (α-ZOL), β-zearalenol (β-ZOL), α-zearalalanol (α-ZAL) and β-zearalalanol (β-ZAL)).

It was also observed that 39%, 7%, and 47% of chicken, pig, and lamb liver samples, respectively, were contaminated by at least two toxins. The lowest co-occurrence frequency was found in pig liver samples with 7% of samples contaminated with four mycotoxins (ZON + α-ZOL + β-ZOL + β-ZAL). In chicken liver samples, 16% showed the combination ZON + α-ZOL + β-ZOL + ZAN, followed by 13% β-ZOL + ZAN, 7% ZON + α-ZOL + β-ZOL + ZAN + β-ZAL and 3% β-ZOL + ZAN + β-ZAL. A total of 13% of samples showed only one mycotoxin or metabolite and 48% of the samples were <LQ. Lamb liver samples showed the highest binary combinations (30%): ZON + ZAN (10%), β-ZOL + ZAN (10%), ZAN + β-ZAL (7%) and β-ZOL + β-ZAL (3%). Only four samples (13%) showed the combination of three mycotoxins (ZON + β-ZOL + ZAN and ZON + β-ZOL + β-ZAL) and one sample (3%) a combination of four mycotoxins (ZON + α-ZOL + ZAN + β-ZAL). A total of 47% of the samples did not show a co-occurrence and 7% of the samples were <LQ. Lastly, pig liver samples had the lowest co-occurrence since only 7% of the samples (2 out of 30 samples) were contaminated by more than one mycotoxin or metabolite (ZON + α-ZOL+β-ZOL + β-ZAL). A total of 53% of the samples did not show co-occurrence and 40% of the samples were <LQ.

3. Conclusions

The validated GC-MS/MS method presented good results in terms of accuracy, sensitivity and robustness for the simultaneous determination of six target mycotoxins in three different type of animal liver (chicken, lamb and pig). The method was suitable for analyzing 91 animal liver samples. The analytical data showed that 69% of analyzed liver samples were contaminated with at least one of the analyzed mycotoxins. β-ZOL was the most detected (42%), with the highest value (max. 152.62 ng/g) observed in chicken liver samples. This can be associated with the fact that chickens are mainly fed with corn or feed rich in corn, and this cereal is the ideal substrate for grown *Fusarium graminearum*, which is the primary producer of ZON [5,7,15,40,44]. In this sense, the implementation of a hazard analysis and critical control point (HACCP) system should be applied throughout the food chain from primary production to final consumer to reduce the presence and production of mycotoxins in feed. Furthermore, control systems to analyze and monitor mycotoxins should be strengthened to prevent exposure and protect animal and human health.

4. Material and Methods

4.1. Standards

Mycotoxin standards and metabolites specifically ZON, α-ZOL, β-ZOL, ZAN, α-ZAL and β-ZAL were obtained from Sigma-Aldrich (St. Louis, MO, USA). Individual stock solutions of all analytes were prepared at identical concentration (1000 mg/L) in methanol. The stock solutions were diluted with acetonitrile to obtain a working standard solutions of 50 mg/L with the six mycotoxins. All standards were stored in darkness and kept at −20 °C until the GC-MS/MS analysis.

4.2. Chemical, Reagents and Other Material

The derivatization reagent composed of BSA (N,O-bis(trimethylsilyl)acetamide) + TMCS (trimethylchlorosilane) + TMSI (N-trimethylsilyimidazole) (3:2:3) was purchased from Supelco (Bellefonte, PA, USA).

Sodium dihydrogen phosphate and disodium hydrogen phosphate, used to prepare phosphate buffer, were acquired from Panreac Quimica S.L.U. (Barcelona, Spain). β-Glucuronidase Type H-1 from Helix pomatia (glucuronidase activity: ≥300,000 units/g solid and sulfatase activity: ≥10,000 units/g solid) was purchased from Sigma–Aldrich (St. Louis, MO, USA).

All solvents, acetonitrile, hexane and methanol (HPLC grade) were purchased from Merck KGaA (Darmstadt, Germany). Anhydrous magnesium sulfate (thin powder) was obtained from Alfa Aesar GmbH & Co (Karlsruhe, Germany); sodium chloride was purchased from Merck and C18-E (50 μm, 65A) was purchased from Phenomenex (Torrance, CA, USA).

4.3. Apparatus

ZON, α-ZOL, β-ZOL, ZAN, α-ZAL and β-ZAL were analyzed by GC-MS/MS. Aliquots of 1 μL of the derivatized extract were injected in splitless mode at 250 °C in programmable temperature vaporization (PTV). A GC system Agilent 7890A coupled with an Agilent 7000A triple quadrupole mass spectrometer with inert electron-impact ion source and an Agilent 7693 autosampler (Agilent Technologies, Palo Alto, CA, USA) were used for MS/MS analysis [44]. An HP-5 MS 30 m × 0.25 mm × 0.25 μm capillary column was used. All analytes eluted within 17 min, reaching the requirement for a high throughout determination.

The MS was operating in electron impact ionization (EI, 70 eV). The source and transfer line temperatures were 230 °C and 280 °C, respectively. The collision gas for MS/MS experiments was nitrogen, and the helium was used as carrier gas at a fixed pressure of 20.3 psi, both at 99.999% purity supplied by Carburos Metálicos S.L. (Barcelona, Spain). The oven temperature program was initially 80 °C, and the temperature increased to 245 °C at 60 °C/min. After a 3-min hold time, the temperature was increased to 260 °C progressively at 3 °C/min and finally to 270 °C at 10 °C/min and then held for 10 min. The analysis was performed with a solvent delay of 3 min in order to prevent instrument damage.

Quantitation data were acquired at SRM mode, with a couple of transition ions, and the mass spectrometer operated in electrospray ionization (EI) mode (Table 1). The transfer line and source temperatures were 280 °C and 230 °C, respectively. The EI energy used was 70 eV as in that region the maximum abundance was observed. The collision energies varied from 5 to 20 eV, depending on the precursor and product ions. The analysis was performed with a filament-multiplier delay of 3.50 min. The collision gas for MS/MS experiments was nitrogen, and the helium was used as quenching gas, both at 99.999% purity supplied by Carburos Metálicos S.L. (Barcelona, Spain). The dwell times also varied from 5 to 35 eV. Data was acquired and processed using the Agilent Masshunter version B.04.00 software.

4.4. Sample Preparation

Livers were analyzed for total ZON, α-ZOL, β-ZOL, ZAN, α-ZAL and β-ZAL concentrations, including their conjugated glucuronides.

For the β-Glucuronidase hydrolysis, the enzymatic hydrolysis method used to deconjugate glucuronides was adapted from [45]. Each sub-sample of tissue (0.5 g) was homogenized in 0.64 mL pH 5.0 ammonium acetate buffer (1 M, BioWorld, Fisher Scientific,

1768, Pittsburgh, PA, USA). Homogenization was carried out using a Polytron PT 10–35 with PTA-10T generator (Kinematica AG, Luzern, Switzerland). After homogenization, 10 μL of β-glucuronidase (Helix pomatia, H-2, Millipore Sigma, Saint Louis, MO, USA) containing 100,000 U of β-glucuronidase/mL was added to the mixture, incubated at 37 °C for 18 h, then allowed to cool to room temperature. Methods for extraction of ZEN and α-ZEL from enzyme-hydrolyzed liver tissue were adapted from standard mycotoxin analysis methods previously described by Mahmoud et al. (2018) [46].

Then, 1.5 mL of acetonitrile was added to 0.5 g of liver and vortexed for 1 min. It was sonicated 10 min at room temperature and centrifuged at 4000 rpm for 3 min at 5 °C. Supernatant was collected in a 15 mL falcon and 150 mg of $MgSO_4$ and 50 mg of C18 were added prior to be vortexed for 1 min, sonicated 10 min at room temperature and centrifuged at 4000 rpm for 3 min at 5 °C. Then, the upper layer was collected in a vial and it was evaporated to dryness under nitrogen flow.

The dry extract was derivatized with 50 mL of BSA + TMCS + TMSI (3:2:3) and left for 30 min at room temperature. After that, it was diluted to 200 μL with hexane and mixed thoroughly on a vortex for 30 s. Then, the diluted derivatized sample was added with 1 mL of phosphate buffer (60 mM, pH 7) to purify the derivate with a liquid–liquid extraction and the upper layer (hexane phase) was transferred to an autosampler vial for the chromatographic analysis.

4.5. Sampling

Livers (n = 91) were purchased from different supermarkets of the Valencian Community (Spain) during the period comprised between October 2021 and February 2022. Chicken liver (n = 31), pork liver (n = 30) and lamb liver (n = 30) were collected and milled separately. Then, 0.5 g of each sample were weighted and kept at −20 °C in dark and dry place until further analysis.

4.6. Method Validation

Commission Decision 2002/657/EC [30] was used as guidelines for the validation studies. All the parameters were evaluated by spiking blank samples. For identification purposes, retention times of mycotoxins in standards and samples were compared at tolerance of ±0.5%. Moreover, in accordance with the 2002/657/EC Decision [30], the relative ion intensity of analytes studied in the standard solution and the spiked samples at the concentration levels used for the calibration curve were compared.

Method performance characteristics such as linearity, LD, LQ, matrix effect, extraction recovery, repeatability and reproducibility were evaluated for all tested mycotoxins.

Author Contributions: Conceptualization, M.A.S., P.L.C. and C.J.G.; methodology, M.A.S., P.L.C. and C.J.G.; validation, M.A.S., P.L.C. and C.J.G.; formal analysis, M.A.S., P.L.C. and C.J.G.; data curation, M.A.S., P.L.C. and C.J.G.; writing, original draft preparation, P.L.C., M.A.S. and C.J.G.; writing—review and editing, P.L.C., M.A.S., C.J.G., I.A. and J.C.M.C.; supervision, C.J.G., I.A. and J.C.M.C. All authors have read and agreed to the published version of the manuscript.

Funding: This research received no external funding.

Institutional Review Board Statement: Not applicable.

Informed Consent Statement: Not applicable.

Data Availability Statement: Not applicable.

Acknowledgments: This project was financially supported by the Spanish Ministry of Science and Innovation PID2019-108070RB-I00ALI.

Conflicts of Interest: The authors declare no conflict of interest.

References

1. Ropejko, K.; Twarużek, M. Zearalenone and Its Metabolites-General Overview, Occurrence, and Toxicity. *Toxins* **2021**, *13*, 35. [CrossRef] [PubMed]
2. Rai, A.; Das, M.; Tripathi, A. Occurrence and toxicity of a fusarium mycotoxin, zearalenone. *Crit. Rev. Food Sci. Nutr.* **2020**, *60*, 2710–2729. [CrossRef] [PubMed]
3. Gil-Serna, J.; Vázquez, C.; González-Jaén, M.T.; Patiño, B. Mycotoxins. Toxicology. In *Encyclopedia of Food Microbiology*; Batt, C.A., Tortorello, M.L., Eds.; Academic Press: Cambridge, MA, USA, 2014; pp. 887–892.
4. Omar, S.S. Prevalence, level and health risk assessment of mycotoxins in the fried poultry eggs from Jordan. *Environ. Res.* **2021**, *200*, 111701. [CrossRef] [PubMed]
5. Ochieng, P.E.; Scippo, M.L.; Kemboi, D.C.; Croubels, S.; Okoth, S.; Kang'ethe, E.K.; Doupovec, B.; Gathumbi, J.K.; Lindahl, J.F.; Antonissen, G. Mycotoxins in Poultry Feed and Feed Ingredients from Sub-Saharan Africa and Their Impact on the Production of Broiler and Layer Chickens: A Review. *Toxins* **2021**, *13*, 633. [CrossRef]
6. Zinedine, A.; Brera, C.; Elakhdari, S.; Catano, C.; Debegnach, F.; Angelini, S.; de Santis, B.; Faid, M.; Benlemlih, M.; Minardi, V.; et al. Natural occurrence of mycotoxins in cereals and spices commercialized in Morocco. *Food Control* **2006**, *17*, 868–874. [CrossRef]
7. Schollenberger, M.; Müller, H.-M.; Rüfle, M.; Suchy, S.; Plank, S.; Drochner, W. Natural Occurrence of 16 *Fusarium* Toxins in Grains and Feedstuffs of Plant Origin from Germany. *Mycopathologia* **2006**, *161*, 43–52. [CrossRef]
8. Iqbal, S.Z.; Rabbani, T.; Asi, M.R.; Jinap, S. Assessment of aflatoxins, ochratoxin A and zearalenone in breakfast cereals. *Food Chem.* **2014**, *157*, 257–262. [CrossRef]
9. European Commission. Regulation of the European Commission (EC) No. 1881/2006 of December 19, 2006, as Amended Fixing Maximum Levels for Certain Contaminants in Foodstuffs (OJ. L. 364/5 of 20.12.2006, Annex "Maximum Levels for Certain Contaminants in foodstuffs". *Off. J. Eur. Union* **2006**, *364*, 5–24.
10. Lu, Q.; Luo, J.Y.; Ruan, H.N.; Wang, C.J.; Yang, M.H. Structure-toxicity relationships, toxicity mechanisms and health risk assessment of food-borne modified deoxynivalenol and zearalenone: A comprehensive review. *Sci. Total Environ.* **2022**, *806*, 151192. [CrossRef]
11. Kinkade, C.W.; Rivera-Núñez, Z.; Gorcyzca, L.; Aleksunes, L.M.; Barrett, E.S. Impact of Fusarium-Derived Mycoestrogens on Female Reproduction: A Systematic Review. *Toxins* **2021**, *13*, 373. [CrossRef]
12. Tian, Y.; Zhang, M.Y.; Zhao, A.H.; Kong, L.; Wang, J.J.; Shen, W.; Li, L. Single-cell transcriptomic profiling provides insights into the toxic effects of Zearalenone exposure on primordial follicle assembly. *Theranostics* **2021**, *11*, 5197–5213. [CrossRef] [PubMed]
13. Dolenšek, T.; Švara, T.; Knific, T.; Gombač, M.; Luzar, B.; Jakovac-Strajn, B. The Influence of *Fusarium* Mycotoxins on the Liver of Gilts and Their Suckling Piglets. *Animals* **2021**, *11*, 2534. [CrossRef] [PubMed]
14. Joint FAO/WHO Expert Committee on Food Additives. *Safety Evaluation of Certain Food Additives and Contaminants: Prepared by the Seventy-Third Meeting of the Joint FAO/WHO Expert Committee on Food Additives (JECFA)*; World Health Organization: Geneva, Switzerland, 2011.
15. EFSA Panel on Contaminants in the Food Chain (CONTAM); Knutsen, H.K.; Alexander, J.; Barregård, L.; Bignami, M.; Brüschweiler, B.; Ceccatelli, S.; Cottrill, B.; Dinovi, M.; Edler, L.; et al. Risks for animal health related to the presence of zearalenone and its modified forms in feed. *EFSA J.* **2017**, *15*, 04851.
16. Carballo, D.; Pallarés, N.; Ferrer, E.; Barba, F.J.; Berrada, H. Assessment of Human Exposure to Deoxynivalenol, Ochratoxin A, Zearalenone and Their Metabolites Biomarker in Urine Samples Using LC-ESI-qTOF. *Toxins* **2021**, *13*, 530. [CrossRef]
17. Schmidt, J.; Cramer, B.; Turner, P.C.; Stoltzfus, R.J.; Humphrey, J.H.; Smith, L.E.; Humpf, H.U. Determination of Urinary Mycotoxin Biomarkers Using a Sensitive Online Solid Phase Extraction-UHPLC-MS/MS Method. *Toxins* **2021**, *13*, 418. [CrossRef]
18. Foerster, C.; Ríos-Gajardo, G.; Gómez, P.; Muñoz, K.; Cortés, S.; Maldonado, C.; Ferreccio, C. Assessment of Mycotoxin Exposure in a Rural County of Chile by Urinary Biomarker Determination. *Toxins* **2021**, *13*, 439. [CrossRef]
19. Pack, E.; Stewart, J.; Rhoads, M.; Knight, J.; de Vita, R.; Clark-Deener, S. Quantification of zearalenone and α-zearalenol in swine liver and reproductive tissues using GC-MS. *Toxicon X* **2020**, *8*, 100058. [CrossRef]
20. Kwaśniewska, K.; Gadzała-Kopciuch, R.; Cendrowski, K. Analytical procedure for the determination of zearalenone in environmental and biological samples. *Crit. Rev. Anal. Chem.* **2015**, *45*, 119–130. [CrossRef]
21. Singh, J.; Mehta, A. Rapid and sensitive detection of mycotoxins by advanced and emerging analytical methods: A review. *Food Sci. Nutr.* **2020**, *8*, 2183–2204. [CrossRef]
22. Kinani, S.; Bouchonnet, S.; Bourcier, S.; Porcher, J.-M.; Aït-Aïssa, S. Study of the chemical derivatization of zearalenone and its metabolites for gas chromatography–mass spectrometry analysis of environmental samples. *J. Chromatogr. A* **2008**, *1190*, 307–315. [CrossRef]
23. Niknejad, F.; Escrivá, L.; Adel Rad, K.B.; Khoshnia, M.; Barba, F.J.; Berrada, H. Biomonitoring of Multiple Mycotoxins in Urine by GC-MS/MS: A Pilot Study on Patients with Esophageal Cancer in Golestan Province, Northeastern Iran. *Toxins* **2021**, *13*, 243. [CrossRef] [PubMed]
24. Alvarez-Ortega, N.; Caballero-Gallardo, K.; Taboada-Alquerque, M.; Franco, J.; Stashenko, E.E.; Juan, C.; Juan-García, A.; Olivero-Verbel, J. Protective Effects of the Hydroethanolic Extract of *Fridericia chica* on Undifferentiated Human Neuroblastoma Cells Exposed to α-Zearalenol (α-ZEL) and β-Zearalenol (β-ZEL). *Toxins* **2021**, *13*, 748. [CrossRef]

25. Pleadin, J.; Mihaljević, Ž.; Barbir, T.; Vulić, A.; Kmetič, I.; Zadravec, M.; Brumen, V.; Mitak, M. Natural incidence of zearalenone in Croatian pig feed, urine and meat in 2014. *Food Addit. Contam. Part B* **2015**, *80*, 277–283. [CrossRef] [PubMed]
26. Gambacorta, L.; Olsen, M.; Solfrizzo, M. Pig Urinary Concentration of Mycotoxins and Metabolites Reflects Regional Differences, Mycotoxin Intake and Feed Contaminations. *Toxins* **2019**, *11*, 378. [CrossRef] [PubMed]
27. Stoev, S.D.; Dutton, M.F.; Njobeh, P.B.; Mosonik, J.S.; Steenkamp, P.A. Mycotoxic nephropathy in Bulgarian pigs and chickens: Complex aetiology and similarity to Balkan Endemic Nephropathy. *Food Addit. Contam. Part A* **2010**, *27*, 72–88. [CrossRef] [PubMed]
28. Curtui, V.G.; Gareis, M.; Usleber, E.; Märtlbauer, E. Survey of Romanian slaughtered pigs for the occurrence of mycotoxins ochratoxins A and B, and zearalenone. *Food Addit. Contam.* **2001**, *18*, 730–738. [CrossRef]
29. Tanaka, T.; Yoneda, A.; Inoue, S.; Sugiura, Y.; Ueno, Y. Simultaneous determination of trichothecene mycotoxins and zearalenone in cereals by gas chromatography–mass spectrometry. *J. Chromatogr. A.* **2000**, *882*, 23–28. [CrossRef]
30. EC (European Commission). Commission Decision 2002/657/EC of 12 August 2002 implementing Council Directive 96/23/EC concerning the performance of analytical methods and the interpretation of results. *Off. J. Eur. Communities* **2002**, *50*, 8–36.
31. Zöllner, P.; Jodlbauer, J.; Kleinova, M.; Kahlbacher, H.; Kuhn, T.; Hochsteiner, W.; Lindner, W. Concentration Levels of Zearalenone and Its Metabolites in Urine, Muscle Tissue, and Liver Samples of Pigs Fed with Mycotoxin-Contaminated Oats. *J. Agric. Food Chem.* **2002**, *50*, 2494–2501. [CrossRef]
32. EFSA. Scientific Opinion on the risks for public health related to the presence of zearalenone in food. *EFSA J.* **2011**, *9*, 2197. [CrossRef]
33. Gajęcka, M.; Sławuta, P.; Nicpoń, J.; Kołacz, R.; Kiełbowicz, Z.; Zielonka, Ł.; Dąbrowski, M.; Szweda, W.; Gajęcki, M.; Nicpoń, J. Zearalenone and its metabolites in the tissues of female wild boars exposed *per os* to mycotoxins. *Toxicon* **2016**, *114*, 1–12. [CrossRef] [PubMed]
34. Mally, A.; Solfrizzo, M.; Degen, G.H. Biomonitoring of the mycotoxin Zearalenone: Current state-of-the art and application to human exposure assessment. *Arch. Toxicol.* **2016**, *90*, 1281–1292. [CrossRef]
35. Iqbal, S.Z.; Nisar, S.; Así, M.R.; Jinap, S. Natural incidence of aflatoxins, ochratoxin A and zearalenone in chicken meat and eggs. *Food Control* **2014**, *43*, 98–103. [CrossRef]
36. Meyer, K.; Usleber, E.; Märtlbauer, E.; Bauer, J. Occurrence of zearalenone, alpha- and beta-zearalenol in bile of breeding sows in relation to reproductive performance. *Berl. Munch. Tierarztl. Wochenschr.* **2000**, *113*, 374–379. [PubMed]
37. Malekinejad, H.; Maas-Bakker, R.; Fink-Gremmels, J. Species differences in the hepatic biotransformation of zearalenone. *Vet. J.* **2006**, *172*, 96–102. [CrossRef] [PubMed]
38. Zinedine, A.; Soriano, J.M.; Molto, J.C.; Mañes, J. Review on the toxicity, occurrence, metabolism, detoxification, regulations and intake of zearalenone: An oestrogenic mycotoxin. *Food Chem. Toxicol.* **2007**, *45*, 1–18. [CrossRef]
39. Mirocha, C.J.; Schauerhamer, B.; Christensen, C.M.; Niku-Paavola, M.L.; Nummi, M. Incidence of zearalenol (Fusarium mycotoxin) in animal feed. *Appl. Environ. Microbiol.* **1979**, *38*, 749–750. [CrossRef]
40. Döll, S.; Dänicke, S.; Ueberschär, K.H.; Valenta, H.; Schnurrbusch, U.; Ganter, M.; Klobasa, F.; Flachowsky, G. Effects of graded levels of Fusarium toxin contaminated maize in diets for female weaned piglets. *Arch. Tierernährung* **2003**, *57*, 311–334.
41. Biehl, M.L.; Prelusky, D.B.; Koritz, G.D.; Hartin, K.E.; Buck, W.B.; Trenholm, H.L. Biliary excretion and enterohepatic cycling of zearalenone in immature pigs. *Toxicol. Appl. Pharmacol.* **1993**, *121*, 152–159. [CrossRef]
42. Jodlbauer, J.; Zöllner, P.; Lindner, W. Determination of zeranol, taleranol, zearalenone, alpha- and beta-zearalenol in urine and tissue by high-performance liquid chromatography-tandem mass spectrometry. *Chromatographia* **2000**, *51*, 681–687. [CrossRef]
43. Miles, C.O.; Erasmuson, A.F.; Wilkins, A.L.; Towers, N.R.; Smith, B.L.; Garthwaite, I.; Scahill, B.G.; Hansen, R.P. Ovine metabolism of zearalenone to alpha-zearalanol (zeranol). *J. Agric. Food Chem.* **1996**, *44*, 3244–3250. [CrossRef]
44. Juan, C.; Oueslati, S.; Mañes, J.; Berrada, H. Multimycotoxin Determination in Tunisian Farm Animal Feed. *J. Food Sci.* **2019**, *84*, 3885–3893. [CrossRef] [PubMed]
45. Lattanzio, V.M.T.; Solfrizzo, M.; de Girolamo, A.; Chulze, S.N.; Torres, A.M.; Visconti, A. LC–MS/MS characterization of the urinary excretion profile of the mycotoxin deoxynivalenol in human and rat. *J. Chromatogr. B* **2011**, *879*, 707–715. [CrossRef]
46. Mahmoud, A.F.; Escrivá, L.; Rodríguez-Carrasco, Y.; Moltó, J.C.; Berrada, H. Determination of trichothecenes in chicken liver using gas chromatography coupled with triple-quadrupole mass spectrometry. *LWT-Food Sci. Technol.* **2018**, *93*, 237–242. [CrossRef]

toxins

Article

Zearalenone Exposure Affects the Keap1–Nrf2 Signaling Pathway and Glucose Nutrient Absorption Related Genes of Porcine Jejunal Epithelial Cells

Qun Cheng [1], Shuzhen Jiang [2], Libo Huang [2], Yuxi Wang [3] and Weiren Yang [2,*]

[1] Department of Animal Sciences and Technology, Qingdao Agricultural University, Qingdao 266109, China
[2] Department of Animal Sciences and Technology, Shandong Agricultural University, Taian 271018, China
[3] Agriculture and Agri-Food Canada, Lethbridge Research and Development Centre, Lethbridge, AB T1J 4B1, Canada
* Correspondence: wryang@sdau.edu.cn

Abstract: This study aims to examine the impact of zearalenone (ZEA) on glucose nutrient absorption and the role of the Kelch-like erythroid cell-derived protein with CNC homology-associated protein 1 (Keap1)–nuclear factor erythroid 2-related factor 2 (Nrf2) signaling pathway in zearalenone-induced oxidative stress of porcine jejunal epithelial cells (IPEC-J2). For 24 and 36 h, the IPEC-J2 cells were exposed to ZEA at concentrations of 0, 10, 20, and 40 (Control, ZEA10, ZEA20, ZEA40) mol/L. With the increase of ZEA concentration and prolongation of the action time, the apoptosis rate and malondialdehyde level and relative expression of sodium-dependent glucose co-transporter 1 (Sglt1), glucose transporter 2 (Glut2), Nrf2, quinone oxidoreductase 1 (Nqo1), and hemeoxygenase 1 (Ho1) at mRNA and protein level, fluorescence intensity of Nrf2 and reactive oxygen species increased significantly ($p < 0.05$), total superoxide dismutase and glutathione peroxidase activities and relative expression of Keap1 at mRNA and protein level, fluorescence intensity of Sglt1 around the cytoplasm and the cell membrane of IPEC-J2 reduced significantly ($p < 0.05$). In conclusion, ZEA can impact glucose absorption by affecting the expression of Sglt1 and Glut2, and ZEA can activate the Keap1-Nrf2 signaling pathway by enhancing Nrf2, Nqo1, and Ho1 expression of IPEC-J2.

Keywords: zearalenone; intestinal porcine jejunal epithelial cells; Keap1-Nrf2 signaling pathway; oxidative stress; glucose nutrient absorption

Key Contribution: The results demonstrated that ZEA increased the expression of Nrf2, Nqo1, ROS, and Ho1 while interfering with the expression of Sglt1 and Glut2, hence impacting IPEC-J2 cells.

Citation: Cheng, Q.; Jiang, S.; Huang, L.; Wang, Y.; Yang, W. Zearalenone Exposure Affects the Keap1–Nrf2 Signaling Pathway and Glucose Nutrient Absorption Related Genes of Porcine Jejunal Epithelial Cells. *Toxins* **2022**, *14*, 793. https://doi.org/10.3390/toxins14110793

Received: 18 June 2022
Accepted: 20 July 2022
Published: 14 November 2022

Publisher's Note: MDPI stays neutral with regard to jurisdictional claims in published maps and institutional affiliations.

Copyright: © 2022 by the authors. Licensee MDPI, Basel, Switzerland. This article is an open access article distributed under the terms and conditions of the Creative Commons Attribution (CC BY) license (https://creativecommons.org/licenses/by/4.0/).

1. Introduction

Zearalenone (ZEA), a mycotoxin with an estrogen-like structure, competes with 17-estradiol for the estrogen receptor in target cells, impairing fertility, the ability to reproduce, and overall health [1]. α-zearalenol (α-ZEL), and β-zearalenol (β-ZEL) are metabolites of zearalenone that coexist in nature in grains and other foods. The harmful consequences of ZEA include carcinogenicity, nephrotoxicity, hepatotoxicity, endocrine disruptors, mutagenesis, and genotoxicity and are connected to alterations in endocrine disruptors and their metabolites (α-ZEL and β-ZEL) [2,3]. Since ZEA has an estrogen-like structure, studies have found that there are many estrogen receptor-positive cells in the intestine, so ZEA may also affect the function and integrity of the intestine [4,5]. Studies have demonstrated that when experimental animals are fed ZEA-contaminated feed, the intestine, which serves as the first line of defense against natural toxins, is exposed to ZEA and becomes the primary target organ for ZEA [6]. ZEA can alter intestinal villi structure [7,8], affect the integrity of porcine intestinal epithelial cells [9], and guide significant changes in gene expression levels of porcine intestinal cells [10]. We have found, in vivo, that ZEA can induce intestinal tissue damage, cause intestinal oxidative stress, and activate the Kelch-like erythroid cell-derived protein with CNC homology-associated protein 1 (Keap1)-nuclear factor erythroid 2-related factor 2 (Nrf2) signal pathway in post-weaning piglets [11–13]. Combining in vitro research is required to further demonstrate the potential contribution of the Keap1-Nrf2 signaling pathway to ZEA-induced intestinal oxidative stress since the internal environment governs the interaction and influence between various tissues and organs.

Glucose is the main energy substance for animal life activities and an intermediate product of animal body metabolism. Essential to its utilization is the intestine's digestion and absorption [14,15]. The two primary transporters in the intestinal absorption of glucose are sodium-dependent glucose co-transporter 1 (Sglt1) and glucose transporter 2 (Glut2), making Sglt1 and Glut2 crucial for maintaining the homeostasis of the intestinal environment [16–18]. The intestinal epithelium absorption of nutrients may be influenced by the oxidative state across the gut lumen [19]. Studies have demonstrated that the degree of oxidative stress affects the stimulatory impact on Sglt1-mediated transport [19,20]. According to the research, mycotoxin lowered the levels of Sglt1 and Glut2 and caused abnormal expression of nutrient transporters in IPEC-J2 cells [21]. However, there are few reports about the barrier function of ZEA on cells and the absorption of glucose nutrients. The major goal of this study is to further investigate the oxidative stress toxicity of ZEA on cells and the effect on cell permeability and the glucose absorption capacity through in vivo experiments, thereby affecting animal health. At the same time, this study will provide new ideas and methods for the study of the toxicity of ZEA to nutrient absorption.

2. Results

2.1. IPEC-J2 Cells' Morphology and Apoptosis

As shown in Figure 1, the morphology of IPEC-J2 cells changed significantly with the extension of time and the increase of ZEA concentration. The results showed that after 24 h and 36 h, ZEA had no significant effect on the cells at low concentrations; when the concentration of ZEA reached 20 μmol/L, the cell viability decreased, and the cell morphology changed from normal paving stone to irregular shape; when the concentration of ZEA reached 80 and 160 μmol/L, the cells were destroyed seriously, and a large number of cells began to die and floated with extremely low activity.

With an increase in ZEA concentration and an extension of the action duration, IPEC-J2 cells' apoptosis rate considerably increased after 24 and 36 h in comparison to the control group (Figure 2, $p < 0.05$). When the ZEA concentration reached 40 μmol/L and the cells were exposed for 36 h, the apoptosis rate reached the highest, indicating that the toxicity of ZEA on IPEC-J2 cells has a time and dose-dependent.

2.2. The TEER of IPEC-J2 Cells

As shown in Figure 3, compared with the control group, the TEER value of IPEC-J2 cells was significantly reduced ($p < 0.05$) after culturing the cells with various concentrations of ZEA for 24 and 36 h. The TEER value of cells in the ZEA40 group is the lowest, and ZEA's influence on TEER value of cells exhibits some time-dose dependence.

2.3. The Relative mRNA and Proteins Expression of Sglt1 and Glut2

The relative expression levels of Sglt1 and Glut2 mRNA and protein in IPEC-J2 cells significantly increased linearly and quadratically with the rise in ZEA concentration at 24 and 36 h, respectively, as compared to the control group (Figures 4 and 5, $p < 0.05$). The relative expression of Sglt1 protein in IPEC-J2 cells showed quadratically ($p < 0.05$) increase, with increasing ZEA concentration, and the expression level of Sglt1 protein increased significantly ($p < 0.05$) only in the ZEA40 group.

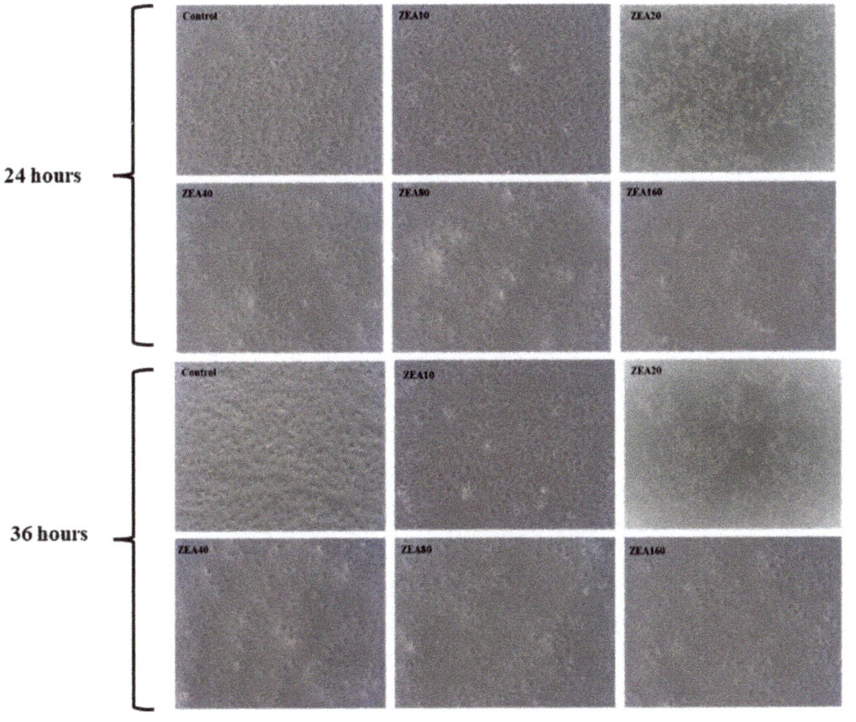

Figure 1. The morphology of intestinal porcine jejunal epithelial cells (IPEC-J2) exposed to ZEA at concentrations of 0 (Control), 10 (ZEA10), 20 (ZEA20), 40 (ZEA40), 80 (ZEA80) and 160 (ZEA160) μmol/L for 24 and 36 h and observed under a light microscope (10×).

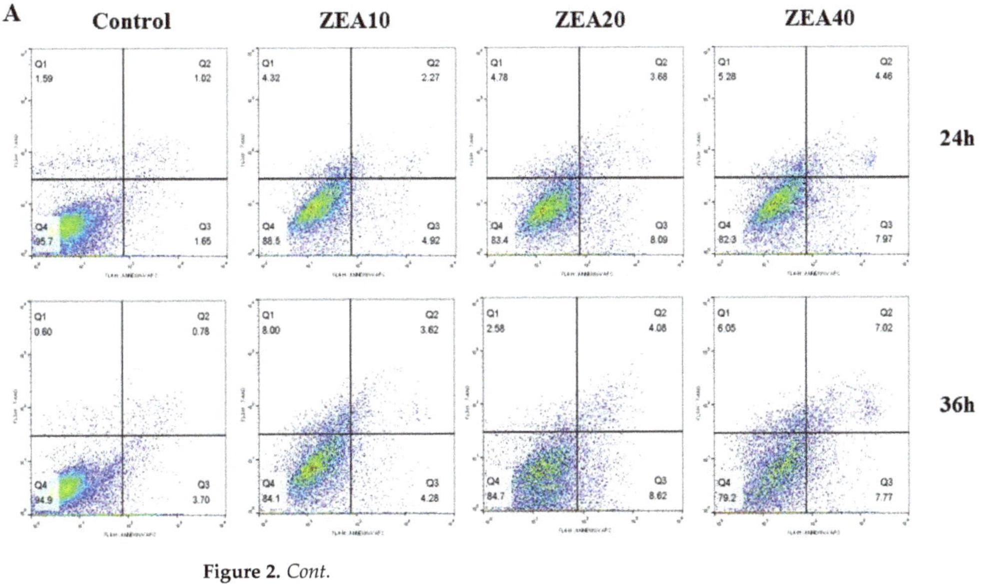

Figure 2. *Cont.*

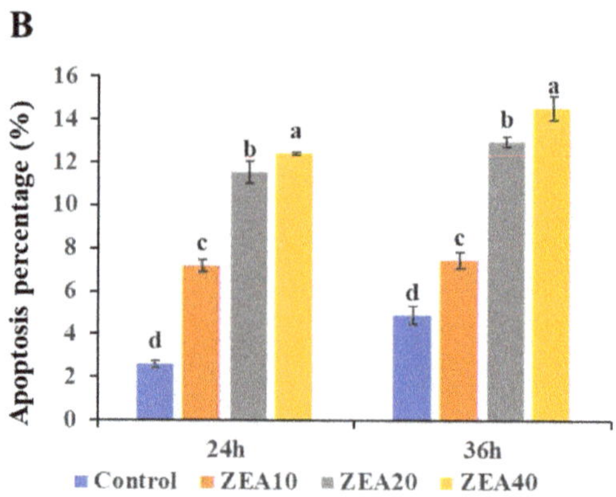

Figure 2. The apoptosis of intestinal porcine jejunal epithelial cells (IPEC-J2) exposed to ZEA at concentrations of 0, 10, 20, and 40 μmol/L (Control, ZEA10, ZEA20, ZEA40) for 24 and 36 h. (**A**) The apoptosis of IPEC-J2 cells was detected by flow cytometry. (**B**) The apoptosis percentage of IPEC-J2 cells. [a–d] The mean values differ significantly ($p < 0.05$).

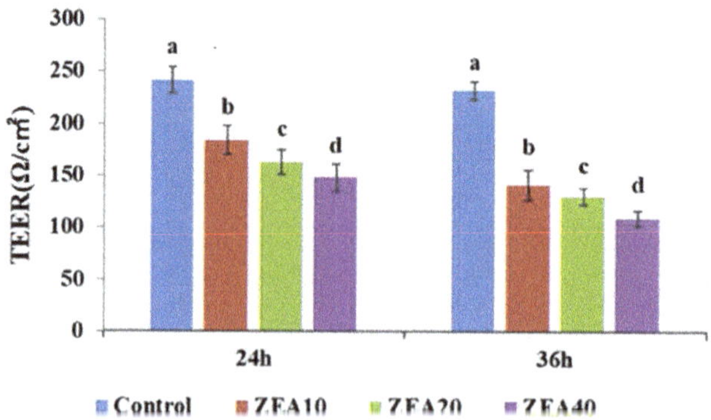

Figure 3. The transepithelial electrical resistance of intestinal porcine jejunal epithelial cells (IPEC-J2) exposed to ZEA at concentrations of 0, 10, 20, and 40 μmol/L (Control, ZEA10, ZEA20, ZEA40) for 24 and 36 h. [a–d] The mean values differ significantly ($p < 0.05$).

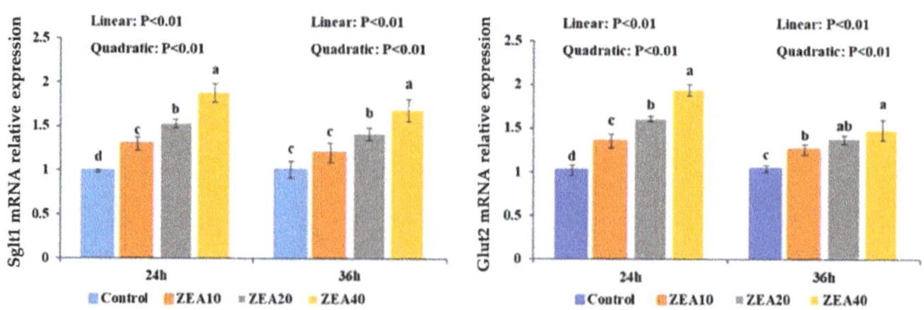

Figure 4. The relative mRNA expression of Sglt1, and Glut2 in intestinal porcine jejunal epithelial cells (IPEC-J2) exposed to ZEA at concentrations of 0, 10, 20, and 40 μmol/L (Control, ZEA10, ZEA20, ZEA40) for 24 and 36 h. [a–d] The mean values differ significantly ($p < 0.05$).

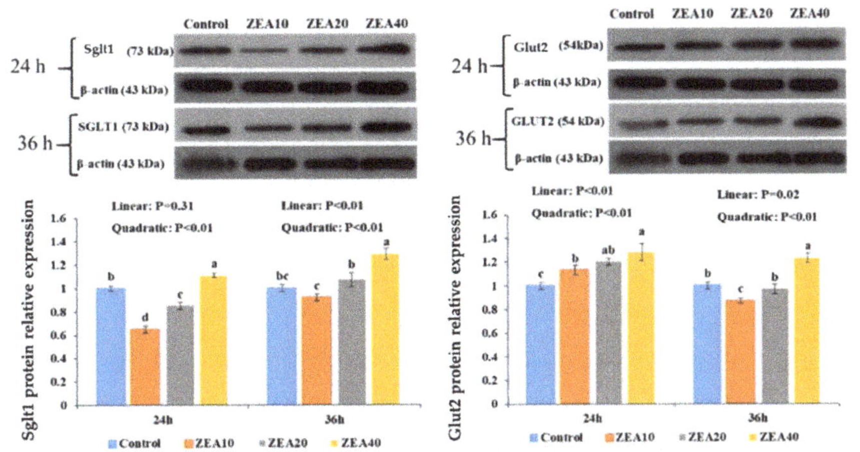

Figure 5. The relative protein expression of Sglt1, and Glut2 in intestinal porcine jejunal epithelial cells (IPEC-J2) exposed to ZEA at concentrations of 0, 10, 20, and 40 μmol/L (Control, ZEA10, ZEA20, ZEA40) for 24 and 36 h. $^{a\text{–}d}$ The mean values differ significantly ($p < 0.05$).

2.4. *The Immunofluorescence Localization of Sglt1 in IPEC-J2 Cells*

Immunofluorescence data obtained at 24 and 36 h showed that a large amount of Sglt1 is mostly dispersed in the cell membrane under normal conditions, and a minor amount is free in the cytoplasm (Figure 6). As the ZEA concentration rises, the fluorescence intensity around the nucleus decreases, indicating that the expression of Sglt1 in the cell membrane decreases, when the ZEA concentration reached 40 μmol/L, the expression of Sglt1 cell membrane was the lowest.

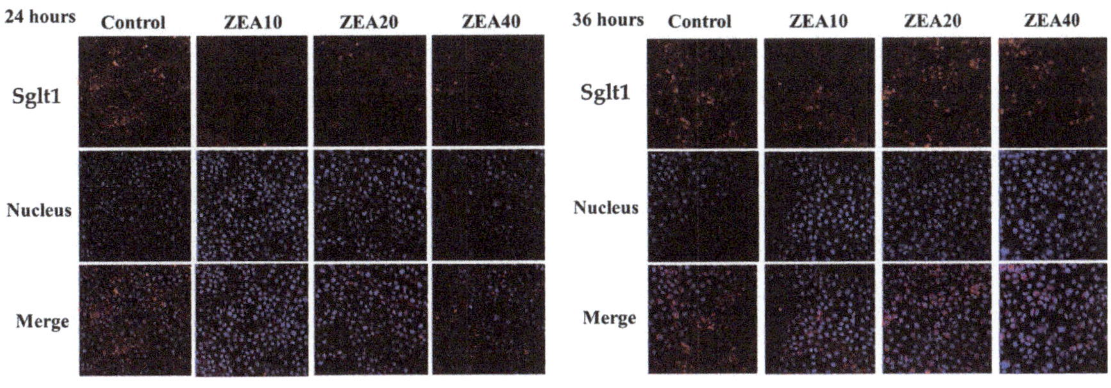

Figure 6. Immunostaining of Sglt1 on intestinal porcine jejunal epithelial cells (IPEC-J2) after exposure to to ZEA at concentrations of 0, 10, 20, and 40 μmol/L (Control, ZEA10, ZEA20, ZEA40) for 24 and 36 h, as detected under a light microscope (40×).

2.5. *The Antioxidant Enzyme Activity of IPEC-J2 Cells*

In comparison to the control group, the T-SOD and GSH-PX activities of IPEC-J2 cells decreased quadratically with increasing ZEA concentration at 24 and 36 h (Figure 7, $p < 0.05$). IPEC-J2 cells exhibited the lowest T-SOD and GSH-PX activity at a ZEA concentration of 10 mol/L. At 24 h and 36 h, the MDA level of the cells in the ZEA treatment group increased linearly and quadratically with the increase of the ZEA concentration ($p < 0.05$). The MDA level was greatest in the ZEA40 group.

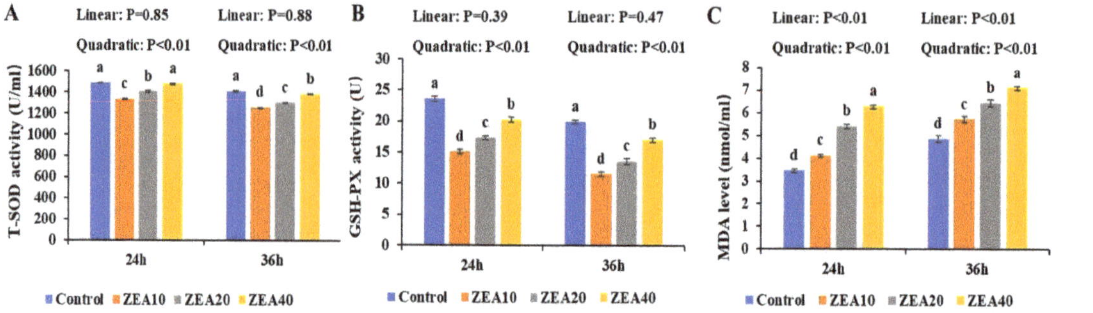

Figure 7. The antioxidant capacity of intestinal porcine jejunal epithelial cells (IPEC-J2) exposed to ZEA at concentrations of 0, 10, 20, and 40 μmol/L (Control, ZEA10, ZEA20, ZEA40) for 24 and 36 h. (**A**) The total superoxide dismutase (T-SOD) activities of IPEC-J2 cells. (**B**) The glu-tathione peroxidase (GSH-PX) activities of IPEC-J2 cells. (**C**) The malondialdehyde (MDA) content of IPEC-J2 cells. [a–d] The mean values differ significantly ($p < 0.05$).

2.6. The Level ROS of IPEC-J2 Cells

Flow cytometer histogram analysis shows that at 24 h and 36 h, the peak value of DCF fluorescence intensity shifted to the right as ZEA concentration increased, and the mean fluorescence intensity (MFI) of ROS in the ZEA treatment group was significantly higher than in the control group (Figure 8, $p < 0.05$). At 36 h, the MFI value of ROS of IPEC-J2 cells in the ZEA40 treatment group was the highest, and the effect of ZEA on the MFI value of cellular ROS showed a certain time-dose dependence (Figure 8 A). At 24 h and 36 h, as the level of ZEA increased, the MFI value of ROS in IPEC-J2 cells increased linearly and quadratically (Figure 8B, $p < 0.05$).

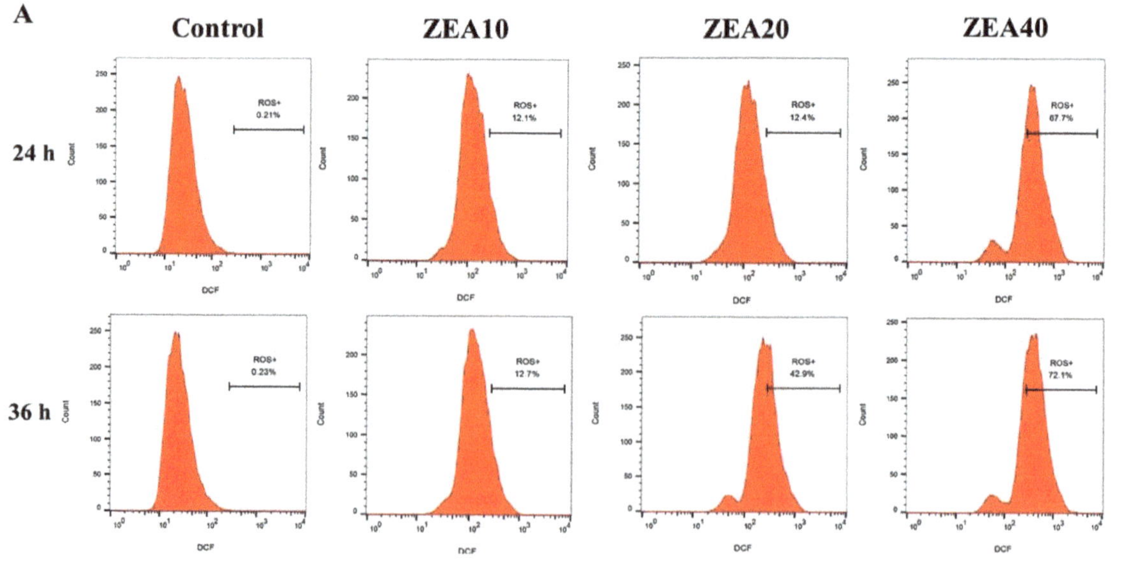

Figure 8. *Cont.*

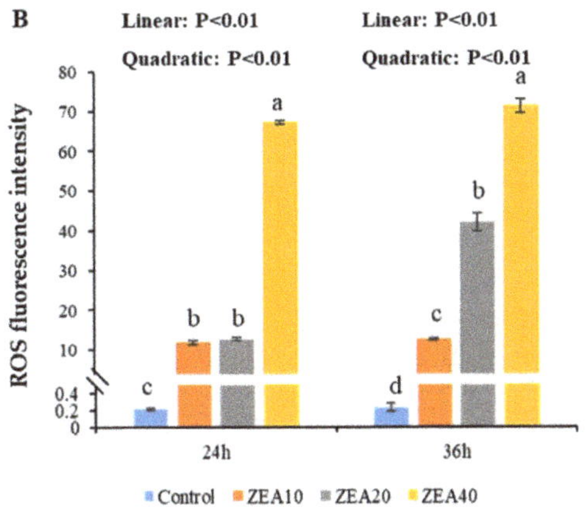

Figure 8. The reactive oxygen species of intestinal porcine jejunal epithelial cells (IPEC-J2) exposed to ZEA at concentrations of 0, 10, 20, and 40 μmol/L (Control, ZEA10, ZEA20, ZEA40) for 24 and 36 h. (**A**) The reactive oxygen species (ROS) of IPEC-J2 cells was detected by flow cytometry. (**B**) The ROS fluorescence intensity of IPEC-J2 cells. [a–d] The mean values differ significantly ($p < 0.05$).

2.7. The Relative Expression of Keap1, Nrf2, Nqo1, and Ho1 mRNA and Protein

At 24 h and 36 h, the relative expression of Keap1 mRNA and protein in IPEC-J2 cells treated with ZEA decreased linearly with increasing ZEA concentration relative to the control group (Figures 9 and 10, $p < 0.05$). The ZEA40 group exhibited the lowest level of expression (Figures 9A and 10A). The relative expression of Nrf2, Nqo1, and Ho1 mRNA and protein increased linearly and quadratically in IPEC-J2 cells at 24 and 36 h ($p < 0.05$). In the ZEA40 group, Nrf2, Nqo1, and Ho1 mRNA and protein expression levels are the highest (Figures 9B–D and 10B–D).

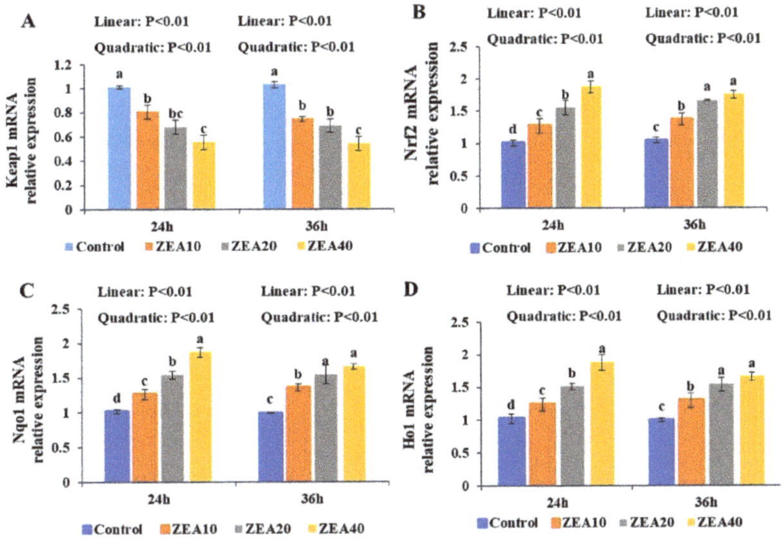

Figure 9. The relative mRNA expression of Keap1, Nrf2, Nqo1, and Ho1 in intestinal porcine jejunal epithelial cells (IPEC-J2) exposed to ZEA at concentrations of 0, 10, 20, and 40 μmol/L (Control, ZEA10, ZEA20, ZEA40) for 24 and 36 h. The Keap1 (**A**), Nrf2 (**B**), Nqo1 (**C**), and Ho1 (**D**) mRNA relative expression of IPEC-J2 cells. [a–d] The mean values differ significantly ($p < 0.05$).

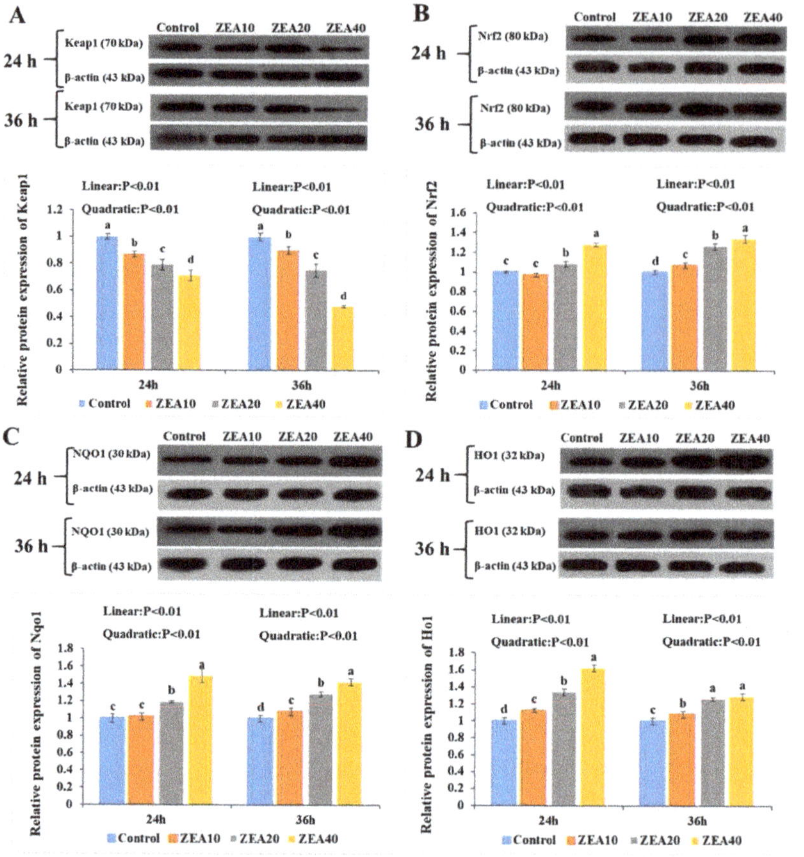

Figure 10. The relative protein expression of Keap1, Nrf2, Nqo1, and Ho1 in intestinal porcine jejunal epithelial cells (IPEC-J2) exposed to ZEA at concentrations of 0, 10, 20, and 40 μmol/L (Control, ZEA10, ZEA20, ZEA40) for 24 and 36 h. The Keap1 (**A**), Nrf2 (**B**), Nqo1 (**C**), and Ho1 (**D**) protein relative expression of IPEC-J2 cells. [a–d] The mean values differ significantly ($p < 0.05$).

2.8. The Immunofluorescence localization of ROS and Nrf2 in the IPEC-J2 Cells

Immunofluorescence results showed that ROS was weakly expressed in the control group IPEC-J2 cells at 24 and 36 h, and mainly distributed in the nucleus and cytoplasm (Figure 11). Fluorescence intensity surrounding the nucleus rose as ZEA concentration increased, indicating that ROS expression increased significantly. At 36 h, when the ZEA concentration reached 40 μmol/L, the cell ROS expression reached the highest.

At 24 h and 36 h, the immunofluorescence results showed that under normal circumstances, Nrf2 was mostly localized in the cytoplasm and a tiny quantity in the nucleus (Figure 12). With the increase of ZEA concentration, the fluorescence intensity around and within the nucleus increased significantly, indicating that Nrf2 began to move to the nucleus, and the expression of Nrf2 increased significantly. At 36 h, when the ZEA concentration reaches 40 μmol/L, the expression of Nrf2 is the highest, and the phenomenon of Nrf2 invading the nucleus was very obvious.

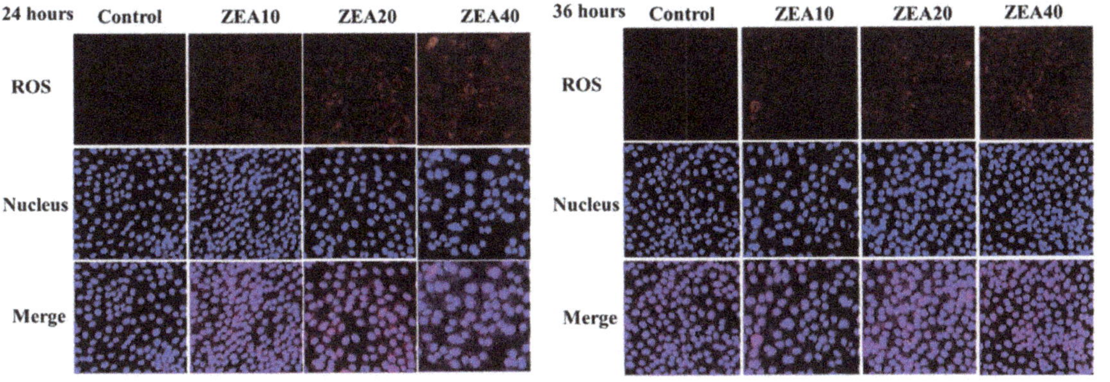

Figure 11. Immunostaining of ROS of intestinal porcine jejunal epithelial cells (IPEC-J2) after exposure to to ZEA at concentrations of 0, 10, 20, and 40 μmol/L (Control, ZEA10, ZEA20, ZEA40) for 24 and 36 h, as detected under a light microscope (40×).

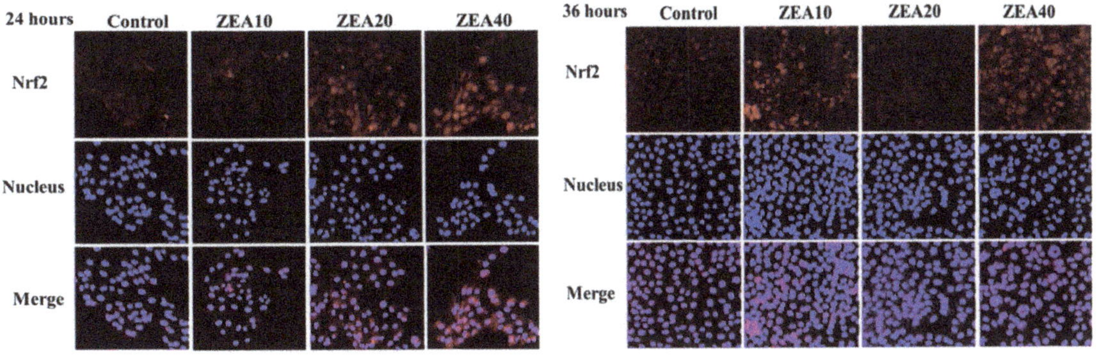

Figure 12. Immunostaining of Nrf2 of intestinal porcine jejunal epithelial cells (IPEC-J2) after exposure to to ZEA at concentrations of 0, 10, 20, and 40 μmol/L (Control, ZEA10, ZEA20, ZEA40) for 24 and 36 h, as detected under a light microscope (40×).

3. Discussion

As we all know, zearalenone has toxic effects and can remain in the body for a long time after being absorbed by animals and humans, which brings huge losses to animal husbandry and poses a threat to human health. Our research confirmed that ZEA can produce oxidative stress in the intestines of weaned pigs, damage the structure and morphology of intestinal villi and affect the healthy development of the intestinal tract. Therefore, we guess that ZEA will have a certain effect on the nutrient absorption of pigs. In order to establish a new theoretical foundation for more effectively addressing the toxic effect of ZEA on pig intestines, the IPEC-J2 cell line was chosen as the cell model in this study to investigate the toxicity of ZEA.

Apoptosis is a basic biological phenomenon of cells, which is activated, expressed and controlled by a succession of genes in the body. When the regulation of apoptosis is unbalanced, excessive apoptosis and death of cells will cause serious damage to the body, which frequently results in the onset of disorders like autoimmune diseases [22,23]. In this study, we found that ZEA significantly reduces the cell viability of IPEC-J2 cells in a time- and dose-dependent manner, causing cell apoptosis and even cell death. In vitro investigations have demonstrated that ZEA and its metabolites can interfere with cell function and even induce apoptosis [24,25]. Some studies have also revealed that ZEA promotes apoptosis in rat germ cells, and after 12 h of ZEA treatment, the most apoptotic cells were detected, and then gradually decreased [26,27]. HepG2 cells were treated with ZEA at concentrations ranging from 7.1–250 M for 24 h, and the findings demonstrated a dose-dependent reduction in cell viability [28]. However, some studies have shown that ZEA (500 μg/L) cannot induce IPEC-J2 cell apoptosis [29], and ZEA (10–40 μM) cannot induce porcine kidney cell apoptosis [30]. This may be because various factors such as different cell types, exposure time, dose and type of mycotoxins, and metabolites affect cell viability [31,32].

The transmembrane resistance (TEER) of monolayer cells is one of the important indicators for studying the integrity of the intestinal tract, and its variations signify alterations in the permeability and integrity of the cell monolayer [33]. Some studies have shown that aflatoxin B1 (AFB1) and ochratoxin A (OTA) can reduce the TEER value of intestinal cells, resulting in increased permeability and a serious impact on the intestinal barrier function [34,35]. Deoxynivalenol (DON) has been shown to have impacts on the TEER of three distinct human intestinal epithelial cell lines, including HT-29, Caco-2, and T48, and it was shown that these effects were dose-dependent [33]. The similar result was also reached by DON's investigation on the TEER value of the swine intestinal epithelial cell lines IPEC-1 and IPEC-J2 [36]. These studies' findings and those of this one are comparable. This study found that ZEA significantly decreased the TEER of IPEC-J2 cells, and that this reduction in TEER was dose- and time-dependent. We analyzed that the decrease of TEER induced by ZEA may be due to the destruction of tight junction, the change of ion transmembrane transport and the oxidative stress induced by mycotoxin, and it may also be due to the oxidative stress induced by ZEA, which needs further research and verification.

The small intestine can be actively transported through the Sglt1 of the small intestinal mucosal epithelial cells to transport the glucose in the intestinal lumen of the small intestine to the epithelial cells, followed by Glut2 facilitating diffusion, which transfers the intracellular glucose to the blood [37]. According to studies, three distinct transporters involved in the intestinal absorption of glucose and fructose have varying susceptibility to oxidative stress. Sglt1 is the most sensitive, followed by Glut2, and Glut5 is the lowest [20]. Notably, an increase in oxidative stress can stimulate and promote Sglt1 and Glut2. Nonetheless, oxidative stress or situations associated to oxidative stress have been demonstrated to have detrimental effects on the Glut transporter [20,38,39]. This study revealed that the protein expression of Sglt1 and Glut2 was significantly reduced in low-dose ZEA group and increased significantly in high-dose ZEA group. Cellular immunofluorescence technology has been widely used in the fields of medicine and biology. It is a common method to detect the intracellular localization of antigen proteins and quantify their expression by using fluorescein-labeled antibodies by using the specific binding reaction between antigens and antibodies [40]. At the same time, the immunofluorescence results showed that Sglt1 expression in the cell membrane was considerably downregulated in ZEA treatment group. It is found that ZEA (10 μg/mL) induces oxidative damage in intestinal cells, impairs nutritional digestion and absorption, and increases the mRNA expression level of Sglt1 [41]. Studies have also found that the combination of 500 g/L ZEA and 40 g/L AFB1 may increase the expression of the Glut2 gene in IPEC-J2 cells [29]. Therefore, we think that different kinds and concentrations of toxins have different effects on the expression of Sglt1 and Glut2 in different cells and may have a certain relationship with the exposure time of toxins. In this study, we found that the up-regulation of Sglt1 and Glut2 mRNA expression levels was inconsistent with protein expression, indicating that low concentrations of ZEA increased the expression of Sglt1 and Glut2 mRNA, but did not promote the simultaneous increase in the transport activity of Sglt1 and Glut2. Some studies have explained this phenomenon. The activity of GLUTS is inhibited by one or more regulatory proteins with very short half-life. When protein synthesis is inhibited, the expression level of regulatory proteins decreases, and the inhibitory effect on GLUTS activity is lifted, resulting in the increase of mRNA expression and activity [41].

Oxidative stress in the organism can be brought on by an imbalance between antioxidant defense and ROS free radical generation [42]. ZEA is believed to be a potent inducer of ROS in the mammalian body, which can induce the massive production of ROS [43,44]. According to some reports, ZEA can induce ROS production and lipid peroxidation, and reduce cell proliferation. The oxidative stress that ZEA causes may be responsible for these cytotoxic effects [9,45,46]. In order to inhibit the increase of ROS and improve the ability of cells to resist oxidative stress damage, cells will activate their own antioxidant defense mechanisms, containing SOD and GSH-PX antioxidant enzymes [42,47,48]. Numerous in vivo and in vitro investigations have demonstrated that ZEA can reduce the activities of the antioxidant enzymes T-SOD and GSH-PX and increase the level of MDA by regulating the antioxidant mechanism in the cell [9,46,49]. The relative expression of ROS mRNA and protein as well as the MDA content in the cells were dramatically raised following ZEA treatment of IPEC-J2 cells for 24 h and 36 h in this investigation, whereas the activities of the antioxidant enzymes T-SOD and GSH-PX were significantly decreased. ZEA promotes a large number of cells in a state of oxidative stress, which is consistent with our previous research results in the pig intestine [11–13]. Some studies also found that the level of ROS in human embryonic kidney cells exposed to ZEA for 1 h did not change significantly, but it increased significantly after 2 h [50]. ZEA can promote cytotoxicity in rat intestinal epithelial cells by decreasing T-SOD and GSH-Px activities and elevating MDA levels [6]. Therefore, previous studies and our own have demonstrated that ROS induced by ZEA injury produces oxidative stress and reduces cellular antioxidant capacity, which may be an important reason for ZEA damage to cell homeostasis and structure, leading to cell death.

Under oxidative stress, the Keap1-Nrf2 signaling pathway is activated, and the conformational change of Keap1 in the cytoplasm shifts the Nrf2 released from the low binding site to the nucleus, triggering the binding of the programmed antioxidant enzyme and the phase II detoxifying enzyme to the ARE. This will produce proteins that protect cells, including antioxidant enzymes and Ho1, Nqo1, etc., so as to improve the body's antioxidant capacity and playing an important role in maintaining cell homeostasis when cells are under oxidative stress [51,52]. The findings of this study showed that ZEA significantly increased the relative expression of Nrf2, Ho1 and Nqo1 mRNA and protein in IPEC-J2 cells, and significantly reduced the relative expression of Keap1 mRNA and protein. Meanwhile, Nrf2 immunofluorescence showed that ZEA significantly induced the increase of Nrf2 gene expression in the cells, and Nrf2 transferred from the cytoplasm to the nucleus. Nrf2 expression reached its maximum level in the ZEA40 treatment group, which was consistent with our previous results in the pig intestine [11–13]. As a result, we think that ZEA can activate the Keap1-Nrf2 signaling pathway of IPEC-J2 cells, and the activation of this pathway increases the ability of IPEC-J2 cells to resist ZEA toxicity. Studies have found that ZEA toxicity can significantly induce differences in gene expression in IPEC-1 cells, and low doses of ZEA (10 μmol/L) significantly upregulate the expression of 70% of genes in IPEC-1cells, including encoding Gpx [9,10]. Interestingly, despite the fact that the expression of Nrf2, Ho1, and Nqo1 genes increases significantly when this route is active, the response is restricted since ZEA-induced ROS also activates a variety of signal pathways such as cell death [53,54]. However, the influence of the Keap1-Nrf2 signaling pathway on the oxidative stress induced by ZEA in IPEC-J2 cells is poorly studied. In order to determine whether the increased expression of Ho1 and Nqo1 genes in IPEC-J2 cells caused by ZEA is controlled by the classic Keap1-Nrf2 signaling pathway alone or by various pathways, we will further use ShRNA to interfere with the expression of Nrf2 in cells to inhibit the activation of the Keap1-Nrf2 signaling pathway for verification experiments.

4. Conclusions

In conclusion, ZEA (10, 20, 40 μmol/L) changes the morphology of IPEC-J2 cells in a certain time- and dose-dependent manner, which changes the permeability of cells, interferes with the expression of Sglt1 and Glut2 genes, and may affect the glucose and nutrient absorption capacity of the cells. ZEA induces oxidative stress in IPEC-J2 cells, which reduces cell antioxidant capacity, and the Keap1-Nrf2 signaling pathway is activated to resist the toxic effects of ZEA during ZEA-induced cellular oxidative stress. However, the link between ZEA's activation of the Keap1-Nrf2 signaling pathway and the mechanism by which ZEA modulates the expression of the Sglt1 and Glut2 genes influences cellular glucose food absorption requires additional investigation.

5. Materials and Methods

5.1. Preparation of IPEC-J2 Cells Culture

The IPEC-J2 cells were obtained commercially (Beina biological Co., Ltd., Beijing, China) The IPEC-J2 cells were cultured with 2 mL of Dulbecco's Modified Eagle Medium (DMEM) high glucose (11995-065, GIBCO Co., Ltd., Shanghai, China) containing 10% fetal bovine serum (FBS, 16000-044, GIBCO Co., Ltd., Shanghai, China) and 1% Penicillin-Streptomycin (P/S, 15140-163, GIBCO Co., Ltd., Shanghai, China). Before being employed in each test, the incubation was carried out at 37 °C in a cell incubator with 5% CO_2.

5.2. Preparation of ZEA Treatment of IPEC-J2 Cells

Commercially available zearalenone (Sigma, Z2125, MO, USA) was dissolved in dimethyl sulfoxide (DMSO) (Sigma, D2650, MO, USA) at a concentration of 20 mmol/L and refrigerated at −20 °C prior to use. The IPEC-J2 cells were seeded onto 6-well plates at a density of 1×10^6 cells per well before aliquots of ZEA solution were added to achieve culture concentrations of 0, 10, 20, and 40 μmol/L (Control, ZEA10, ZEA20, ZEA40), respectively. After mixing the cultures, they were incubated at 37 °C for 24 and 36 h, after which the cells were collected for further testing.

5.3. Determination of Apoptosis in IPEC-J2 Cells

The cell samples were obtained, twice-washed with 2-mL of cold PBS, and re-suspended into 100 μL 1×Binding Buffer. The collected IPEC-J2 cells were then stained for 15 min at 25 °C in the dark with PI (Beyotime, C1052-2, Shanghai, China) and FITC Annexin V (Beyotime, C1052-3, Shanghai, China). After the reaction, add 400 μL of 1 × Binding Buffer to each tube and mix well, and then examined using a BD FACSCaliburTM flow cytometer (FACSCalibur, BD, San Jose, CA, USA) with Flowjo 7.6 software.

5.4. Determination of Cell Transepithelial Electrical Resistance (TEER) of IPEC-J2 Cells

The cells collected and diluted to 5×10^3 μL^{-1}. Then 100 μL cell suspension was inoculated in the chamber of 12-hole Transwell plate (membrane area 1.12 cm^2), then 400 μL complete medium was added, and 1.5 mL complete medium was added in the hole under the chamber. After placing the Transwell plate in the incubator for one day, a compact monolayer cell was formed, and then the complete medium was replaced with ZEA different treatment groups of medium, each treatment was repeated for three times. After 24 h and 36 h of treatment, the cells in each group were measured by cell resistance meter (Millipore Mers00002 Millicell ERS, MA, USA). Three points in different directions were selected for each hole.

5.5. Determination of Antioxidant Enzyme Activity

T-SOD A001-1, GSH-PX A005 and MDA A003 detection kits (Nanjing Aoqing Biotechnology Co., Ltd., Nanjing, China) were used to assess the cell samples for malondialdehyde (MDA) content and total superoxide dismutase (T-SOD) and glu-tathione peroxidase (GSH-PX) activities [55].

5.6. Determinations of Relative mRNA Expressions

The total RNA from the cells was extracted using RNAiso Plus (Applied TaKaRa, Dalian, China) per the manufacturer's instructions. Using an Eppendorf Bio-photometer (DS-11, Denovix, Wilmington, DE, USA) with an absorbance ratio of 260/208 nm, the purity and concentration of the RNA were assessed. A reverse transcription system kit (PrimeScriptTM RT Master Mix, RR036A, Applied TaKaRa, Dalian, China) was used to convert total RNA to cDNA, and the resulting cDNA was split into two subsamples.

The cDNA subsample was utilized for quantitative real-time PCR analysis. The qRT-PCR study employed a total volume of 20 μL of the PCR reaction mixture made up of 10 μL SYBRY Premix Ex Taq II, 0.4 μL DyeII (SYBRY Premix Ex Taq-TIi RNaseH Plus, DRR420A; Applied TaKaRa), 0.4 μL of both forward and reverse primers, and 2 μL cDNA (<100 ng). Sangon Biological Engineering Technology and Services Co. Ltd. (Shanghai, China) designed all of the primers, and the Beijing Genomics Institute synthesized them (BGI, Beijing, China). An initial denaturation phase at 95 °C for 30 s was followed by 43 cycles at 95 °C for 5 s, 60 °C for 34 s, 95 °C for 15 s, 60 °C for 60 s, and 95 °C for 15 s in the optimized qRT-PCR procedure. An AB 7500 Real Time PCR System was used to carry out the qRT-PCR experiments (Applied Biosystems, Foster City, CA, USA). Using the $2^{-\Delta\Delta Ct}$ method, the relative expression level of Sglt1, Glut2, Keap1, Nrf2, Nqo1, Ho1, and β-actin mRNA was determined [56]. For each sample, the analysis was done three times. Table 1 presents the primer sequences and production lengths.

Table 1. Sequence of primers for real-time PCR.

Target Gene	Primer Sequences (5′ to 3′)	Accession No.
Nrf2	F: GAGTTAGATAGTGCCCCTGGAA R: ACTGGAGCACTATTACCCTGAG	XM_005671981.3
Keap1	F: GTGTGTGCTCCATGTCATGAAT R: CTCCCCAAAGTGCATGTAGATG	NM_001114671.1
Ho1	F: AGGTCCTCAAGAAGATTGCTCA R: CATCTCCAGAGTGTTCATTCGG	NM_001004027.1
Nqo1	F: AAAAGCACTGATCATACTGGCC R: TTCTGGAGATGACGGGATTGAA	NM_ 001159613.1
Sglt1	F: CGTCCATCTTTAACAGCAGCAG R: GCATGTAGATGAAGAGCTGCC	NM_001012297.1
Glut2	F: ACCGACAGCCTATTCTAGTAGC R: AGGAAAACAGAGAGAGCAGTGA	NM_001097417.1
β-actin	F: AGATCACTCCCCCAATGACAG R: AGAGCAAGAGAGGCATCCTG	XM_003124280.5

5.7. Determination of Sglt1, ROS, and Nrf2 Distribution in IPEC-J2 Cells

The IPEC-J2 were initially seeded on microslides, which were then treated with ZEA as stated above, fixed with 4% paraformaldehyde for 1 h, then penetrated with 0.5% Triton X-100 for 10 min at room temperature. The resulting cells underwent the following processing steps: washing with PBS three times for 5 min each, blocking with 10% FBS for 1 h, Nrf2 (1:500, ab89443, Abcam, Cambridge, UK), reactive oxygen species (ROS, 1:200, ab236409, Abcam, Cambridge, UK), Sglt1 (1:150, ab247121, Abcam, Cambridge, UK) incubation at 4 °C overnight, three washings with PBS, mixing with goat anti-rabbit lgG that has been Alexa Fluor 555-labeled (1:200, ab150079, Abcam, Cambridge, UK) at 37 °C in the dark for 1 h, and washing with PBS. The appropriate Hoechst 33342 (C1022, Beyotime, Shanghai, China) was then added to the previously prepared cells, stirred for 5 min, and then rinsed with PBS. Under a confocal microscope, the treated samples were analyzed (FLUOVIEW FV3000, Olympus, Japan).

5.8. Determination of Protein Expression

The cell samples were then washed with 2 mL of PBS and centrifuged once more (1200× g, 5 min). Following the manufacturer's instructions, they were extracted with lysate containing PMSF (1 mmol/L, Beyotime, Shanghai, China). Using a BCA protein assay kit (Beyotime, Shanghai, China), the total protein content of the extract was assessed before it was subjected to western blotting to detect the protein expression of relevant mRNA, as indicated below. Each sample was loaded with 60 μg of protein and put through 1.5 h of electrophoresis on polyacrylamide gels. After then, the bands that had been separated were moved to immobilon-p transfer membranes (So-larbio, Beijing, China). These membranes were first blocked in 10% skim milk powder for 2 h, then washed three times with Tris Buffered Saline Tween (TBST, pH 7.6), and finally incubated overnight at 4 °C with monoclonal mouse antibody β-actin (1:1500, SC-47778, Santa Cruz, CA, USA), polyclonal rabbit antibody Nrf2 (1:1000, ab92946, Abcam, Cambridge, UK), polyclonal rabbit antibody Keap1 (1:1000, ab196346, Abcam, UK), Nqo1 (1:500, ab2346, Abcam, UK), Ho1 (1:500, ab13248, Abcam, Cambridge, UK), Sglt1 (1:1000, bs-1128R, BIOSS, Beijing, China), Glut2 (1:500, bs-0351R, BIOSS, Beijing, China). Following the primary incubation, the membrane was washed with TBST three times for a total of 5 min each time, and then it was subjected to a secondary incubation with goat anti-rabbit lgG (1:5000, Thermo Pierce 31210, Thermo Fisher Scientific, MA, USA) and goat anti-mouse lgG (1:5000, Thermo Pierce 31160, Thermo Fisher Scientific, MA, USA) in diluted secondary antibody dilution buffer (Beyotime, Shanghai, China) at room temperature for 2 h. After 30 min of washing with TBST, the membranes were submerged in a high-sensitivity luminescence reagent (BeyoECL Plus; Beyotime Biotechnology). After that, the membranes were exposed to film with a Fusion FX imaging system and processed with a FusionCapt Advance FX7 software program (Beijing Oriental Science and Technology Development Co., Ltd., Beijing, China). Image-Pro Plus 6.0 was utilized to measure the protein concentration (Media Cy-bernetics, Inc., Rockville, MD, USA).

5.9. Statistical Analysis

All data were analyzed using the generalized linear model (GLM) method of SAS 9.2 for one-way analysis of variance (SAS Inst. Inc., Cary, NC, USA). Initially, the data were analyzed using a totally random design, with the treatment as the fixed effect and each cell as the random component. For the purpose of determining linear and quadratic responses to the ZEA concentrations, orthogonal polynomial contrasts were utilized. Duncan's multiple range tests were utilized in order to evaluate the significance of changes between treatments. The threshold for determining significance was set at $p < 0.05$.

Author Contributions: Conceptualization, W.Y.; data curation, Q.C.; formal analysis, L.H. and S.J.; funding acquisition, W.Y.; methodology, W.Y., L.H. and S.J.; writing—original draft, Q.C.; writing—review and editing, S.J. and Y.W. All authors have read and agreed to the published version of the manuscript.

Funding: This research was financed in part by the Natural Science Foundation of Shandong Province (grant no. ZR2021MC048); Founds of Shandong Agriculture Research System in Shandong Province (grant no. SDAIT-08-04 and SDAIT-08-05); Major Innovative Projects of Shandong Province (grant no. 2019JZZY020609).

Institutional Review Board Statement: The authors confirm that the ethical policies of the journal, as noted on the journal's author guidelines page, have been adhered to. All animal care and experimental procedures were in accordance with the guidelines for the care and use of laboratoryanimals prescribed by the Ministry of Agriculture of China.

Informed Consent Statement: Not applicable.

Data Availability Statement: Not applicable.

Acknowledgments: Financial supports of the natural science fund, the major innovation fund and the pig industry fund of Shandong Province.

Conflicts of Interest: We certify that there are no conflict of interest with any financial organization regarding the material discussed in the manuscript.

References

1. Frizzell, C.; Ndossi, D.; Verhaegen, S.; Dahl, E.; Eriksen, G.; Sřrlie, M.; Ropstad, E.; Muller, M.; Elliott, C.T.; Connolly, L. Endocrine disrupting effects of zearalenone, alpha- and beta-zearalenol at the level of nuclear receptor binding and steroidogenesis. *Toxicol. Lett.* **2011**, *206*, 210–217. [CrossRef] [PubMed]
2. Agahi, F.; Juan, C.; Font, G.; Juan-García, A. In silico methods for metabolomic and toxicity prediction of zearalenone, α-zearalenone and β-zearalenone. *Food Chem. Toxicol.* **2020**, *146*, 111818. [CrossRef] [PubMed]

3. Oueslati, S.; Berrada, H.; Juan-García, A.; Mañes, J.; Juan, C. Multiple mycotoxin determination on Tunisian cereals-based food and evaluation of the population exposure. *Food Anal. Method.* **2020**, *13*, 1271–1281. [CrossRef]
4. Kawano, N.; Koji, T.; Hishikawa, Y.; Murase, K.; Murata, I.; Kohno, S. Identification and localization of estrogen receptor alpha- and beta-positive cells in adult male and female mouse intestine at various estrogen levels. *Histochem. Cell Biol.* **2004**, *121*, 399–405. [CrossRef]
5. Wada-Hiraike, O.; Imamov, O.; Hiraike, H.; Hultenby, K.; Schwend, T.; Omoto, Y.; Warner, M.; Gustafsson, J.A. Role of estrogen receptor beta in colonic epithelium. *Proc. Natl. Acad. Sci. USA* **2006**, *103*, 2959–2964. [CrossRef]
6. Long, M.; Chen, X.; Wang, N.; Wang, M.; Pan, J.; Tong, J.; Li, P.; Yang, S.; He, J. Proanthocyanidins Protect Epithelial Cells from Zearalenone-Induced Apoptosis via Inhibition of Endoplasmic Reticulum Stress-Induced Apoptosis Pathways in Mouse Small Intestines. *Molecules* **2018**, *23*, 1508. [CrossRef]
7. Gerez, J.R.; Pinton, P.; Callu, P.; Grosjean, F.; Oswald, I.P.; Bracarense, A.P. Deoxynivalenol alone or incombination with nivalenol and zearalenone induce systemic histological changes in pigs. *Exp. Toxicol. Pathol.* **2015**, *67*, 89–98. [CrossRef]
8. Liu, M.; Gao, R.; Meng, Q.; Zhang, Y.; Bi, C.; Shan, A. Toxic effects of maternal zearalenone exposure on intestinal oxidative stress, barrier function, immunological and morphological changes in rats. *PLoS ONE* **2014**, *9*, e106412. [CrossRef]
9. Marin, D.E.; Pistol, G.C.; Neagoe, I.V.; Calin, L.; Taranu, I. Effects of zearalenone on oxidative stress and inflammation in weanling piglets. *Food Chem. Toxicol.* **2015**, *58*, 408–415. [CrossRef]
10. Taranu, I.; Braicu, C.; Marin, D.E.; Pistol, G.C.; Motiu, M.; Balacescu, L.; Neagoe, I.B.; Burlacu, R. Exposure to zearalenone-mycotoxin alters in vitro porcine intestinal epithelial cells by differential gene expression. *Toxicol. Lett.* **2015**, *232*, 310–325. [CrossRef]
11. Cheng, Q.; Jiang, S.Z.; Huang, L.B.; Ge, J.S.; Wang, Y.X.; Yang, W.R. Zearalenone induced oxidative stress in the jejunum in postweaning gilts through modulation of the Keap1–Nrf2 signaling pathway and relevant genes. *J. Anim. Sci.* **2019**, *97*, 1722–1733. [CrossRef] [PubMed]
12. Cheng, Q.; Jiang, S.Z.; Huang, L.B.; Wang, Y.X.; Yang, W.R.; Yang, Z.B.; Ge, J.S. Effects of zearalenone-induced oxidative stress and Keap1–Nrf2 signaling pathway-related gene expression in the ileum and mesenteric lymph nodes of post-weaning gilts. *Toxicology* **2020**, *429*, 152337. [CrossRef] [PubMed]
13. Cheng, Q.; Jiang, S.Z.; Huang, L.B.; Yang, W.R.; Yang, Z.B. Zearalenone regulates key factors of the Kelch-like erythroid cell-derived protein with CNC homology-associated protein 1-nuclear factor erythroid 2-related factor 2 signaling pathway in duodenum of post-weaning gilts. *Anim. Biosci.* **2021**, *34*, 1403–1414. [CrossRef]
14. Klinger, S.; Lange, P.; Brandt, E.; Hustedt, K.; Schröder, B.; Breves, G.; Herrmann, J. Degree of SGLT1 phosphorylation is associated with but does not determine segment-specific glucose transport features in the porcine small intestines. *Physiol. Rep.* **2018**, *6*, e13562. [CrossRef]
15. Röder, P.V.; Geillinger, K.E.; Zietek, T.S.; Thorens, B.; Koepsell, H.; Daniel, H. The Role of SGLT1 and GLUT2 in Intestinal Glucose Transport and Sensing. *PLoS ONE* **2014**, *9*, e89977. [CrossRef]
16. Farrell, T.L.; Ellam, S.L.; Forrelli, T.; Williamson, G. Attenuation of glucose transport across Caco-2 cell monolayers by a polyphenol-rich herbal extract: Interactions with SGLT1 and GLUT2 transporters. *BioFactors* **2013**, *39*, 448–456. [CrossRef]
17. Grefner, N.M.; Gromova, L.V.; Gruzdkov, A.A.; Komissarchik, Y.Y. Interaction of glucose transporters SGLT1 and GLUT2 with cytoskeleton in enterocytes and Caco2 cells during hexose absorption. *Cell Tissue Biol.* **2015**, *9*, 45–52. [CrossRef]
18. Thorens, B. GLUT2, glucose sensing and glucose homeostasis. *Diabetologia* **2014**, *58*, 221–232. [CrossRef]
19. Ana, F.; Monteiro, R.; Pestana, D.; Freitas, V.D.; Mateus, N.; Azevedo, I.; Calhau, C. Intestinal oxidative state can alter nutrient and drug bioavailability. *Oxidative Med. Cell. Longev.* **2009**, *2*, 322–327. [CrossRef]
20. Nelson, A.; Cláudia, S.; Fátima, M. The effect of oxidative stress upon intestinal sugar transport: An in vitro study using human intestinal epithelial (Caco-2) cells. *Toxicol. Res.* **2018**, *7*, 1236–1246. [CrossRef]
21. Chen, Z.; Zhou, L.; Yuan, Q.; Chen, H.; Lei, H.; Su, J. Effect of fumonisin B 1 on oxidative stress and gene expression alteration of nutrient transporters in porcine intestinal cells. *J. Biochem. Mol. Toxicol.* **2021**, *35*, e22706. [CrossRef] [PubMed]
22. Nasu, K.; Nishida, M.; Kawano, Y.; Tsuno, A.; Abe, W.; Yuge, A. Aberrant expression of apoptosis-related molecules in endometriosis: A possible mechanism underlying the pathogenesis of endometriosis. *Reprod. Sci.* **2011**, *18*, 206–218. [CrossRef]
23. Fadeel, B.; Orrenius, S.; Fadeel, B.; Orrenius, S. Apoptosis: A basic biological phenomenon with wide-ranging implications in human disease. *J. Intern. Med.* **2006**, *258*, 479–517. [CrossRef] [PubMed]
24. Li, Y.; Zhang, B.; Huang, K.; He, X.; Luo, Y.; Liang, R.; Luo, H.; Shen, X.L.; Xu, W. Mitochondrial proteomic analysis reveals the molecular mechanisms underlying reproductive toxicity of zearalenone in MLTC-1 cells. *Toxicology* **2014**, *324*, 55–67. [CrossRef] [PubMed]
25. Wang, H.W.; Wang, J.Q.; Zheng, B.Q.; Li, S.L.; Zhang, Y.D.; Li, F.D.; Zheng, N. Cytotoxicity induced by ochratoxin A, zearalenone, and α-zearalenol: Effects of individual and combined treatment. *Food Chem. Toxicol.* **2014**, *71*, 17–24. [CrossRef]
26. Kim, I.H.; Son, H.Y.; Cho, S.W.; Ha, C.S.; Kang, B.H. Zearalenone induces male germ cell apoptosis in rats. *Toxicol. Lett.* **2003**, *138*, 185–192. [CrossRef]
27. Jee, Y.; Noh, E.-M.; Cho, E.-S.; Son, H.-Y. Involvement of the Fas and Fas ligand in testicular germ cell apoptosis by zearalenone in rat. *J. Vet. Sci.* **2010**, *11*, 115. [CrossRef]
28. Karaman, E.F.; Zeybel, M.; Ozden, S. Evaluation of the epigenetic alterations and gene expression levels of HepG2 cells exposed to zearalenone and α-zearalenol. *Toxicol. Lett.* **2020**, *326*, 52–60. [CrossRef]

29. Huang, W.; Chang, J.; Wang, P.; Liu, C.; Yin, Q.; Song, A.D.; Gao, T.Z.; Dang, X.W.; Lu, F.S. Effect of Compound Probiotics and Mycotoxin Degradation Enzymes on Alleviating Cytotoxicity of Swine Jejunal Epithelial Cells Induced by Aflatoxin B1 and Zearalenone. *Toxins* **2019**, *11*, 12. [CrossRef]
30. Lei, M.Y.; Zhang, N.Y.; Qi, D.S. In vitro investigation of individual and combined cytotoxic effects of aflatoxin B1 and other selected mycotoxins on the cell line porcine kidney 15. *Exp. Toxicol. Pathol.* **2013**, *65*, 1149–1157. [CrossRef]
31. Tatay, E.; Meca, G.; Font, G.; Ruiz, M.J. Interactive effects of zearalenone and its metabolites on cytotoxicity and metabolization in ovarian CHO-K1 cells. *Toxicol. Vitr.* **2014**, *28*, 95–103. [CrossRef] [PubMed]
32. Zain, M.E. Impact of mycotoxins on humans and animals. *J. Saudi. Chem. Soc.* **2011**, *15*, 129–144. [CrossRef]
33. Pinton, P.; Nougayrède, J.P.; Del Rio, J.C.; Moreno, C.; Marin, D.E.; Ferrier, L.; Bracarense, A.P.; Kolf-Clauw, M.; Oswald, I.P. The food contaminant deoxynivalenol, decreases intestinal barrier permeability and reduces claudin expression. *Toxicol. Appl. Pharmacol.* **2009**, *237*, 41–48. [CrossRef] [PubMed]
34. Romero, A.; Ares, I.; Ramos, E.; Castellano, V.; Martínez, M.; Martínez-Larrañaga, M.R.; Anadón, A.; Martínez, M.A. Mycotoxins modify the barrier function of Caco-2 cells through differential gene expression of specific claudin isoforms: Protective effect of illite mineral clay. *Toxicology* **2016**, *353*, 21–33. [CrossRef]
35. Mclaughlin, J.; Padfield, P.J.; Burt, J.P.H.; O'Neill, C.A. Ochratoxin A increases permeability through tight junctions by removal of specific claudin isoforms. *J. Physiol.-Cell Physiol.* **2004**, *287*, C1412–C1417. [CrossRef]
36. Diesing, A.; Nossol, C.; Dänicke, S.; Walk, N.; Post, A.; Kahlert, S.; Rothkötter, H. Vulnerability of polarized intestinal porcine epithelial cells to mycotoxin deoxynivalenol depends on the route of application. *PLoS ONE* **2011**, *6*, e17472. [CrossRef]
37. Kishida, K.; Iida, T.; Yamada, T.; Toyoda, Y. d-Allose is absorbed via sodium-dependent glucose cotransporter 1 (SGLT1) in the rat small intestine. *Metab. Open* **2021**, *11*, 100112. [CrossRef]
38. Arya, A.; Al-Obaidi, M.M.; Karim, R.B.; Taha, H.; Khan, A.K.; Shahid, N.; Sayem, A.S.; Looi, C.Y.; Mustafa, M.R.; Mohd, M.A.; et al. Extract of Woodfordia fruticose flowers ameliorates hyperglycemia, oxidative stress and improves beta-cell function in streptozotocin-nicotinamide induced diabetic rats. *J. Ethnopharmacol.* **2015**, *175*, 229–240. [CrossRef]
39. Debray-Garcia, Y.; Sanchez, E.I.; Rodriguez-Munoz, R.; Venegas, M.A.; Velazquez, J.; Reyes, J.L. Diabetes and exposure to peritoneal dialysis solutions alter tight junction proteins and glucose transporters of rat peritoneal mesothelial cells. *Life Sci.* **2016**, *161*, 78–89. [CrossRef]
40. Yeong, J.; Tan, T.; Chow, Z.L.; Cheng, Q.; Lee, B.; Seet, A.; Lim, J.X.; Lim, J.C.T.; Ong, C.C.H.; Thike, A.A.; et al. Multiplex immunohistochemistry/immunofluorescence (mIHC/IF) for PD-L1 testing in triple-negative breast cancer: A translational assay compared with conventional IHC. *J. Clin. Pathol.* **2020**, *73*, 557–562. [CrossRef]
41. Maresca, M.; Mahfoud, R.; Garmy, N.; Fantini, J. The mycotoxin deoxynivalenol affects nutrient absorption in human intestinal epithelial cells. *J. Nutr.* **2002**, *9*, 2723–2731. [CrossRef] [PubMed]
42. Mallebrera, B.; Juan-García, A.; Font, G.; Ruiz, M.J. Mechanisms of beauvericin toxicity and antioxidant cellular defense. *Toxicol. Lett.* **2016**, *246*, 28–34. [CrossRef] [PubMed]
43. Jia, Z.; Liu, M.; Qu, Z.; Zhang, Y.; Yin, S.; Shan, A. Toxic effects of zearalenone on oxidative stress, inflammatory cytokines, biochemical and pathological changes induced by this toxin in the kidney of pregnant rats. *Environ. Toxicol. Pharmacol.* **2014**, *37*, 580–591. [CrossRef]
44. Ben Salem, I.; Prola, A.; Boussabbeh, M.; Guilbert, A.; Bacha, H.; Abid-Essefi, S.; Lemaire, C. Crocin and Quercetin protect HCT116 and HEK293 cells from Zearalenone-induced apoptosis by reducing endoplasmic reticulum stress. *Cell Stress Chaperones* **2015**, *20*, 927–938. [CrossRef] [PubMed]
45. Marin, D.E.; Taranu, I.; Pistol, G.; Stancu, M. Effects of zearalenone and its metabolites on the swine epithelial intestinal cell line: IPEC 1. *Proc. Nutr. Soc.* **2013**, *72*, 85–89. [CrossRef]
46. Qin, X.; Cao, M.; Lai, F.; Yang, F.G.W.; Zhang, X.; Cheng, S.; Sun, X.; Qin, G.; Shen, W.; Li, L. Oxidative stress induced by zearalenone in porcine granulosa cells and its rescue by curcumin in vitro. *PLoS ONE* **2015**, *10*, e0127551. [CrossRef]
47. Mallebrera, B.; Font, G.; Ruiz, M.J. Disturbance of antioxidant capacity produced by beauvericin in CHO-K1 cells. *Toxicol. Lett.* **2014**, *226*, 337–342. [CrossRef]
48. Zhou, J.; Ao, X.; Lei, Y.; Ji, C.; Ma, Q. Bacillus subtilis ansb01g culture alleviates oxidative stress and cell apoptosis induced by dietary zearalenone in first-parity gestation sows. *Anim. Nutr.* **2020**, *6*, 372–378. [CrossRef]
49. Liu, Y.; Wang, W.J. Aflatoxin B1 impairs mitochondrial functions, activates ROS generation, induces apoptosis, and involves in Nrf2 signal pathway in primary broiler hepatocytes. *Anim. Sci. J.* **2016**, *87*, 1490–1500. [CrossRef]
50. Gao, F.; Jiang, L.P.; Chen, M.; Geng, C.Y.; Yang, G.; Ji, F.; Zhong, L.F.; Liu, X.F. Genotoxic effects induced by zearalenone in a human embryonic kidney cell line. *Mutat. Res.-Gen. Tox. En.* **2013**, *755*, 6–10. [CrossRef]
51. Sun, X.; Ou, Z.; Chen, R.; Niu, X.; Chen, D.; Kang, R.; Tang, D.L. Activation of the p62-keap1-nrf2 pathway protects against ferroptosis in hepatocellular carcinoma cells. *Hepatology* **2016**, *63*, 173–184. [CrossRef] [PubMed]
52. Li, L.; Liu, J.; Nie, S.; Ding, L.; Wang, L.; Liu, J.; Liu, W.; Zhang, T. Direct inhibition of keap1–nrf2 interaction by egg-derived peptides dkk and ddw revealed by molecular docking and fluorescence polarization. *RSC Adv.* **2017**, *7*, 34963–34971. [CrossRef]
53. Jin, J.; Xiong, T.; Hou, X.; Sun, X.; Liao, J.; Huang, Z.; Huang, M.; Zhao, Z. Role of Nrf2 activation and NF-κB inhibition in valproic acid induced hepatotoxicity and in diammonium glycyrrhizinate induced protection in mice. *Food Chem. Toxicol.* **2014**, *73*, 95–104. [CrossRef] [PubMed]

54. Valko, M.; Leibfritz, D.; Moncol, J.; Cronin, M.T.D.; Mazur, M.; Telser, J. Free radicals and antioxidants in normal physiological functions and human disease. *Int. J. Biochem. Cell Biol.* **2007**, *39*, 44–84. [CrossRef] [PubMed]
55. Jiang, S.Z.; Yang, Z.B.; Yang, W.R.; Wang, S.J.; Liu, F.X.; Johnston, L.A.; Chi, F.; Wang, Y. Effect of purified zearalenone with or without modified montmorillonite on nutrient availability, genital organs and serum hormones in post-weaning piglets. *Livest. Sci.* **2012**, *144*, 110–118. [CrossRef]
56. Livak, K.J.; Schmittgen, T.D. Analysis of relative gene expression data using realtime quantitative pcr and the 2(-delta delta c(t)) method. *Methods* **2001**, *25*, 402–408. [CrossRef] [PubMed]

toxins MDPI

Review

Aflatoxins in Feed: Types, Metabolism, Health Consequences in Swine and Mitigation Strategies

Roua Gabriela Popescu, Andreea Luminița Rădulescu, Sergiu Emil Georgescu * and Anca Dinischiotu

Department of Biochemistry and Molecular Biology, Faculty of Biology, University of Bucharest, Splaiul Independentei No. 91–95, 050095 Bucharest, Romania
* Correspondence: sergiu.georgescu@bio.unibuc.ro

Abstract: Feeding farm animals with aflatoxin-contaminated feed can cause various severe toxic effects, leading to increased susceptibility to infectious diseases and increased mortality, weight loss, poor performance and reduced reproductive capability. Following ingestion of contaminated foodstuffs, aflatoxins are metabolized and biotransformed differently in animals. Swine metabolism is not effective in detoxifying and excreting aflatoxins, meaning the risk of aflatoxicosis is increased. Thus, it is of great importance to elucidate the metabolism and all metabolic pathways associated with this mycotoxin. The damage induced by AFB1 in cells and tissues consists of inhibition of cell proliferation, carcinogenicity, immunosuppression, mutagenicity, oxidative stress, lipid peroxidation and DNA damage, leading to pathological lesions in the liver, spleen, lymph node, kidney, uterus, heart, and lungs of swine. At present, it is a challenging task and of serious concern to completely remove aflatoxins and their metabolites from feedstuff; thus, the aim of this study was a literature review on the deleterious effects of aflatoxins on swine metabolism, as well as alternatives that contribute to the detoxification or amelioration of aflatoxin-induced effects in farm animal feed.

Keywords: mycotoxin; aflatoxin; toxicity; metabolism; swine; decontamination

Key Contribution: Understanding aflatoxin metabolism, effects on swine health, as well as alternative procedures that contribute to the detoxification or amelioration of aflatoxin-induced effects in farm animal feed.

Citation: Popescu, R.G.; Rădulescu, A.L.; Georgescu, S.E.; Dinischiotu, A. Aflatoxins in Feed: Types, Metabolism, Health Consequences in Swine and Mitigation Strategies. *Toxins* **2022**, *14*, 853. https://doi.org/10.3390/toxins14120853

Received: 31 October 2022
Accepted: 1 December 2022
Published: 3 December 2022

Publisher's Note: MDPI stays neutral with regard to jurisdictional claims in published maps and institutional affiliations.

Copyright: © 2022 by the authors. Licensee MDPI, Basel, Switzerland. This article is an open access article distributed under the terms and conditions of the Creative Commons Attribution (CC BY) license (https://creativecommons.org/licenses/by/4.0/).

1. Introduction

Mycotoxins are toxins produced by certain fungal species. They are classified into five main groups (Figure 1), with specific chemical structures, that occur frequently in foods and feeds, i.e., trichothecenes, zearalenone, ochratoxins, fumonisins and aflatoxins. At the same time, fungi that produce mycotoxins are divided into two groups: those that invade before grain harvesting, a group commonly called field fungi, and those that grow only after harvesting, called storage fungi. Among the field fungi, several types of mycotoxin-producing species can be distinguished. The most important are i. *Fusarium graminearum* (deoxynivalenol, nivalenol), normally developed on the field plants; ii. *Fusarium moniliforme* (fumonisins), and sometimes *Aspergillus flavus* (aflatoxin), present in the case of senile or stressed plants; iii. *Penicillium verrucosum* (ochratoxin) and *A. flavus* (aflatoxin) that colonize the plant prior to harvesting, and subsequently predispose the crop to mycotoxin contamination. Mycotoxins are spread in animal feed, cereal crops, vegetables, and animal products. Feeding stuffs for farmed animals are considered as having the highest levels of mycotoxins [1–6].

Aflatoxins are a group of secondary metabolites that are produced by several *Aspergillus* species with increased toxicity and carcinogenic potential. Pigs, poultry and cattle are the most important farm animals affected by aflatoxicosis. The most potent toxicant is AFB1 [7].

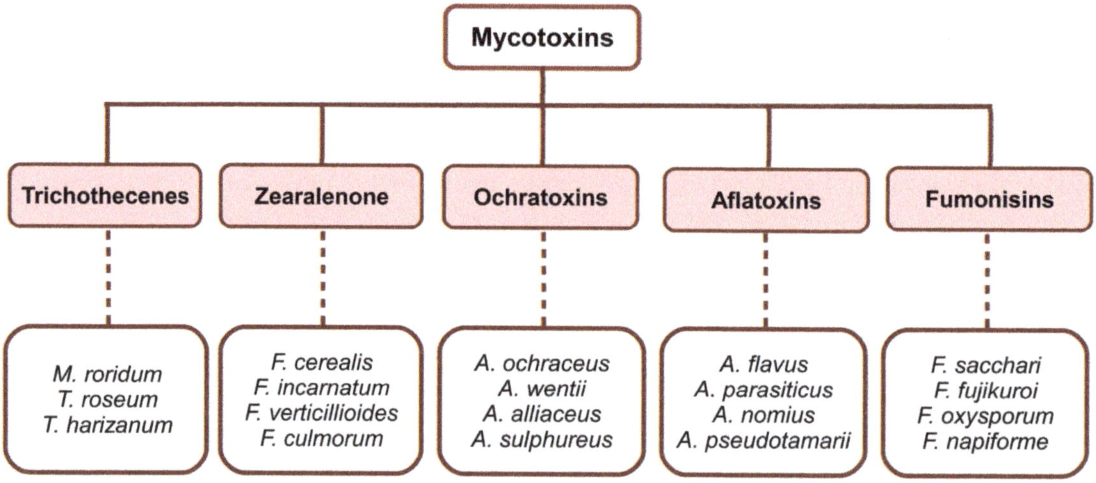

Figure 1. Classification of mycotoxins and the main producing species. Adapted after [8–13]. This image was made in OpenOffice Draw software.

Until 1985, the Food and Agriculture Organization reported that approximately 25% of the world's agricultural production is contaminated with mycotoxins [14]. Taking into consideration the predicted climate change in southeastern Europe, increased cereal contamination with AFB1 and OTA is expected [15]. Contamination with aflatoxins is most predominant in the regions of Africa and Asia, due to climatic conditions that favor the development of aflatoxigenic strains in both field and storage conditions [16,17]. The risks of aflatoxin-contaminated feed depend largely on the age and physiologic status of farm animals.

The main purpose of this review is to create an overview of aflatoxin metabolism, its effects on swine health, as well as alternative procedures that contribute to the detoxification or amelioration of aflatoxin-induced effects in farm animal feed.

2. Types of Aflatoxins

Mycotoxins are natural compounds of low molecular weight, up to 500 Da; aflatoxins are considered the most toxic, responsible for a significant decline in agriculture. They represent the most abundant groups found in foodstuffs, oilseeds, cereals, and dairy products [6,18]. All types of aflatoxins are derived from fungal species belonging to the genus *Aspergillus* and are considered among the most harmful mycotoxins for both animals and humans [19–23].

Aflatoxins are colorless to pale yellow crystalline substances, freely soluble in moderately polar solvents such as chloroform, methanol, dimethyl sulfoxide, with a water solubility of 10–20 μg/mL. In conditions such as under ultraviolet light in the presence of oxygen, extremes of $pH < 3$ or $pH > 10$ and oxidizing agents, aflatoxins are unstable. For example, ammonization at high temperatures results in the opening of the lactone ring, generating the decarboxylation of an aflatoxin molecule, an irreversible reaction. Some important physical and chemical properties of aflatoxins are given in Table 1 [20,24–29].

Currently, over 20 types of aflatoxins are known and among the best known are B1, B2, G1, G2, M1, M2, aflatoxicol and aflatoxin Q1 (Figure 2). Some of these forms are derivatives or metabolites of animal metabolism. For example, aflatoxin M1 and aflatoxin M2 are the metabolites of aflatoxin B1 and aflatoxin B2 which are found in the milk of lactating mammals fed with aflatoxin-contaminated feed [20,29,30].

Table 1. Physical and chemical properties of major aflatoxins. Adapted after [29,31–34].

Aflatoxin Type	Molecular Formula	Molecular Weight (g /mol)	Melting Point (°C)	Fluorescence	
				λ Excitation (nm)	λ Emission (nm)
B1 [29]	$C_{17}H_{12}O_6$	312	268–269	223	425
B2 [29]	$C_{17}H_{14}O_6$	314	286–289	265	425
G1 [29]	$C_{17}H_{12}O_7$	328	244–246	243	450
G2 [29]	$C_{17}H_{14}O_7$	330	237–240	265	450
M1 [33]	$C_{17}H_{12}O_7$	328	299	365	435
M2 [34]	$C_{17}H_{14}O_7$	330	293	360	450
Aflatoxicol [32]	$C_{17}H_{14}O_6$	314	225	325	425
Aflatoxin Q1 [31]	$C_{17}H_{12}O_7$	328	250	365	466

2.1. *Aflatoxins B1 and B2*

Aflatoxin B1 (AFB1) is the most potent carcinogenic mycotoxin naturally produced by Aspergillus species such as A. flavus, A. parasiticus, A. nomius, A. bombycis, A. arachidicola, A. minisclerotigenes, A. ochraceoroseus, A. pseudotamarii and A. rambellii, and it exerts harmful effects on humans and animals. The sensitivity degree and toxicity of AFB1 vary significantly between species, due to differences in its biotransformation. Some animals are considered extremely susceptible to AFB1, especially turkeys, rats, pigs, sheep, and dogs, whereas others such as monkeys, mice and chickens are considered resistant. The LD_{50} values for aflatoxin B1 are variable, depending on species and sex, with values ranging from 9 to 60 mg of AFB1 per kg of body weight [20,30,35–38].

Aflatoxin B2 (AFB2) is a blue-fluorescent, toxic secondary metabolite produced by the same species as AFB1, such as *A. arachidicola, A. flavus, A. minisclerotigenes, A. nomius* and *A. parasiticus.* This metabolite can be synthesized through multiple sequences that begin with a [2+3]-cycloaddition between quinone and 2,3-dihydrofuran [20,39–41].

2.2. *Aflatoxins G1 and G2*

Aflatoxin G1 (AFG1) and aflatoxin G2 (AFG2) are toxins produced by species of the common soil fungi, *A. parasiticus, A. nominus, A. bombyccis, A. arachidicola* and *A. flavus.* The presence of AFG1 is associated with toxicity and hepato-carcinogenicity in human and animal populations, while AFG2 has much lower activity [20,30,42,43].

2.3. *Aflatoxins M1 and M2*

The aflatoxins M1 (AFM1) and M2 (AFM2) are mammalian bio-conversion products or 4-hydroxy derivatives of AFB1 and AFB2, respectively, produced by *A. flavus* and *A. parasiticus.* After entering the body of humans or animals, AFB1 and AFB2 are metabolized by the hepatic microsomal mixed function oxidase system (cytochrome P450) to a reactive epoxide intermediate, but they can be also hydroxylated to the less harmful aflatoxins M1 and M2. In the case of an animal that ingests feed contaminated with AFB1, a percentage between 0.5% and 5% of the toxin ingested is biotransformed in the liver into AFM1. Milk, cheese, and other dairy products contain residues of AFM1 and AFM2 that should not exceed the limit of 50 ng per kg in Europe, 500 ng per kg in the USA, and 100 ng per kg in Iran [20,23,30,44–47] for human consumption.

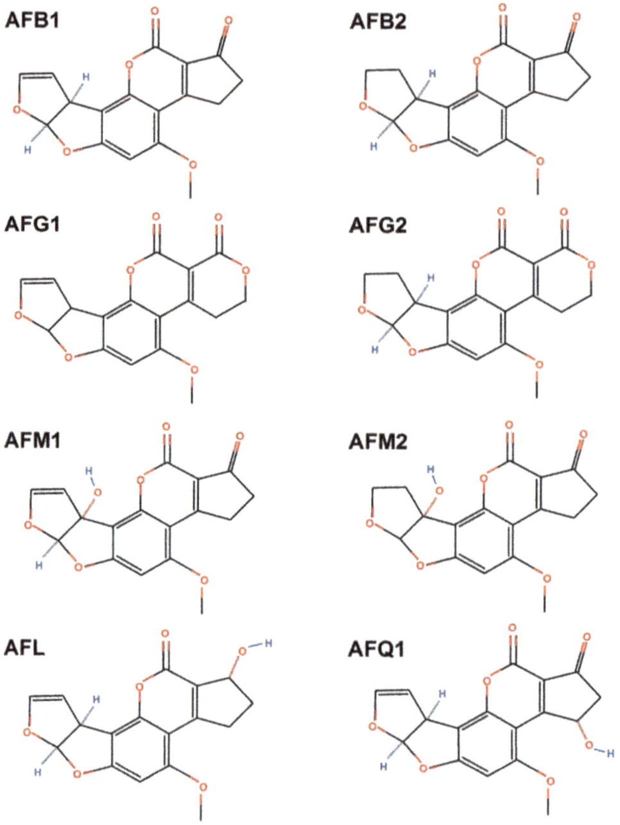

Figure 2. Chemical structures of aflatoxin B1 (AFB1), aflatoxin B2 (AFB2), aflatoxin G1 (AFG1), aflatoxin G2 (AFG2), aflatoxin M1 (AFM1), aflatoxin M2 (AFM2), aflatoxicol (AFL) and aflatoxin Q1 (AFQ1). This image was made in OpenOffice Draw software, v 4.1.9.

2.4. *Aflatoxicol*

The first report on natural contamination of food with aflatoxicol (AFL) appeared in 1984 [48]. AFL is one of the metabolites of AFB1, formed by the selective reduction of cyclopentanone carbonyl of AFB1, and has two stereoisomers (AFL1 /AFL-A /Ro and AFL2 or AFL-B) which differ by the orientation of the hydroxyl group in the cyclopentene ring. Both AFL forms are produced by the biological reduction catalyzed by enzymes present in fungi, such as: *Tetrahymena pyriformis, Trichoderma viride, Dactylium dendroides, Streptococcus lactis, Absidia repens, Mucor griseocyanus, Aspergillus niger, Mucor ambiguus, Tetrahymena pyriformis* and *Rhizopus spp*. Although AFL is eighteen times less toxic than AFB1, it was shown that AFL is carcinogenic and a potent frameshift mutagen [32,49–52].

2.5. *Aflatoxin Q1*

Aflatoxin Q1 (AFQ1) is a monohydroxylated derivative of AFB1, being one of the major AFB1 metabolites which appear after incubation of microsomal fraction from the mammalian liver with AFB1. The microsomal fraction is rich in CYP3A4 and other CYP450 enzymes which are responsible for the activation of AFB1 into the epoxide form, and for conversion into a less toxic detoxification metabolite, AFQ1. Initially it was found in the urine of rhesus monkeys orally exposed to AFB1. On the other hand, Yourtee et al. [53] showed that AFQ1 might be a major metabolite in the detoxification pathway of the native mycotoxin. AFQ1 is approximately eighteen times less toxic and approximately eighty-three times less mutagenic than AFB1 [30,53–55].

3. Aflatoxins' Metabolism: Biochemical, Molecular and Cell Signaling Aspects

After ingestion of contaminated food, aflatoxins are absorbed in the intestine; following their distribution, metabolism and excretion, the liver is the first and main organ affected (Figure 3). They also accumulate in muscle. P450 cytochromes play an important role in phase I biotransformation of xenobiotics, especially those belonging to families 1 and 3 [56]. In mammals, the enzymes with the highest levels of protein expression, and involved in the conversion of aflatoxins, are CYP1A2 and CYP3A4. The metabolite resulting from the oxidation reaction can bind to DNA, causing genotoxicity, and proteins generating cytotoxicity. For example, AFB1 binds to guanine residues of nucleic acids, resulting in AFB1 adducts that can lead to transversion of guanine–cytosine (GC) to thymine–adenine (TA) and implicitly to irreversible DNA damage. Binding of AFB1 to proteins is irreversible, the most well-known adduct being ADB1-lysine in albumin. In the first stage of metabolic oxidation in the liver, an epoxy reactive intermediate (e.g., AFB1-8,9-epoxide) is formed or this is hydrolyzed to a less toxic form, AFM1 [57,58].

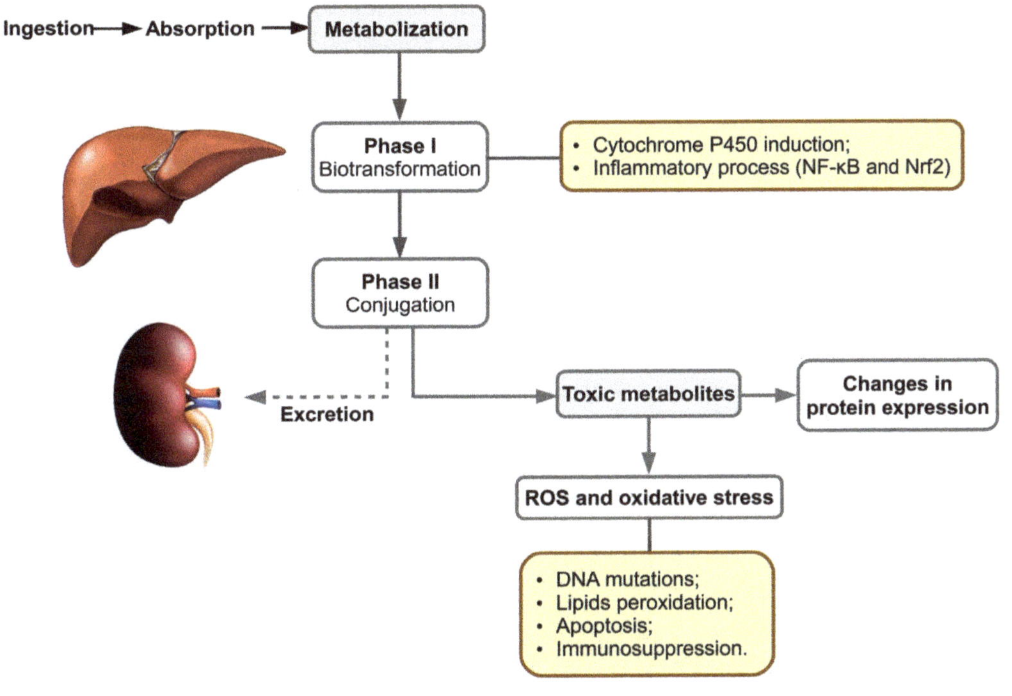

Figure 3. The adverse cellular effects of mycotoxins and their metabolites. Adapted after [56,59–61]. This image was made in OpenOffice Draw software, v 4.1.9.

The cytochrome P450 superfamily consists of enzymes involved in xenobiotic metabolism and endogenous compound oxidation; thus, Phase I enzymes catalyze the reactions of hydroxylation, sulphoxidation, epoxidation, N-, O- and S-dealkylation, oxidative aromatic hydroxylation, desulfuration, denitrosation, and dehalogenation aiming for the addition of functional polar group(s). In porcine hepatic tissue, the CYP450 proteins expressed are represented by CYP2A19 (34%), CYP2D25 (25,5%), CYP2C49 (11.2%), CYP2E1 (8.1%), CYP3A39 (8,1%), CYP3A29 (5,8%), CYP2C33 (5%) and CYP1A2 (2.3% of the total liver CYPs, respectively) [62–68].

Phase II of metabolism implicates conjugation reactions of metabolites previously formed [69] with glucuronic acid and sulfate especially. Subsequently, the epoxide metabolite generated in phase I may be detoxified in phase II by glutathione conjugation, through hydrolysis by an epoxide hydrolase to AFB1-8,9-dihydrodiol, or by reduction to a less toxic metabolite such as AFM1 or AFQ1 [43,70–73]. The resulting metabolites are excreted through the biliary pathway, followed by the urinary pathway.

By RNA-seq technology it was proved that in vitro exposure of bovine fetal hepatocyte cell line (BFH12) to AFB1 affected the cells' transcriptome. Gap junction protein beta 2 and Follistatin genes—the latter being involved in proliferation and colony expansion of progenitor populations of hepatocytes—as well as those of ornithine decarboxylase and A-Raf proto-oncogene have been upregulated. Instead, genes that codify for tumor suppressors, such as those of collagen type XVIII alpha 1 chain (COL18A1), collagen type 1 alpha 2 chain (COL1A2), as well as that for natriuretic peptide receptor 3 have been downregulated. The treatment with this mycotoxin also upregulated the following CYP isoforms: CYP26B1, CYP3A4, CYP27B1 and downregulated CYP1A1, CYP1B1, CYP19A1, CYP36A1, CYP4B1 [74].

The same study from Pauletto et al. [74] revealed that all analyzed glutathione-S-transferase genes, except those for omega 1 and pi1 isoforms, have been downregulated. The gene sets for TNF-α signaling via NF-kB, oxidative phosphorylation, DNA repair, inflammatory response, KRAS signaling, p53 pathway, PI3K-Akt-mTOR signaling, apoptosis and hypoxia have been upregulated by AFB1 treatment of BFH12 cells. In the same conditions, other gene sets for epithelial–mesenchymal transition, bile acid metabolism, estrogen response and heme metabolism have been downregulated.

Recently, based on transcriptomic data and post translational analyses, it was postulated that Toll-Like Receptor (TLR2) activation is involved in AFB1-induced inflammation and oxidative stress in BFH12 cells [75]. Moreover, in a chicken hepatocarcinoma cell line (LMH) exposed to AFB1 differentially, expression analysis revealed that 1006 genes have been upregulated and 791 downregulated, compared with the control treatment. The mRNA expression of CYP27A1, CYP1A4, FABP2, PPARα and GSTT1 were significantly decreased by this mycotoxin treatment, whereas genes responsible for focal adhesion and MAPK pathways were upregulated compared with control ones [76].

Previously it was noticed that in HepG2 cells treated with AFB1, increases in the expressions of miR-34A and miR-33a-5p led to an important decrease of β-catenin, c-myc and cyclin D1 levels in the Wnt signaling pathway, generating an important risk of hepatocellular carcinoma [77,78]. The exposure to this mycotoxin also inhibits protein synthesis and due to this, enzymes' levels of different metabolic pathways are affected [79].

Recently, it was proved that AFB1 exposure of cells affects the respiratory chain, generating reactive oxygen species (ROS). If these are not counteracted by the antioxidant enzymatic and non-enzymatic systems, oxidative stress occurs [80]. The excess of ROS attacks polyunsaturated fatty acids from glycerophospholipids, generating end products of lipid peroxidation, as well as DNA and proteins. Lipid peroxidation and oxidative damage to DNA play a major role in the toxicity of aflatoxins.

4. Aflatoxin's Toxicity in Swine

The monogastric animals are more susceptible to AFB1, compared to the ruminants, since bacteria from the rumen section of the stomach can metabolize mycotoxins [81]. The pig's caloric need is supplied by carbohydrates and fats in great extent. Cereal grains represent a source of carbohydrates, and they are included in swine ratio in up to 85% of the ingredients [82]. Maize, wheat, barley and oat are used in pigs' nutrition, and represent a common source of mycotoxins in feed.

In recent years, the domestic production and industrial swine industry have been heavily affected by viral infections, such as African swine fever and mycotoxin contamination [83,84]. Mycotoxins with the greatest economic impact on swine breeders are aflatoxins, zearalenone, deoxynivalenol, trichothecenes, T-2 toxin and ochratoxins [85–87]. Growing pigs are highly susceptible to mycotoxins. One of the main difficulties encountered in controlling mycotoxins is that more than one type of mycotoxin is present in a batch of fodder, or cereal for pigs, at the same time. Thus, feeding pigs contaminated feed with several types of mycotoxins, even if they are in concentrations at the minimum recommended by the European Union, can cause numerous negative effects in animals due to cumulative toxic effects. The most common symptoms of mycotoxicosis in swine are the refusal to eat, decrease in growth rate, reproductive disorders and decreased immune status [88–92]. In the process of breeding and producing pigs about 60–70% of the costs are due to feed [93].

Luthy et al. [94] showed that in pigs approximately 20% of the radioactive AFB1 dose was excreted in the urine during nine days. AFM1, the metabolite of AFB1, was found in the range of 80–420 pg/mL in the urine of pigs fed with 26.48 μg AFB1/kg of body weight for 42 days [94–99].

Exposure of swine to aflatoxins can cause a variety of chronic or acute syndromes depending on the type of aflatoxin and level of consumption; aflatoxins can generate increased susceptibility to infectious diseases and increased mortality, weight loss and poor performance, reduced reproductive capability, changes in clinical biochemical patterns, and suppressed immune function [100].

The maximum tolerable levels of aflatoxins in pig diets depend on age. According to the FDA, the regulatory limits for swine aflatoxin B1 are <20 ppb for piglets, <100 ppb for specimens used for reproduction and <200 ppb AFB1 for those in the finishing period [101,102]. Aflatoxin exposure generates in pigs: a low growth rate, poor conversion of food, increased mortality, impaired coagulation of blood and kidney function, changes in the immune response, increased susceptibility to disease and decreased resistance to stress [103,104].

The liver is the organ most affected by the ingestion of aflatoxins because it receives and concentrates all compounds carried by the bloodstream. Extremely high concentrations of aflatoxins (over 1000 ppb, previously reported in Ugandan crops, mainly maize, peanuts and cassava) cause hepatitis, hepatic necrosis, increased clotting time, and finally the death of animals caused by severe hemorrhage. In a lighter, subacute form, aflatoxicosis causes hepatic lipidosis, portal fibrosis and liver tumors [105,106].

Additionally, the resulting aflatoxin metabolites can be transmitted from lactating sows to nursing pigs, via milk, consequently contaminating the piglets which are more sensitive to stunted growth; thus, this may cause up to a 20% mortality in piglets, characterized by enterocolitis, diarrhea, and a suppressed immune system which leads to decreased resistance to infectious diseases. Prodanov-Radulović et al. [91] reported the presence of AFM1 in the milk of nursing sows consuming diets containing AFB1. Furthermore, Weaver [107] showed that the concentration of AFM1 was about 1.7 times higher in colostrum, than milk of nursing sows, because AFM1 binds to milk casein and therefore is transferred to the piglets [29,72,91,108–111].

The diseases caused by the consumption of aflatoxins are known as aflatoxicosis. Swine metabolism is not effective in detoxifying and excreting aflatoxins, which increases the risk of aflatoxicosis. The main biological effects of aflatoxins in suckling piglets, growing, and finished and breeding pigs are carcinogenicity, immunosuppression, mutagenicity, teratogenicity, decreased feed efficiency and poor weight gain, impaired liver and altered serum biochemical parameters. Severe effects in swine can lead to acute hepatitis, systemic hemorrhages, nephrosis and death [71,72,112]. Some authors have shown that swine fed with low levels of aflatoxins presented signs of pulmonary edema, reduced feed consumption and body weight gain, as well as a decrease in the activity of enzymes that catalyze the oxidative decarboxylation, total serum proteins, total leukocyte count and blood pressure [72,113–115].

Another toxic effect in swine exposed to aflatoxins is the alteration of the inflammatory response, known as immunotoxicity. In weanling pigs fed for 28 days with low doses of aflatoxins [116], reduced synthesis of pro-inflammatory cytokines and an increase in anti-inflammatory ones were noticed.

Immunomodulatory effects of AFB1 have also been proven in swine. Studies conducted by Meissonnier et al. highlighted impaired lymphocyte activation and increased cytokine expression (TNF-α, IL-1β, IL-6, IL-10, and IFN-γ) in pigs vaccinated with ovalbumin, after dietary AFB1 exposure [117,118]. In contrast, Marin et al. [118] showed that aflatoxins did not exert any effect on regulatory cytokines produced by either the Th1 (IL-2) or the Th2 (IL-4) subset of lymphocytes.

AFB1 is a very strong inhibitor of lymphocyte proliferation. Stec et al. [119] showed that a concentration of 0.02 μg AFB1/mL reduced up to 50% of lymphocytes, isolated from peripheral blood taken from 7-week-old pigs after a 72-hour exposure period, suggesting that AFB1 is a very strong inhibitor of in vitro lymphocyte proliferation in pigs [116,119,120].

The effects on sperm motility or on the reproductive performance of gilts depend on aflatoxin doses also. The maturation rates of oocytes decreased significantly in the case of acute exposure to 50 μM AFB1, probably because most oocytes have been arrested at the germinal vesicle breakdown or meiosis I stage, resulting in early oocyte apoptosis and increased Bak, Bax, Bcl-xl mRNA levels. This could suggest that AFB1 disrupts porcine oocyte maturation through the modulation of epigenetic modifications, oxidative stress, excessive autophagy and apoptosis [121,122].

In summary, aflatoxins induce pathological lesions in the liver, spleen, lymph node, kidney, uterus, heart and lungs of swine. Severe toxicity causes collapse and death within several hours, acute toxicity causes death within 12 h, and with subacute toxicity death occurs after about 20 days [123–128].

5. Methods to Reduce Aflatoxins' Toxicity

The need for solutions to ameliorate the effects of mycotoxins on food-producing swine prompts increased research in this area. Currently, there are few national and international studies that focus on the effects of aflatoxins at the hepato-nephrotoxic level in swine. Considering this, studies regarding detoxification methods and the influence of certain feed additives on the toxicity of aflatoxins, in swine liver and kidneys, are of great importance.

Aflatoxin decontamination procedures have been developed to inactivate or remove it from feed stuffs, without leaving any chemical residues. These must be cost-effective to keep the final market price reasonable. The methods used for decontamination of aflatoxins can be divided into biological, chemical and physical methods. All these methods must ensure that the degradation process maintains the nutritive value of feed and will not introduce one or more toxic substances. Prevention is the most desirable method of reducing aflatoxin contamination but needs much more improvement in terms of agricultural storage methods, practices in harvesting and handling of crops. Therefore, the recognition of problems caused by mycotoxins in food and feed is the first step to prevention, which will allow farmers to produce good quality food for the animals [29,72,129–131].

The general chemical methods used against aflatoxins are based on chemical agents that deactivate and degrade aflatoxins by oxidation and/or hydrolysis of the lactone ring from the polyketide backbone of aflatoxins, or by oxidation of the double bond of the terminal furan ring. However, the use of these agents is limited due to the problems associated with their residues [29,131,132].

Physical methods involve the separation of contaminated fractions, removal, or inactivation of aflatoxins by physical means, such as heating, cooking, roasting, and radiation. Due to the limited solubility of aflatoxins in water, these procedures are regarded as being unfeasible and economically inefficient. Therefore, decontaminating products contaminated with mold requires a multi-step process that involves mechanical sorting and washing. Jalili [133] mentioned that processing methods such as boiling, roasting, baking, and steaming in maize products destroyed aflatoxins to a considerable extent of 50–70%.

Adsorption is another physical method for aflatoxin decontamination and involves the binding of a toxic compound, to the adsorbent compound, during digestion in the gastrointestinal tract of farm animals. Examples of adsorbents are active carbon, diatomaceous earth, alumina clay, alumina bentonite, montmorillonite; sodium and calcium aluminum silicates, mainly zeolite; phyllosilicates and hydrated sodium calcium aluminosilicate; complex carbohydrates such as cellulose, and the polysaccharides present at the cellular walls of yeasts and bacteria e.g., glucomannans and peptidoglycans; and synthetic polymers of cholestyramine, polyvinyl pyrrolidone, and its derivatives [29,131,133,134].

Efficient drying of farmed feed is an effective measure against fungal growth and aflatoxin production. The correct way of drying is the best manner of avoiding fungal growth and mycotoxin production in grain after harvest. When natural drying in the sun is not possible, most of the time because climate conditions do not allow this, in order to reduce or prevent the production of most mycotoxins, drying should take place as soon as possible after harvest, and as rapidly as feasible, or otherwise it is necessary to use a form of mechanical drying [129,135].

Biological decontamination of aflatoxins is another strategy in which the degradation is achieved, by using modified strains of *Aspergillus* to reduce aflatoxin contamination by competitive inhibition, or by using genetically modified plants. For example, in Africa, Central America and Asia, the populations experience high levels of exposure to dietary aflatoxin from maize, which is an important part of the human diet in these locations. One of the strategies used in these regions involves the use of transgenic maize (Bt corn) in order to control mycotoxin contamination. The second approach involves the use of a food supplement (NovaSil clay) in order to absorb aflatoxins in the gastrointestinal tract and, therefore, reduce the toxin bioavailability. The third method is based on a modified strain of *A. flavus* that does not produce aflatoxins. Another alternative for biological decontamination is the addition of antioxidant compounds in animal feed, in order to reduce the toxic effects of aflatoxins, or to inhibit the growth of aflatoxin-producing fungal species. Examples of antioxidant compounds are chlorophyll and its derivatives, selenium, medicinal herbs and plant extracts [71,136,137].

The presence of polyphenolic compounds in feed, especially representatives of the flavonoid group, can attenuate the mycotoxin-induced inflammatory process by modulating the activities of NF-κB and Nrf2 [138,139]. However, the potential anti-inflammatory effects of polyphenols have, so far, been less investigated in farm animals.

Currently, wine production is one of the main agricultural activities around the world, which is accompanied by the generation of large amounts of waste and by-products rich in antioxidant compounds [140]. Examples of such compounds are stilbens (resveratrol), anthocyans, flavones, flavonones and isoflavones [59].

A strategy for reducing exposure to mycotoxins in animals includes supplementation of feed products with detoxifying additives, which allows for counteraction of their toxic effects [29,141–144].

6. Conclusions

Aflatoxins produced by various fungi during the pre- and post-harvest stages of various food and feed, cause adverse effects in different animals and negative economic impacts worldwide. Significant advances have been achieved in our understanding of aflatoxins' metabolism. Swine are particularly sensitive to aflatoxin exposure due to ineffective detoxification and excretion. The major challenge of ongoing and future research will remain the identification of members of metabolic pathways that link aflatoxin toxicity in swine, to the perturbations of cell metabolism and oxidative stress. Current methods cannot completely remove aflatoxin metabolites from swine diets. Therefore, it is desirable to prevent the contamination of feed by aflatoxins, which is achievable by using different procedures, including feed storage in dry areas and improved management techniques, in order to develop strategies that contribute to the detoxification or amelioration of aflatoxin-induced effects in farm animals, in an efficient and cost-effective manner.

Author Contributions: R.G.P. and A.D. conceived the manuscript; R.G.P., A.L.R. and A.D. wrote the paper. All authors participated in the preparation of the manuscript. All authors have read and agreed to the published version of the manuscript.

Funding: This work was supported by the project 41PFE/30.12.2021, financed by the Ministry of Research, Innovation and Digitalization through Program 1—Development of the National R&D System, Subprogram 1.2. Institutional performance—Financing projects for excellence in RDI.

Institutional Review Board Statement: Not applicable.

Informed Consent Statement: Not applicable.

Data Availability Statement: Not applicable.

Acknowledgments: This work was supported by the project 41PFE/30.12.2021, financed by the Ministry of Research, Innovation and Digitalization through Program 1—Development of the National R&D System, Subprogram 1.2. Institutional performance—Financing projects for excellence in RDI.

Conflicts of Interest: The authors declare no conflict of interest.

References

1. Arroyo-Manzanares, N.; Huertas-Pérez, J.F.; García-Campaña, A.M.; Gámiz-Gracia, L. Aflatoxins in animal feeds: A straightforward and cost-effective analytical method. *Food Control* **2015**, *54*, 74–78. [CrossRef]
2. Pinotti, L.; Ottoboni, M.; Giromini, C.; Dell'Orto, V.; Cheli, F. Mycotoxin contamination in the EU feed supply chain: A focus on Cereal Byproducts. *Toxins* **2016**, *8*, 45. [CrossRef] [PubMed]
3. Senthilkumar, T.; Jayas, D.S.; White, N.D.G.; Fields, P.G.; Gräfenhan, T. Near-Infrared (NIR) hyperspectral imaging: Theory and applications to detect fungal infection and mycotoxin contamination in food products. *Indian J. Entomol.* **2016**, *78*, 91. [CrossRef]
4. Sobral, M.M.C.; Faria, M.A.; Cunha, S.C.; Ferreira, I.M.P.L.V.O. Toxicological interactions between mycotoxins from ubiquitous fungi: Impact on hepatic and intestinal human epithelial cells. *Chemosphere* **2018**, *202*, 538–548. [CrossRef]
5. Tola, M.; Kebede, B. Occurrence, importance and control of mycotoxins: A review. *Cogent Food Agric.* **2016**, *2*, 1191103. [CrossRef]
6. Alshannaq, A.; Yu, J.H. Occurrence, toxicity, and analysis of major mycotoxins in food. *Int. J. Environ. Res. Public Health* **2017**, *14*, 632. [CrossRef] [PubMed]
7. Pleadin, J.; Kovačević, D.; Perković, I. Impact of casing damaging on aflatoxin B 1 concentration during the ripening of dry-fermented meat sausages. *J. Immunoass. Immunochem.* **2015**, *36*, 655–666. [CrossRef] [PubMed]
8. Rheeder, J.P.; Marasas, W.F.O.; Vismer, H.F. Production of Fumonisin Analogs by. *Society* **2002**, *68*, 2101–2105. [CrossRef]
9. Ehrlich, K.C.; Kobbeman, K.; Montalbano, B.G.; Cotty, P.J. Aflatoxin-producing Aspergillus species from Thailand. *Int. J. Food Microbiol.* **2007**, *114*, 153–159. [CrossRef]
10. McCormick, S.P.; Stanley, A.M.; Stover, N.A.; Alexander, N.J. Trichothecenes: From simple to complex mycotoxins. *Toxins* **2011**, *3*, 802–814. [CrossRef]
11. Fakruddin, M.; Chowdhury, A.; Hossain, M.N.; Ahmed, M.M. Characterization of aflatoxin producing Aspergillus flavus from food and feed samples. *Springerplus* **2015**, *4*, 1–6. [CrossRef] [PubMed]
12. Wang, Y.; Wang, L.; Liu, F.; Wang, Q.; Selvaraj, J.N.; Xing, F.; Zhao, Y.; Liu, Y. Ochratoxin A producing fungi, biosynthetic pathway and regulatory mechanisms. *Toxins* **2016**, *8*, 83. [CrossRef] [PubMed]
13. Wentzel, J.F.; Lombard, M.J.; Du Plessis, L.H.; Zandberg, L. Evaluation of the cytotoxic properties, gene expression profiles and secondary signalling responses of cultured cells exposed to fumonisin B1, deoxynivalenol and zearalenone mycotoxins. *Arch. Toxicol.* **2017**, *91*, 2265–2282. [CrossRef] [PubMed]
14. Eskola, M.; Kos, G.; Elliott, C.T.; Hajšlová, J.; Mayar, S.; Krska, R. Worldwide contamination of food-crops with mycotoxins: Validity of the widely cited 'FAO estimate' of 25%. *Crit. Rev. Food Sci. Nutr.* **2020**, *60*, 2773–2789. [CrossRef] [PubMed]
15. Gagiu, V.; Mateescu, E.; Armeanu, I.; Dobre, A.A.; Smeu, I.; Cucu, M.E.; Oprea, O.A.; Iorga, E.; Belc, N. Post-harvest contamination with mycotoxins in the context of the geographic and agroclimatic conditions in Romania. *Toxins* **2018**, *10*, 533. [CrossRef] [PubMed]
16. Shabeer, S.; Asad, S.; Jamal, A.; Ali, A. Aflatoxin Contamination, Its Impact and Management Strategies: An Updated Review. *Toxins* **2022**, *14*, 307. [CrossRef] [PubMed]
17. Kolawole, O.; Siri-Anusornsak, W.; Petchkongkaw, A.; Meneely, J.; Elliott, C. The Efficacy of Additives for the Mitigation of Aflatoxins in Animal Feed: A Systematic Review and Network Meta-Analysis. *Toxins* **2022**, *14*, 707. [CrossRef]
18. Jallow, A.; Xie, H.; Tang, X.; Qi, Z.; Li, P. Worldwide aflatoxin contamination of agricultural products and foods: From occurrence to control. *Compr. Rev. Food Sci. Food Saf.* **2021**, *20*, 2332–2381. [CrossRef]
19. Filazi, A.; Sireli, U.T. Chapter 7. Occurrence of aflatoxins in food. In *Aflatoxins: Recent Advances and Future Prospects*; Mehdi, R.-A., Ed.; InTech Open Access: London, UK, 2012; pp. 143–170.
20. Feddern, V.; Dors, G.C.; Tavernari, F.; Mazzuco, H.; Cunha, J.A.; Krabbe, E.L.; Scheuermann, G.N. Aflatoxins: Importance on animal nutrition. *Aflatoxins Recent Adv. Futur. Prospect. InTech Open Access Croat.* **2013**, 171–195. [CrossRef]
21. Seetha, A.; Munthali, W.; Msere, H.W.; Swai, E.; Muzanila, Y.; Sichone, E.; Tsusaka, T.W.; Rathore, A.; Okori, P. Occurrence of aflatoxins and its management in diverse cropping systems of central Tanzania. *Mycotoxin Res.* **2017**, *33*, 323–331. [CrossRef]

22. Ismail, A.; Gonçalves, B.L.; de Neeff, D.V.; Ponzilacqua, B.; Coppa, C.F.S.C.; Hintzsche, H.; Sajid, M.; Cruz, A.G.; Corassin, C.H.; Oliveira, C.A.F. Aflatoxin in foodstuffs: Occurrence and recent advances in decontamination. *Food Res. Int.* **2018**, *113*, 74–85. [CrossRef] [PubMed]
23. Negash, D. A Review of Aflatoxin: Occurrence, Prevention, and Gaps in Both Food and Feed Safety. *J. Appl. Microbiol. Res.* **2018**, *1*, 1–35. [CrossRef]
24. Okoth, S. *Improving the Evidence Base on Aflatoxin Contamination and Exposure in Africa*; CTA Working Paper; 16/13; CTA: Wageningen, The Netherland, 2016; pp. 1–113. Available online: https://hdl.handle.net/10568/90118 (accessed on 23 September 2022).
25. MdQuadri, S.H.; Niranjan, M.; Chaluvaraju, K.; Shantaram, U.; Enamul, H. An Overview on Chemistry, Toxicity, Analysis and Control of Aflatoxins. *Int. J. Chem. Life Sci.* **2017**, *2*, 1071–1078.
26. Sailaja, O.; Krishnaven, G.; Manoranjani, M. Identification and High-performance Liquid Chromatography Quantification of Aflatoxins in Red Chili. *Asian J. Pharm.* **2017**, *2017*, 933–937.
27. Akeberegn, D.; Alemneh, T.; Zewudie, D. Effects of Aflatoxin Contamination in Milk: A Review. *MRJMBS* **2018**, *6*, 118–128.
28. Sedova, I.; Kiseleva, M.; Tutelyan, V. Mycotoxins in Tea: Occurrence, Methods of Determination and Risk Evaluation. *Toxins* **2018**, *10*, 444. [CrossRef]
29. Vijaya Kumar, V. Aflatoxins: Properties, Toxicity and Detoxification. *Nutr. Food Sci. Int. J.* **2018**, *6*, 555696. [CrossRef]
30. Bbosa, G.S.; Kitya, D.; Odda, J.; Ogwal-Okeng, J. Aflatoxins metabolism. *Health* **2013**, *5*, 14–34. [CrossRef]
31. Franco, C.M.; Fente, C.A.; Vázquez, B.I.; Cepeda, A.; Mahuzier, G.; Prognon, P. Interaction between cyclodextrins and aflatoxins Q1, M1 and P1. Fluorescence and chromatographic studies. *J. Chromatogr. A* **1998**, *815*, 21–29. [CrossRef]
32. Carvajal, M.; Rojo, F.; Méndez, I.; Bolños, A. Aflatoxin B1and its interconverting metabolite aflatoxicol in milk: The situation in Mexico. *Food Addit. Contam.* **2003**, *20*, 1077–1086. [CrossRef]
33. Behfar, A.; Khorasgani, Z.N.; Alemzadeh, Z.; Goudarzi, M.; Ebrahimi, R.; Tarhani, N. Determination of Aflatoxin M1 Levels in Produced Pasteurized Milk in Ahvaz City by Using HPLC. *Jundishapur J. Nat. Pharm. Prod.* **2012**, *7*, 80–84. [CrossRef] [PubMed]
34. Lee, D.; Lee, K.G. Analysis of aflatoxin M1 and M2 in commercial dairy products using high-performance liquid chromatography with a fluorescence detector. *Food Control.* **2015**, *50*, 467–471. [CrossRef]
35. Marai, I.F.M.; Asker, A.A. Aflatoxins in rabbit production: Hazards and control. *Trop. Subtrop. Agroecosystems* **2008**, *8*, 1–28.
36. Rawal, S.; Kim, J.E.; Coulombe, R. Aflatoxin B1 in poultry: Toxicology, metabolism and prevention. *Res. Vet. Sci.* **2010**, *89*, 325–331. [CrossRef] [PubMed]
37. Omar, M.H.E.-D. (Ed.) Mycotoxins-Induced Oxidative Stress and Disease. In *Mycotoxin and Food Safety in Developing Countries*; InTech Open Access: London, UK, 2013; pp. 63–92. [CrossRef]
38. Ishikawa, A.T.; Hirooka, E.Y.; e Silva, P.L.A.; Bracarense, A.P.F.R.L.; Da Costa, K.K.M.; Akagi, C.Y.; Kawamura, O.; Da Costa, M.C.; Itano, E.N. Impact of a single oral acute dose of aflatoxin b1on liver function/cytokines and the lymphoproliferative response in C57BL/6 mice. *Toxins* **2017**, *9*, 374. [CrossRef] [PubMed]
39. Chang, S.B.; Abdel Kader, M.M.; Wick, E.L.; Wogan, G.N. Aflatoxin B2: Chemical identity and biological activity. *Science* **1963**, *142*, 1191–1192. [CrossRef]
40. Yabe, K.; Ando, Y.; Hamasaki, T. Biosynthetic Relationship among aflatoxins B1, B2, G1 and G2. *Appl. Environ. Microbiol.* **1988**, *54*, 2101–2106. [CrossRef]
41. Zhou, G.; Corey, E.J. Short, enantioselective total synthesis of aflatoxin B2 using an asymmetric [3+2]-cycloaddition step. *J. Am. Chem. Soc.* **2005**, *127*, 11958–11959. [CrossRef]
42. Wong, J.J.; Hsieh, D.P.H. Mutagenicity of aflatoxins related to their metabolism and carcinogenic potential. *Proc. Natl. Acad. Sci. USA* **1976**, *73*, 2241–2244. [CrossRef]
43. Bennett, J.W.; Klich, M. Mycotoxins. *Clin. Microbiol. Rev.* **2003**, *16*, 497–516. [CrossRef]
44. Galvano, F.; Galofaro, V.; Galvano, G. Occurrence and Stability of Aflatoxin M 1 in Milk and Milk Products: A Worldwide Review. *J. Food Prot.* **1996**, *59*, 1079–1090. [CrossRef]
45. Yu, J. Current understanding on aflatoxin biosynthesis and future perspective in reducing aflatoxin contamination. *Toxins* **2012**, *4*, 1024–1057. [CrossRef] [PubMed]
46. Ahlberg, S.; Grace, D.; Kiarie, G.; Kirino, Y.; Lindahl, J. A risk assessment of Aflatoxin M1 exposure in low and mid-income dairy consumers in Kenya. *Toxins* **2018**, *10*, 348. [CrossRef] [PubMed]
47. Puga-Torres, B.; Salazar, D.; Cachiguango, M.; Cisneros, G.; Gómez-Bravo, C. Determination of aflatoxin M1 in raw milk from different provinces of Ecuador. *Toxins* **2020**, *12*, 498. [CrossRef] [PubMed]
48. Righetti, L.; Rolli, E.; Dellafiora, L.; Galaverna, G.; Suman, M.; Bruni, R.; Dall'Asta, C. Thinking Out of the Box: On the Ability of Zea mays L. to Biotrasform Aflatoxin B1 Into Its Modified Forms. *Front. Plant Sci.* **2021**, *11*, 1–11. [CrossRef]
49. Salhab, A.S.; Edwards, G.S. Comparative in Vitro Metabolism of Aflatoxicol by Liver Preparations from Animals and Humans. *Cancer Res.* **1976**, *37*, 1016–1021. [CrossRef]
50. Schoenhard, G.L.; Hendricks, J.D.; Nixon, J.E.; Lee, D.J.; Wales, J.H.; Sinnhuber, R.O.; Pawlowski, N.E. Aflatoxicol-induced hepatocellular carcinoma in Rainbow Trout (Salmogairdneri) and the synergistic effects of cyclopropenoid fatty acids. *Cancer Res.* **1981**, *41*, 1011–1014.
51. Nakazato, M.; Morozumi, S.; Saito, K.; Fujinuma, K.; Nishima, T.; Kasai, N. Interconversion of aflatoxin B1 and aflatoxicol by several fungi. *Appl. Environ. Microbiol.* **1990**, *56*, 1465–1470. [CrossRef] [PubMed]

52. Karabulut, S.; Paytakov, G.; Leszczynski, J. Reduction of aflatoxin B1 to aflatoxicol: A comprehensive DFT study provides clues to its toxicity. *J. Sci. Food Agric.* **2014**, *94*, 3134–3140. [CrossRef]
53. Yourtee, D.M.; Bean, T.A.; Kirk-Yourtee, C.L. Human aflatoxin B1metabolism: An investigation of the importance of aflatoxin Q1as a metabolite of hepatic post-mitochondrial fraction. *Toxicol. Lett.* **1987**, *38*, 213–224. [CrossRef]
54. Hendricks, J.D.; Sinnhuber, R.O.; Nixon, J.E.; Wales, J.H.; Masri, M.S.; Hsieh, D.P. Carcinogenic response of rainbow trout (Salmo gairdneri) to aflatoxin Q1 and synergistic effect of cyclopropenoid fatty acids. *J. Natl. Cancer Inst.* **1980**, *64*, 523–528. [PubMed]
55. Fan, T.S.L.; Zhang, G.S.; Chu, F.S. Production and characterization of antibody against aflatoxin Q1. *Appl. Environ. Microbiol.* **1986**, *47*, 526–532. [CrossRef] [PubMed]
56. Antonissen, G.; Devreese, M.; De Baere, S.; Martel, A.; Van Immerseel, F.; Croubels, S. Impact of Fusarium mycotoxins on hepatic and intestinal mRNA expression of cytochrome P450 enzymes and drug transporters, and on the pharmacokinetics of oral enrofloxacin in broiler chickens. *Food Chem. Toxicol.* **2017**, *101*, 75–83. [CrossRef] [PubMed]
57. Allocati, N.; Masulli, M.; Di Ilio, C.; Federici, L. Glutathione transferases: Substrates, inihibitors and pro-drugs in cancer and neurodegenerative diseases. *Oncogenesis* **2018**, *7*, 1–15. [CrossRef]
58. Carvajal-Moreno, M. Metabolic Changes of Aflatoxin B1 to become an Active Carcinogen and the Control of this Toxin. *Immunome Res.* **2015**, *11*, 1. [CrossRef]
59. Mulero, J.; Martínez, G.; Oliva, J.; Cermeño, S.; Cayuela, J.M.; Zafrilla, P.; Martínez-Cachá, A.; Barba, A. Phenolic compounds and antioxidant activity of red wine made from grapes treated with different fungicides. *Food Chem.* **2015**, *180*, 25–31. [CrossRef]
60. Jarolim, K.; Del Favero, G.; Pahlke, G.; Dostal, V.; Zimmermann, K.; Heiss, E.; Ellmer, D.; Stark, T.D.; Hofmann, T.; Marko, D. Activation of the Nrf2-ARE pathway by the Alternaria alternata mycotoxins altertoxin I and II. *Arch. Toxicol.* **2017**, *91*, 203–216. [CrossRef] [PubMed]
61. Wen, J.; Mu, P.; Deng, Y. Mycotoxins: Cytotoxicity and biotransformation in animal cells. *Toxicol. Res.* **2016**, *5*, 377–387. [CrossRef]
62. Burkina, V.; Rasmussen, M.K.; Oliinychenko, Y.; Zamaratskaia, G. Porcine cytochrome 2A19 and 2E1. *Basic Clin. Pharmacol. Toxicol.* **2019**, *124*, 32–39. [CrossRef] [PubMed]
63. Nannelli, A.; Chirulli, V.; Longo, V.; Gervasi, P.G. Expression and induction by rifampicin of CAR- and PXR-regulated CYP2B and CYP3A in liver, kidney and airways of pig. *Toxicology* **2008**, *252*, 105–112. [CrossRef]
64. Yao, M.; Dai, M.; Liu, Z.; Huang, L.; Chen, D.; Wang, Y.; Peng, D.; Wang, X.; Liu, Z.; Yuan, Z. Comparison of the substrate kinetics of pig CYP3A29 with pig liver microsomes and human CYP3A4. *Biosci. Rep.* **2011**, *31*, 211–220. [CrossRef] [PubMed]
65. Puccinelli, E.; Gervasi, P.G.; La Marca, M.; Beffy, P.; Longo, V. Expression and inducibility by phenobarbital of CYP2C33, CYP2C42, CYP2C49, CYP2B22, and CYP3As in porcine liver, kidney, small intestine, and nasal tissues. *Xenobiotica* **2010**, *40*, 525–535. [CrossRef] [PubMed]
66. Shang, W.; Nuffer, J.H.; Muñiz-Papandrea, V.A.; Colón, W.; Siegel, R.W.; Dordick, J.S. Cytochrome c on silica nanoparticles: Influence of nanoparticle size on protein structure, stability, and activity. *Small* **2009**, *5*, 470–476. [CrossRef] [PubMed]
67. Kojima, M.; Morozumi, T. Cloning of six full-length cDNAs encoding pig cytochrome P450 enzymes and gene expression of these enzymes in the liver and kidney. *J. Health Sci.* **2004**, *50*, 518–529. [CrossRef]
68. Achour, B.; Barber, J.; Rostami-Hodjegan, A. Correction to "Cytochrome P450 pig liver pie: Determination of individual cytochrome P450 isoform contents in microsomes from two pig livers using liquid chromatography in conjunction with mass spectroscopy" (Drug Metabolism and Disposition (2011) 39, (2130-2134)). *Drug Metab. Dispos.* **2012**, *40*, 227. [CrossRef]
69. Lehman-McKeeman, L.D.; Ruepp, S.U. *Biochemical and Molecular Basis of Toxicity*, 3rd ed.; Elsevier Inc.: Amsterdam, The Netherlands, 2018; ISBN 9780128098424.
70. Gallagher, E.P.; Kunze, K.L.; Stapleton, P.L.; Eaton, D.L. The kinetics of aflatoxin B1oxidation by human cDNA-expressed and human liver microsomal cytochromes P450 1A2 and 3A4. *Toxicol. Appl. Pharmacol.* **1996**, *141*, 595–606. [CrossRef]
71. Devreese, M.; De Backer, P.; Croubels, S. Overview of the most important mycotoxins for the pig and poultry husbandry Overzicht van de meest belangrijke mycotoxines voor de varkens-en pluimveehouderij. *Vlaams Diergeneeskd. Tijdschr.* **2013**, *82*, 171–180. [CrossRef]
72. Dhama, K.; Singh, K.P. Aflatoxins- Hazard to Livestock and Poultry Production: A Review. *Vet. Immunol. Immunopathol.* **2007**, *9*, 1–15.
73. Diaz, G.J.; Murcia, H.W.; Cepeda, S.M.; Boermans, H.J. The role of selected cytochrome P450 enzymes on the bioactivation of aflatoxin B1 by duck liver microsomes. *Avian Pathol.* **2010**, *39*, 279–285. [CrossRef]
74. Pauletto, M.; Tolosi, R.; Giantin, M.; Guerra, G.; Barbarossa, A.; Zaghini, A.; Dacasto, M. Insights into Aflatoxin B1 Toxicity in Cattle: An in vitro whole-transcriptomic approach. *Toxins* **2020**, *12*, 429. [CrossRef]
75. Iori, S.; Pauletto, M.; Bassan, I.; Bonsembiante, F.; Gelain, M.E.; Bardhi, A.; Barbarossa, A.; Zaghini, A.; Dacasto, M.; Giantin, M. Deepening the Whole Transcriptomics of Bovine Liver Cells Exposed to AFB1: A Spotlight on Toll-like Receptor 2. *Toxins* **2022**, *14*, 504. [CrossRef] [PubMed]
76. Choi, S.Y.; Kim, T.H.; Hong, M.W.; Park, T.S.; Lee, H.; Lee, S.J. Transcriptomic alterations induced by aflatoxin B1 and ochratoxin A in LMH cell line. *Poult. Sci.* **2020**, *99*, 5265–5274. [CrossRef] [PubMed]
77. Fang, Y.; Feng, Y.; Wu, T.; Srinivas, S.; Yang, W.; Fan, J.; Yang, C.; Wang, S. Aflatoxin B1 Negatively Regulates Wnt/β-Catenin Signaling Pathway through Activating miR-33a. *PLoS ONE* **2013**, *8*, 1–12. [CrossRef]

78. Zhu, L.; Gao, J.; Huang, K.; Luo, Y.; Zhang, B.; Xu, W. miR-34a screened by miRNA profiling negatively regulates Wnt/β-catenin signaling pathway in Aflatoxin B1 induced hepatotoxicity. *Sci. Rep.* **2015**, *5*, 1–13. [CrossRef] [PubMed]
79. Caloni, F.; Cortinovis, C. Toxicological effects of aflatoxins in horses. *Vet. J.* **2011**, *188*, 270–273. [CrossRef] [PubMed]
80. Ma, J.; Liu, Y.; Guo, Y.; Ma, Q.; Ji, C.; Zhao, L. Transcriptional profiling of aflatoxin b1-induced oxidative stress and inflammatory response in macrophages. *Toxins* **2021**, *13*, 401. [CrossRef]
81. Li, C.; Liu, X.; Wu, J.; Ji, X.; Xu, Q. Research progress in toxicological effects and mechanism of aflatoxin B1 toxin. *PeerJ* **2022**, *10*, e13850. [CrossRef] [PubMed]
82. Velayudhan, D.E.; Kim, I.H.; Nyachoti, C.M. Characterization of dietary energy in swine feed and feed ingredients: A review of recent research results. *Asian-Australasian J. Anim. Sci.* **2015**, *28*, 1–13. [CrossRef]
83. Sánchez-Cordón, P.J.; Montoya, M.; Reis, A.L.; Dixon, L.K. African swine fever: A re-emerging viral disease threatening the global pig industry. *Vet. J.* **2018**, *233*, 41–48. [CrossRef]
84. Munkvold, G.P.; Arias, S.; Taschl, I.; Gruber-Dorninger, C. *Mycotoxins in Corn: Occurrence, Impacts, and Management*, 3rd ed.; Elsevier Inc.: Amsterdam, The Netherlands, 2019; ISBN 978-0-12-811971-6.
85. Diekman, M.A.; Coffey, M.T.; Purkhiser, E.D.; Reeves, D.E.; Young, L.G. Mycotoxins and Swine Performance. *Nutrition* **1914**, *6*, 1–6.
86. Guerre, P. Worldwide mycotoxins exposure in pig and poultry feed formulations. *Toxins* **2016**, *8*, 350. [CrossRef]
87. Kong, C.; Park, C.S.; Kim, B.G. Evaluation of a mycotoxin adsorbent in swine diets containing barley naturally contaminated with Fusarium mycotoxins. *Rev. Colomb. Ciencias Pecu.* **2016**, *29*, 169–177. [CrossRef]
88. Tiemann, U.; Dänicke, S. In vivo and in vitro effects of the mycotoxins zearalenone and deoxynivalenol on different non-reproductive and reproductive organs in female pigs: A review. *Food Addit. Contam.* **2007**, *24*, 306–314. [CrossRef] [PubMed]
89. Kanora, A.; Maes, D. The role of mycotoxins in pig reproduction: A review. *Vet. Med.* **2009**, *54*, 565–576. [CrossRef]
90. Chaytor, A.C.; Hansen, J.A.; Van Heugten, E.; See, M.T.; Kim, S.W. Occurrence and decontamination of mycotoxins in swine feed. *Asian-Australasian J. Anim. Sci.* **2011**, *24*, 723–738. [CrossRef]
91. Prodanov-Radulovic, J.; Dosen, R.; Stojanov, I.; Pusic, I.; Zivkov-Balos, M.; Ratajac, R. Influence of mycotoxin zearalenone on the swine reproductive failure. *Zb. Matice Srp. Za Prir. Nauk.* **2013**, *124*, 121–129. [CrossRef]
92. Li, X.; Zhao, L.; Fan, Y.; Jia, Y.; Sun, L.; Ma, S.; Ji, C.; Ma, Q.; Zhang, J. Occurrence of mycotoxins in feed ingredients and complete feeds obtained from the Beijing region of China. *J. Anim. Sci. Biotechnol.* **2014**, *5*, 1–8. [CrossRef] [PubMed]
93. Choe, J.; Moyo, K.M.; Park, K.; Jeong, J.; Kim, H.; Ryu, Y.; Kim, J.; Kim, J.; Lee, S.; Go, G. Meat Quality Traits of Pigs Finished on Food Waste. *Korean J. Food Sci. Anim. Resour.* **2017**, *37*, 690–697. [CrossRef]
94. Luthy, J.; Zweifel, U.; Schlatter, C. Metabolism [14C] Aflatoxin B, in Pigs. *Food Cosmet. Toxicol.* **1980**, *18*, 253–256. [CrossRef]
95. Patterson, D.S.P. Metabolism as a factor in determining the toxic action of the aflatoxins in different animal species. *Food Cosmet. Toxicol.* **1973**, *11*, 287–294. [CrossRef]
96. Tang, D.; Sauceda, J.C.; Lin, Z.; Ott, S.; Basova, E.; Goryacheva, I.; Biselli, S.; Lin, J.; Niessner, R.; Knopp, D. Magnetic nanogold microspheres-based lateral-flow immunodipstick for rapid detection of aflatoxin B2 in food. *Biosens. Bioelectron.* **2009**, *25*, 514–518. [CrossRef] [PubMed]
97. Furtado, R.M.; Pearson, A.M.; Hogberg, M.G.; Miller, E.R.; Gray, J.I.; Aust, S.D. Withdrawal Time Required for Clearance of Aflatoxins from Pig Tissues. *J. Agric. Food Chem.* **1982**, *30*, 101–106. [CrossRef] [PubMed]
98. Hsieh, D.P.H.; Wong, J.J. Pharmacokinetics and excretionof aÑatoxins. In *The Toxicology of Afatoxins*; Eaton, D.L., Groopman, J.D., Eds.; Academic Press Inc.: San Diego, CA, USA, 1994; pp. 73–88.
99. Thieu, N.Q.; Pettersson, H. Zearalenone, deoxynivalenol and aflatoxin B1and their metabolites in pig urine as biomarkers for mycotoxin exposure. *Mycotoxin Res.* **2009**, *25*, 59–66. [CrossRef] [PubMed]
100. AFSSA. Review of mycotoxin-detoxifying agents used as feed additives: Mode of action, efficacy and feed/food safety. *EFSA Support. Publ.* **2009**, *6*, 1–192.
101. Food and Drug Administration. CPG Sec. 683.100 Action Levels for Aflatoxins in Animal Feeds. 2015. Available online: https://www.fda.gov/iceci/compliancemanuals/ (accessed on 23 September 2022).
102. Korley Kortei, N.; Akomeah Agyekum, A.; Akuamoa, F.; Baffour, V.K.; Wiisibie Alidu, H. Risk assessment and exposure to levels of naturally occurring aflatoxins in some packaged cereals and cereal based foods consumed in Accra, Ghana. *Toxicol. Rep.* **2019**, *6*, 34–41. [CrossRef]
103. Lee, H.S.; Lindahl, J.; Nguyen-Viet, H.; Khong, N.V.; Nghia, V.B.; Xuan, H.N.; Grace, D. An investigation into aflatoxin M1in slaughtered fattening pigs and awareness of aflatoxins in Vietnam. *BMC Vet. Res.* **2017**, *13*, 1–7. [CrossRef] [PubMed]
104. Mupunga, I.; Mngqawa, P.; Katerere, D.R. Peanuts, aflatoxins and undernutrition in children in Sub-Saharan Africa. *Nutrients* **2017**, *9*, 1287. [CrossRef]
105. Zain, M.E. Impact of mycotoxins on humans and animals. *J. Saudi Chem. Soc.* **2011**, *15*, 129–144. [CrossRef]
106. Nakavuma, J.L.; Kirabo, A.; Bogere, P.; Nabulime, M.M.; Kaaya, A.N.; Gnonlonfin, B. Awareness of mycotoxins and occurrence of aflatoxins in poultry feeds and feed ingredients in selected regions of Uganda. *Int. J. Food Contam.* **2020**, *7*, 1–10. [CrossRef]
107. Weaver, A.C. The Impact of Mycotoxins on Growth and Health of Swine. Ph.D. Dissertation, North Carolina State University, Raleigh, NC, USA, 2013; p. 211.
108. Newbern, P.M.; Butler, W.H. Acute and Chronic Effects of Aflatoxin on the Liver of Domestic and Laboratory Animals: A Review. *Cancer Res.* **1969**, *29*, 236–250.

109. Silvotti, L.; Petterino, C.; Bonomi, A.; Cabassi, E. Immunotoxicological effects on piglets of feeding sows diets containing aflatoxins. *Vet. Rec.* **1997**, *141*, 469–472. [CrossRef] [PubMed]
110. Barbiroli, A.; Bonomi, F.; Benedetti, S.; Mannino, S.; Monti, L.; Cattaneo, T.; Iametti, S. Binding of Aflatoxin M1 to Different Protein Fractions in Ovine and Caprine Milk. *J. Dairy Sci.* **2007**, *90*, 532–540. [CrossRef] [PubMed]
111. Do, J.H.; Choi, D.-K. Aflatoxins: Detection, Toxicity, and Biosynthesis. *Biotechnol. Bioprocess Eng.* **2007**, *12*, 585–593. [CrossRef]
112. Jw, J.; Nm, W.; Hm, I.; Ss, A. Short Communication Aflatoxicosis Associated with Swine Stillbirth in the Aflatoxicosis Associated with Swine Stillbirth in the Piggery Farm University of Agriculture Makurdi. *CTEB* **2018**, *13*, 13–16. [CrossRef]
113. Wilfred, E.G.; Dungworth, D.L.; Moulton, J.E. Pathologic Effects of Aflatoxin in Pigs. *Vet. Pathol.* **1968**, *5*, 370–384. [CrossRef]
114. Dilkin, P.; Zorzete, P.; Mallmann, C.A.; Gomes, J.D.F.; Utiyama, C.E.; Oetting, L.L.; Corrêa, B. Toxicological effects of chronic low doses of aflatoxin B1 and fumonisin B1-containing Fusarium moniliforme culture material in weaned piglets. *Food Chem. Toxicol.* **2003**, *41*, 1345–1353. [CrossRef]
115. Obuseh, F.A.; Jolly, P.E.; Jiang, Y.; Shuaib, F.M.B.; Waterbor, J.; Ellis, W.O.; Piyathilake, C.J.; Desmond, R.A.; Afriyie-Gyawu, E.; Phillips, T.D. Aflatoxin B1 albumin adducts in plasma and aflatoxin M1 in urine are associated with plasma concentrations of vitamins A and E. *Int. J. Vitam. Nutr. Res.* **2010**, *80*, 355–368. [CrossRef]
116. Marin, D.E.; Taranu, I.; Bunaciu, R.P.; Pascale, F.; Tudor, D.S.; Avram, N.; Sarca, M.; Cureu, I.; Criste, R.D.; Suta, V.; et al. Changes in performance, blood parameters, humoral and cellular immune responses in weanling piglets exposed to low doses of aflatoxin. *J. Anim. Sci.* **2002**, *80*, 1250–1257. [CrossRef]
117. Meissonnier, G.M.; Pinton, P.; Laffitte, J.; Cossalter, A.M.; Gong, Y.Y.; Wild, C.P.; Bertin, G.; Galtier, P.; Oswald, I.P. Immunotoxicity of aflatoxin B1: Impairment of the cell-mediated response to vaccine antigen and modulation of cytokine expression. *Toxicol. Appl. Pharmacol.* **2008**, *231*, 142–149. [CrossRef]
118. Rushing, B.R.; Selim, M.I. Aflatoxin B1: A review on metabolism, toxicity, occurrence in food, occupational exposure, and detoxification methods. *Food Chem. Toxicol.* **2019**, *124*, 81–100. [CrossRef]
119. Stec, J.A.N.; Mudzki, J.Ż.; Rachubik, J.Ł.A.W.; Szczotka, M. Effects of aflatoxin B1, ochratoxin A, patulin, citrinin, and zearalenone on the in vitro proliferation of pig blood lymphocytes. *Bull. Vet. Inst. Pulawy* **2009**, *53*, 129–134.
120. Pierron, A.; Alassane-Kpembi, I.; Oswald, I.P. Impact of mycotoxin on immune response and consequences for pig health. *Anim. Nutr.* **2016**, *2*, 63–68. [CrossRef] [PubMed]
121. Hintz, H.F.; Heitman, H.; Booth, A.N.; Gagne, W.E. Effects of aflatoxin on reproduction in swine. *Proc. Soc. Exp. Biol. Med.* **1967**, *126*, 146–148. [CrossRef] [PubMed]
122. Liu, J.; Wang, Q.C.; Han, J.; Xiong, B.; Sun, S.C. Aflatoxin B1 is toxic to porcine oocyte maturation. *Mutagenesis* **2015**, *30*, 527–535. [CrossRef] [PubMed]
123. Miller, D.M.; Stuart, B.P.; Crowell, W.A. Experimental aflatoxicosis in swine: Morphological and clinical pathological results. *Can. J. Comp. Med.* **1981**, *45*, 343–351. [PubMed]
124. Ketterer, P.J.; Blaney, B.J.; Moore, C.J.; McInnes, I.S.; Cook, P.W. Field cases of aflatoxicosis in pigs. *Aust. Vet. J.* **1982**, *59*, 113–118. [CrossRef] [PubMed]
125. Abidin, Z.; Khatoon, A.; Numan, M. Mycotoxins in broilers: Pathological alterations induced by aflatoxins and ochratoxins, diagnosis and determination, treatment and control of mycotoxicosis. *Worlds Poult. Sci. J.* **2011**, *67*, 485–496. [CrossRef]
126. Shivasharanappa, G.Y.; Mundas, S.; Rao, D.G.K.; Tikare, V.; Shridhar, N.B. Histopathological Changes in Pigs Exposed to Aflatoxin B1 During Pregnancy. *Indian J. Anim. Res.* **2013**, *47*, 386–391.
127. Monson, M.; Coulombe, R.; Reed, K. Aflatoxicosis: Lessons from Toxicity and Responses to Aflatoxin B1 in Poultry. *Agriculture* **2015**, *5*, 742–777. [CrossRef]
128. Liew, W.-P.-P.; Mohd-Redzwan, S. Mycotoxin: Its Impact on Gut Health and Microbiota. *Front. Cell. Infect. Microbiol.* **2018**, *8*, 60. [CrossRef]
129. Freitas, B.V.; Mota, M.M.; Del Santo, T.A.; Afonso, E.R.; Silva, C.C.; Utimi, N.B.P.; Barbosa, L.C.G.S.; Vilela, F.G.; Araujo, L.F. Mycotoxicosis in swine: A review. *J. Anim. Vet. Adv.* **2012**, *2*, 174–181.
130. Varga, J.; Péteri, Z.; Tábori, K.; Téren, J.; Vágvölgyi, C. Degradation of ochratoxin A and other mycotoxins by Rhizopus isolates. *Int. J. Food Microbiol.* **2005**, *99*, 321–328. [CrossRef] [PubMed]
131. Rezaei, R.; Knabe, D.A.; Wu, G.; Dixon, J.B.; Barrientos Velázquez, A.L.; Deng, Y. Impacts of Aflatoxins on Swine Nutrition and Possible Measures of Amelioration. In *Aflatoxin Control: Safeguarding Animal Feed with Calcium Smectite*; ACSESS: Hoboken, NJ, USA, 2014; pp. 54–67. [CrossRef]
132. Yu, J.; Chang, P.-K.; Ehrlich, K.C.; Cary, J.W.; Bhatnagar, D.; Cleveland, T.E.; Payne, G.A.; Linz, J.E.; Woloshuk, C.P.; Bennett, W. Clustered patway genes in aflatoxins biosynthesis. *Appl. Environ. Microbiol.* **2004**, *70*, 1253–1262. [CrossRef] [PubMed]
133. Jalili, M. A Review on Aflatoxins Reduction in Food. *Iran. J. Health Saf. Environ.* **2015**, *3*, 445–459. [CrossRef]
134. Khadem, A.A.; Sharifi, S.D.; Barati, M.; Borji, M. Evaluation of the effectiveness of yeast, zeolite and active charcoal as aflatoxin absorbents in broiler diets. *Glob. Vet.* **2012**, *8*, 426–432.
135. Chulze, S.N. Strategies to reduce mycotoxin levels in maize during storage: A review. *Food Addit. Contam.-Part A Chem. Anal. Control. Expo. Risk Assess.* **2010**, *27*, 651–657. [CrossRef]
136. Juglal, S.; Govinden, R.; Odhav, B. Spice Oils for the Control of Co-Occurring Mycotoxin-Producing Fungi. *J. Food Prot.* **2002**, *65*, 683–687. [CrossRef]

137. Roze, L.V.; Hong, S.-Y.; Linz, J.E. Aflatoxin Biosynthesis: Current Frontiers. *Annu. Rev. Food Sci. Technol.* **2013**, *4*, 293–311. [CrossRef]
138. Marin, D.E.; Pistol, G.C.; Gras, M.A.; Palade, M.L.; Taranu, I. Comparative effect of ochratoxin A on inflammation and oxidative stress parameters in gut and kidney of piglets. *Regul. Toxicol. Pharmacol.* **2017**, *89*, 224–231. [CrossRef]
139. Muhammad, I.; Wang, X.; Li, S.; Li, R.; Zhang, X. Curcumin confers hepatoprotection against AFB1-induced toxicity via activating autophagy and ameliorating inflammation involving Nrf2/HO-1 signaling pathway. *Mol. Biol. Rep.* **2018**, *45*, 1775–1785. [CrossRef]
140. Barba, F.J.; Zhu, Z.; Koubaa, M.; Sant'Ana, A.S.; Orlien, V. Green alternative methods for the extraction of antioxidant bioactive compounds from winery wastes and by-products: A review. *Trends Food Sci. Technol.* **2016**, *49*, 96–109. [CrossRef]
141. Mohajeri, M.; Behnam, B.; Cicero, A.F.G.; Sahebkar, A. Protective effects of curcumin against aflatoxicosis: A comprehensive review. *J. Cell. Physiol.* **2017**, *233*, 3552–3577. [CrossRef] [PubMed]
142. Sayyari, A.; Fæste, C.K.; Hansen, U.; Uhlig, S.; Framstad, T.; Schatzmayr, D.; Sivertsen, T. Effects and biotransformation of the mycotoxin deoxynivalenol in growing pigs fed with naturally contaminated pelleted grains with and without the addition of Coriobacteriaceum DSM 11798. *Food Addit. Contam.-Part A Chem. Anal. Control. Expo. Risk Assess.* **2018**, *35*, 1394–1409. [CrossRef] [PubMed]
143. Popescu, R.G.; Bulgaru, C.; Untea, A.; Vlassa, M.; Filip, M.; Hermenean, A.; Marin, D.; Țăranu, I.; Georgescu, S.E.; Dinischiotu, A. The Effectiveness of Dietary Byproduct Antioxidants on Induced CYP Genes Expression and Histological Alteration in Piglets Liver and Kidney Fed with Aflatoxin B1 and Ochratoxin A. *Toxins* **2021**, *13*, 148. [CrossRef] [PubMed]
144. Popescu, R.G.; Avramescu, S.; Marin, D.E.; Țăranu, I.; Georgescu, S.E.; Dinischiotu, A. The reduction of the combined effects of aflatoxin and ochratoxin a in piglet livers and kidneys by dietary antioxidants. *Toxins* **2021**, *13*, 648. [CrossRef]

toxins

Article

Deepening the Whole Transcriptomics of Bovine Liver Cells Exposed to AFB1: A Spotlight on Toll-like Receptor 2

Silvia Iori [1], Marianna Pauletto [1], Irene Bassan [1], Federico Bonsembiante [1,2], Maria Elena Gelain [1], Anisa Bardhi [3], Andrea Barbarossa [3], Anna Zaghini [3], Mauro Dacasto [1] and Mery Giantin [1,*]

[1] Department of Comparative Biomedicine and Food Science, University of Padua, Viale dell'Università 16, Legnaro, 35020 Padua, Italy; silvia.iori@phd.unipd.it (S.I.); marianna.pauletto@unipd.it (M.P.); irene.bassan@unipd.it (I.B.); federico.bonsembiante@unipd.it (F.B.); mariaelena.gelain@unipd.it (M.E.G.); mauro.dacasto@unipd.it (M.D.)

[2] Department of Animal Medicine, Production and Health, University of Padua, Viale dell'Università 16, Legnaro, 35020 Padua, Italy

[3] Department of Veterinary Medical Sciences, Alma Mater Studiorum University of Bologna, Via Tolara di Sopra 50, Ozzano dell'Emilia, 40064 Bologna, Italy; anisa.bardhi@unibo.it (A.B.); andrea.barbarossa@unibo.it (A.B.); anna.zaghini@unibo.it (A.Z.)

* Correspondence: mery.giantin@unipd.it; Tel.: +39-049-827-2946

Abstract: Aflatoxin B1 (AFB1) is a food contaminant metabolized mostly in the liver and leading to hepatic damage. Livestock species are differently susceptible to AFB1, but the underlying mechanisms of toxicity have not yet been fully investigated, especially in ruminants. Thus, the aim of the present study was to better characterize the molecular mechanism by which AFB1 exerts hepatotoxicity in cattle. The bovine fetal hepatocyte cell line (BFH12) was exposed for 48 h to three different AFB1 concentrations (0.9 μM, 1.8 μM and 3.6 μM). Whole-transcriptomic changes were measured by RNA-*seq* analysis, showing significant differences in the expression of genes mainly involved in inflammatory response, oxidative stress, drug metabolism, apoptosis and cancer. As a confirmatory step, post-translational investigations on genes of interest were implemented. Cell death associated with necrosis rather than apoptosis events was noted. As far as the toxicity mechanism is concerned, a molecular pathway linking inflammatory response and oxidative stress was postulated. Toll-Like Receptor 2 (TLR2) activation, consequent to AFB1 exposure, triggers an intracellular signaling cascade involving a kinase (p38β MAPK), which in turn allows the nuclear translocation of the activator protein-1 (AP-1) and NF-κB, finally leading to the release of pro-inflammatory cytokines. Furthermore, a p38β MAPK negative role in cytoprotective genes regulation was postulated. Overall, our investigations improved the actual knowledge on the molecular effects of this worldwide relevant natural toxin in cattle.

Keywords: aflatoxin B1; bovine; liver; RNA-*seq*; oxidative stress; inflammatory response; toll-like receptor 2

Key Contribution: The present study provided new experimental evidence on the mechanistic toxicology of AFB1 in cattle liver, highlighting the major role played by oxidative stress and inflammatory processes in the exploitation of AFB1 hepatotoxicity.

Citation: Iori, S.; Pauletto, M.; Bassan, I.; Bonsembiante, F.; Gelain, M.E.; Bardhi, A.; Barbarossa, A.; Zaghini, A.; Dacasto, M.; Giantin, M. Deepening the Whole Transcriptomics of Bovine Liver Cells Exposed to AFB1: A Spotlight on Toll-like Receptor 2. *Toxins* **2022**, *14*, 504. https://doi.org/10.3390/toxins14070504

Received: 22 June 2022
Accepted: 16 July 2022
Published: 20 July 2022

Publisher's Note: MDPI stays neutral with regard to jurisdictional claims in published maps and institutional affiliations.

Copyright: © 2022 by the authors. Licensee MDPI, Basel, Switzerland. This article is an open access article distributed under the terms and conditions of the Creative Commons Attribution (CC BY) license (https://creativecommons.org/licenses/by/4.0/).

1. Introduction

Aflatoxin B1 (AFB1) is a widespread mycotoxin produced by fungal species belonging to Aspergillus family (i.e., *A. flavus* and *A. parasiticus*). AFB1 occurrence is reported in several food and feed commodities, such as corn, wheat, peanuts, milk and eggs, posing a great potential risk to human and animal health [1]. As a matter of fact, aflatoxins (AFs) B1, B2, G1 and G2 represent a group of mycotoxins with hepatotoxic, genotoxic and carcinogenic effects [2]. Along this line, the International Agency for Research on Cancer

(IARC) classified AFs as carcinogenic to humans (Group 1) [3]. Among them, AFB1 is considered as the most toxic AF, as well as one of the most potent liver carcinogen [4]. AFB1 contamination mainly affects agricultural crops from tropical and sub-tropical areas. However, greenhouse emission, global warming and precipitations also favour the spread of crop fungal infections in temperate regions. In this scenario, AFB1 crops contamination will likely rise in the next years, thereby becoming increasingly important in the European area [5,6].

The consumption of feed contaminated with AFB1 may affect health, as well as the performance of farm animals. In non-ruminant species such as poultry, the exposure to AFB1 negatively affects production traits like weight, egg production and hatchability [7–9]. In poultry and pigs, liver damage and immunosuppression were also reported, making them more vulnerable to diseases [10]. In cattle, AFs negatively affect production traits (e.g., milk, beef), growth and reproduction, as well as rumen metabolism [11,12].

AFB1 bioactivation and detoxification occur mostly in the liver by hepatic cytochrome P450s (CYPs), which belong to the phase I metabolizing enzymes, and by glutathione S-transferases (GSTs), a phase II family of conjugative enzymes, respectively [13]. In humans as well as other animal species, AFB1 is bioactivated by the CYP1A and CYP3A4 isozymes, giving rise to metabolic derivatives such as AFB1-8,9-exo-epoxide (AFBO). AFBO may form adducts with DNA, RNA and proteins, leading to toxicity, impairment of transcriptional and translational processes, together with carcinogenesis initiation [14].

AFB1 may also be converted in other toxic metabolites, such as the hydroxylated carcinogenic derivative Aflatoxin M1 (AFM1) and aflatoxicol (AFL), and relatively nontoxic metabolites like aflatoxin P1 (AFP1), aflatoxin Q1 (AFQ1) and aflatoxin B2a (AFB2a) [15]. The amount and nature of AFB1 metabolites depend on the CYP isozymes involved, which in turn differ among species, as well as to differences in concentration of hepatic GST isozymes [16]. In this respect, cattle are known to be more susceptible than horses or sheep [17,18]. Several aflatoxicosis outbreaks were reported in cattle, with consequent significant economic losses [19]. AFB1 toxicity depends also on other factors, such as age, duration and dose of exposure [20].

Given all the above-mentioned AFB1 adverse effects, it is undeniable that understanding the mechanistic toxicology of AFB1 in farm animals is a fundamental step in contributing to the management of the associated risk. Whole-transcriptome analysis is a powerful approach helping to reach this goal, providing a deeper look at the molecular pathways involved in hepatic AFB1 response [21]. As far as we know, only few RNA sequencing (RNA-*seq*) studies were conducted on livestock species. Specifically, in-vivo studies on the liver of poultry (i.e., turkey, ducklings) exposed to AFB1 and showing several hepatic disfunctions revealed the dysregulation of genes associated with drug metabolism, carcinogenesis, apoptosis, cell cycle, lipid metabolism and oxidation-reduction reactions [22,23].

Despite the cattle susceptibility to the toxic effects of AFB1, scarce information is available on the molecular mechanism governing AFB1 hepatotoxicity in cattle.

Only recently we demonstrated the occurrence of great transcriptional perturbations linked to inflammation, cellular damage and apoptosis pathways on a bovine fetal hepatocyte cell line (BFH12) exposed to 3.6 μM AFB1 [24]. Since this cell line is of fetal origin, thus characterized by a different drug metabolizing gene expression pattern compared to the adult one (e.g., weak CYP1A1 expression), we used 1 nM 3,3′,4,4′,5-pentachlorobiphenyl (PCB126) to increase CYP1A1 mRNA expression. Indeed, despite PCB126 pretreatment modulated the expression of a very low number of genes (n = 8), it affected both AFB1 metabolic pattern profile and the cytotoxicity rate [24].

In order to disclose AFB1 dose-dependent molecular effects without the influence of PCB126 pre-treatment, in the present study BFH12 cells were exposed for 48 h to increasing AFB1 sub-cytotoxic concentrations (i.e., 0.9 μM, 1.8 μM and 3.6 μM). Transcriptomic changes were measured by RNA-*seq* analysis. Then, confirmatory post-translational investigations on target proteins were executed. The expression of several genes related to inflammation and oxidative stress, together with cancer and drug metabolism, resulted dysregulated in a dose-dependent manner by the mycotoxin. Moreover, we postulated a putative pathway linking oxidative stress and inflammation in the response of cattle liver to AFB1 exposure.

Overall, the results obtained in the present study allowed a deeper characterization of AFB1 mechanistic toxicology in cattle liver, highlighting signaling pathways that could be promising targets in the treatment of aflatoxicosis.

2. Results

2.1. AFB1 Cytotoxicity

AFB1 cytotoxicity was estimated using two different metabolic assays, WST-1 and the CellTiter-Glo^{TM}. Both tests showed a dose-dependent increase of AFB1 cytotoxicity (Figure 1). In particular, the percentage of dead cells obtained with WST-1 assay was 20.90%, 39.63% and 67.21% for AFB1 0.9 µM, 1.8 µM and 3.6 µM, respectively. Conversely, CellTiter-Glo^{TM} assay showed overall lower cytotoxicity values (20.72%, 35.66% and 49.18% for AFB1 0.9 µM, 1.8 µM and 3.6 µM, respectively). The cytotoxicity obtained with the highest concentration approximated the AFB1 IC_{50} estimated in our previous publication [24].

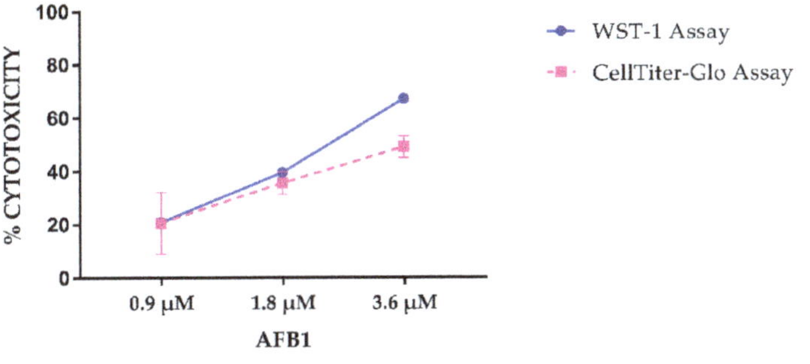

Figure 1. Cytotoxicity evaluation of increasing AFB1 concentrations in BFH12 cells using WST-1 and CellTiter-Glo^{TM} assays. Data are expressed as the mean percentage of dead cells relative to that of cells exposed to the vehicle only (0.1%) DMSO) ± mean standard error (SEM).

2.2. LC-MS/MS Quantification of AFB1, AFM1 and AFL

BFH12 cells metabolized AFB1, as they released AFM1 and AFL into the cell medium. Notably, the amount of such derivatives, as well as of AFB1, measured into the medium by means of LC-MS/MS, increased following a dose-dependent manner (Table 1).

Table 1. Concentration (µM) of AFB1, AFM1 and AFL measured in the cell medium after 48 h of incubation with increasing AFB1 concentrations (i.e., 0.9 µM, 1.8 µM and 3.6 µM). Data are expressed as the mean concentration ± standard deviation (SD) of four independent cell culture experiments.

	AFB1 (µM)	AFM1 (µM)	AFL (µM)
AFB1 0.9 µM	0.800 ± 0.037	0.002 ± 0.001	0.054 ± 0.004
AFB1 1.8 µM	1.170 ± 0.057	0.003 ± 0.001	0.074 ± 0.014
AFB1 3.6 µM	2.812 ± 0.183	0.006 ± 0.001	0.165 ± 0.020

2.3. Quantitation of Apoptosis and Necrosis by Annexin V and Propidium Iodide

For the assessment of the cell death induced by AFB1, Annexin V/propidium iodide (PI) assay was used. The results obtained are shown in Figure 2. Overall, the cell death was primarily associated to necrosis; in this respect, statistically significant differences ($p < 0.001$) were obtained in all pair-wise comparisons between treated and control cells. Conversely, no statistically significant differences were obtained considering apoptotic rates.

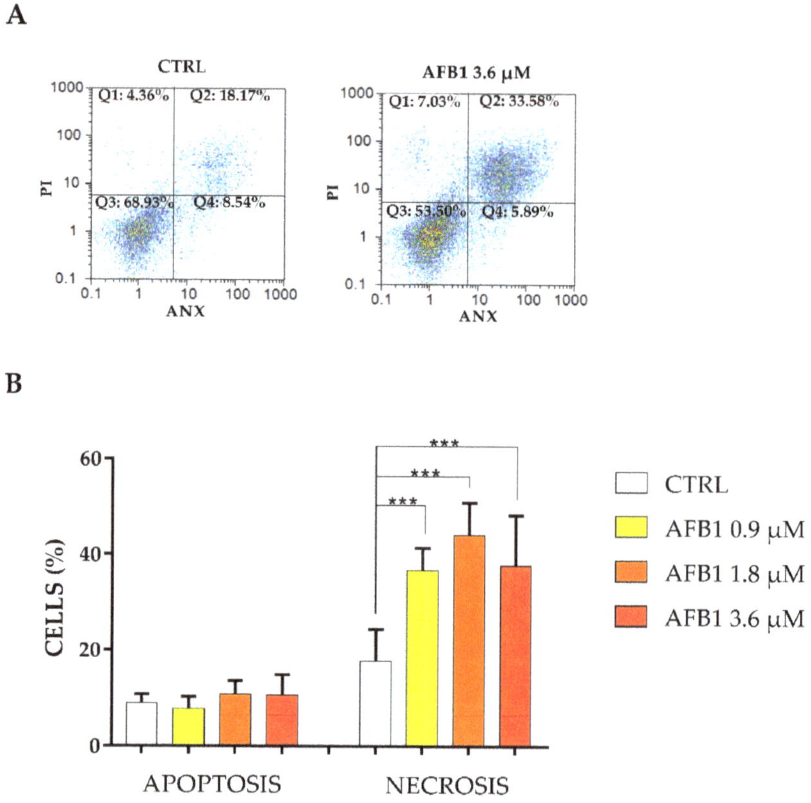

Figure 2. Assessment of apoptotic and necrotic cell rates by means of flow cytometry. (**A**) Scatter-plots of BFH12 cells exposed to 0.1% DMSO only (CTRL) or to 3.6 μM AFB1. ANX = Annexin V, PI = Propidium Iodide. Q1 and Q3 squares represent naked nuclei (negative to annexin V and positive to PI) and alive cells (negative to both annexin V and PI), respectively. Q2 square reports necrotic cells (positive to both annexin V and PI), and apoptotic cells are visible in Q4 square (positive to annexin V and negative to PI). The image is representative of six biological replicates. (**B**) Data are expressed as the mean ± SEM of six biological replicates (i.e., independent cell cultures), each one analyzed in triplicate. Statistical analysis: one-way ANOVA followed by Bonferroni's multiple comparisons test; ***: $p < 0.001$, treated cells vs. control cells.

2.4. Differential Gene Expression Analysis

A total of 23,770,398 raw reads were obtained and deposited in GeneBank under the BioProject accession number ID PRJNA847423. All samples passed quality control measures for raw sequenced reads. After trimming, an average of about 24 million reads per sample were retained, with ~99% of reads mapping to the *B. taurus* reference genome. Numbers of raw reads passing the filters and reads mapping to the cow genome are provided in Table S1. The plot MDS (Figure S1) provided an unsupervised clustering of samples. The first dimension (*x* axis) clearly separated each experimental condition (i.e., CTRL, 0.9 μM AFB1, 1.8 μM AFB1 and 3.6 μM AFB1). Transcriptional changes in response to each AFB1 concentration were assessed through pair-wise comparisons between control and treated samples. In 0.9 μM AFB1-treated cells, a total of 986 DEGs was found, while 2309 and 4200 DEGs were highlighted in 1.8 μM AFB1- and 3.6 μM AFB1-treated cells, respectively (Figure S2). The whole list of DEGs resulting from each comparison is reported in Table S2. A Venn diagram was then constructed to visualize unique and shared DEGs after AFB1 treatments (Figure 3).

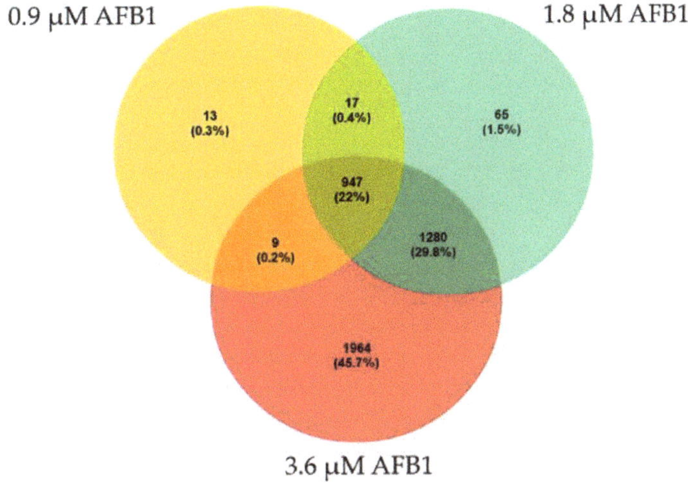

Figure 3. Venn diagram reporting number and percentages of unique and in common DEGs among the different AFB1 treatment conditions (i.e., 0.9 μM, 1.8 μM and 3.6 μM).

2.5. Functional and Gene Set Enrichment Analysis of DEGs

The functional Kyoto Encyclopedia of Genes and Genomes (KEGG) pathway analysis was performed on DEGs modulated by AFB1 treatments. For each condition, the output of the analysis was reported in a dotplot with an adjusted *p*-value, highlighting 11, 10 and 17 enriched pathways for 0.9 μM, 1.8 μM and 3.6 μM AFB1 treatments, respectively (Figure 4, Table S3). Several pathways linked to inflammatory processes and immunity were found to be significantly over-represented (e.g., "Cytokine-cytokine receptor interaction", "TNF signaling pathway" and "Complement and coagulation cascades"). Likewise, a number of pathways associated with pathogens infections (e.g., "Viral protein interaction with cytokine and cytokine receptor", "Malaria", "Staphylococcus aureus infection", "Systemic lupus erythematosus", "Ameobiasis" and "African trypanosomiasis") were enriched; they were mainly represented by genes involved in inflammatory and immune response (e.g., interleukins, complement components, chemokines). Going deeper, several genes such as the prostaglandin-endoperoxide synthase 2 (PTGS2/COX2), the interleukin 6 (IL6), the CD40 and CD44 molecules, the toll-like receptor 2 (TLR2), the Fos proto-oncogene (FOS) and p38 MAPK family member β (MAPK11/p38β) increased their mRNA expression in a dose-dependent way as a consequence of the exposure. Conversely, the expression of few genes, e.g., the complement factor H (CFH), the C-X-C motif chemokine ligand 9 (CXCL9), 10 (CXCL10) and 11 (CXCL11), as well as the interleukin 33 (IL33) displayed a strong dose-related downregulation after AFB1 treatments.

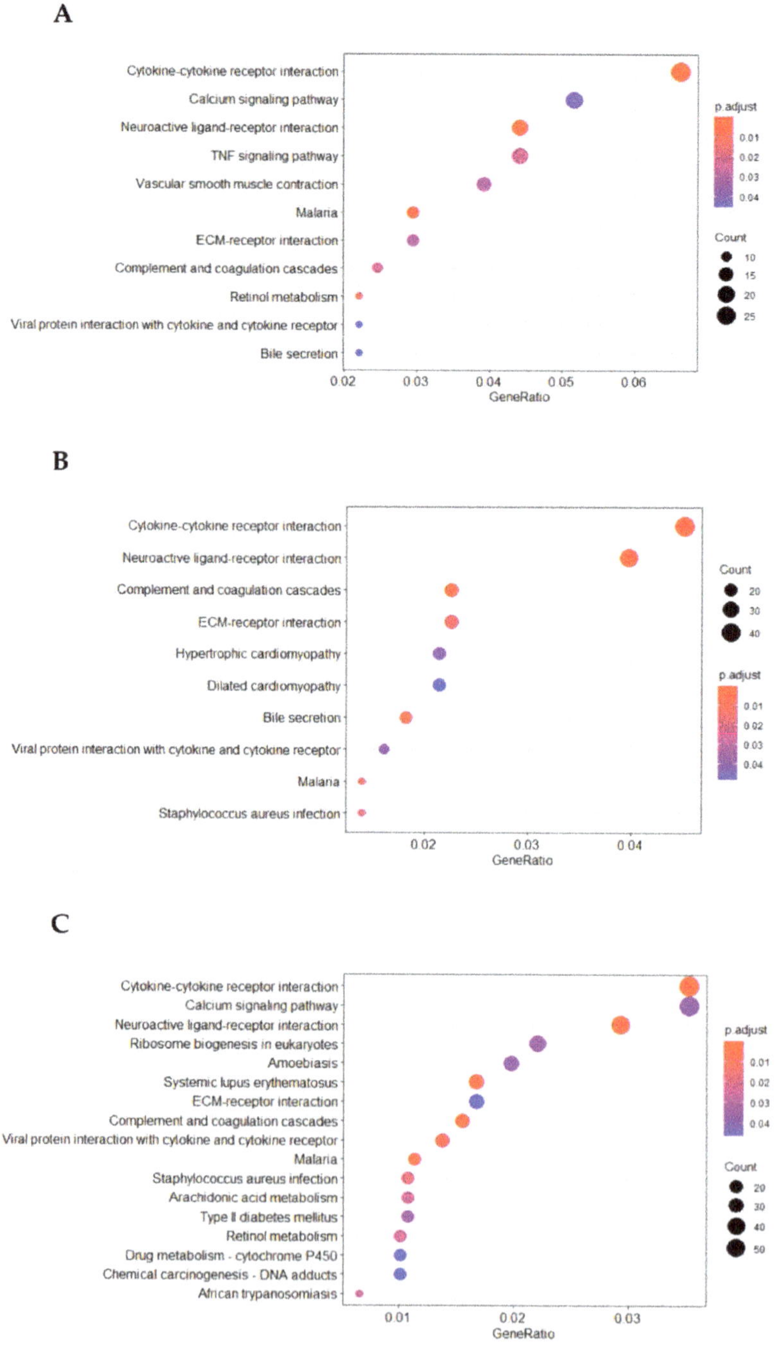

Figure 4. KEGG enrichment analysis of DEGs in 0.9 µM AFB1 (**A**), 1.8 µM AFB1 (**B**) and 3.6 µM AFB1 (**C**) treatments vs. control. Gene ratio is the percentage of DEGs over the total number of genes in a given pathway. Count (dot size) represents the number of DEGs enriched in a certain pathway.

The KEGG pathways related to drug metabolism (e.g., "Retinol metabolism" and "Drug metabolism-cytochrome P450") were also over-represented. Worthy of note, the expression of several CYP family members resulted dysregulated by AFB1 treatment. More specifically, CYP1A1 showed a dose-dependent increase in its mRNA expression, while CYP2B6 gene expression resulted inhibited in a dose-dependent way. The aldehyde dehydrogenase 1 family members A1 (ALDH1A1) was also affected by AFB1, as its mRNA levels decreased as a consequence of the mycotoxin concentration; the same effect was observed for the UDP-glucuronosyltransferases family 1 member A1 (UGT1A1), the microsomal glutathione S-transferase 1 (MSGT1) and the flavin containing dimethylaniline monoxygenase 2 (FMO2).

KEGG enrichment analysis pointed out also the terms "Chemical carcinogenesis—DNA adducts", "ECM-receptor interaction" and "Calcium signaling pathway", which are all related to carcinogenesis processes. The list of genes whose expression resulted upregulated in parallel with the AFB1 concentration used includes, for example, lamin subunit gamma 2 (LAMC2), thrombospondin 1 (THBS1), cartilage oligomeric matrix protein (COMP), endothelin 1 (EDN1), secreted phosphoprotein 1 (PNS1) and neurotensin (NTS). Conversely, few genes such as collagen type IV alpha 5 (COL4A5) and 6 (COL4A6) chains, tenascin XB (TNXB), inositol 1,4,5-trisphosphate receptor type 2 (ITPR2) and calcium voltage-gated channel subunit alpha1 E (CACNA1E) showed an opposite behavior.

As far as KEGG GSEA analysis is considered, 23, 28 and 34 gene sets (GSs) were enriched in the cells incubated with 0.9 μM, 1.8 μM and 3.6 μM AFB1, respectively (Table S4). GSs shared among all AFB1 treatments were consistent with those highlighted with KEGG over-representation test. In fact, most pathways were related to inflammatory response and response to stress conditions (i.e., "NF-kappa B signalling pathway", "P53 signalling pathway", "Cell cycle", "Salmonella infection", "Toxoplasmosis" and "Epstein-Barr virus infection).

Additionally, the GS "Apoptosis" was reported. Looking deeply, the expression of genes such as the TNF Receptor Superfamily Member 10A (TNFRSF10A) and 10B (TNFRSF10A), together with the Growth Arrest and DNA Damage Inducible Alpha (GADD45A) and the Cell Death Inducing DFFA Like Effector B (CIDEB), showed an up-regulation which was AFB1 dose-dependent. Moreover, the mRNA expression of the TNF Receptor Superfamily Member 10D (TNFRSF10D) and the Bcl-2-like protein 1 (BCL2L1) resulted in dysregulation, being inhibited by increasing concentrations of AFB1.

The log fold change (LogFC) of each gene mentioned above was reported in Table S5.

2.6. Protein-Protein Interaction (PPI) Network

To investigate the molecular regulatory network of AFB1 at the protein level, the output of STRING database derived from the analysis of the 947 DEGs shared among all treatments (i.e., 0.9 μM AFB1, 1.8 μM AFB1 and 3.6 μM AFB1), was imported in Cytoscape software. The PPI network consisted of 738 nodes and 2193 edges. The top ten hub genes were implied in pathways related to inflammatory (IL6, CD44, PTGS2, FOS and the Secreted Phosphoprotein 1-SPP1) and carcinogenesis (Cyclin D1-CCND1, EDN1, Androgen Receptor-AR, G Protein Subunit Gamma 7-GNG7 and Connective tissue growth factor-CTGF) processes. According to the MCODE algorithm, 35 clusters were detected, and the module with the highest computed score (9.357) is shown in Figure 5. The KEGG pathway enrichment analysis of the module showed that the proteins were mainly associated with the inflammatory response (e.g., "TNF signaling pathway", "Toll-like receptor signaling pathway", "IL-17 signaling pathways" and "NF-kappa B signaling pathway") as well as the carcinogenesis process (e.g., "PI3K-Akt signaling pathway", "Pathways in cancer", "Proteoglycans in cancer" and "Bladder cancer") (Figure S4).

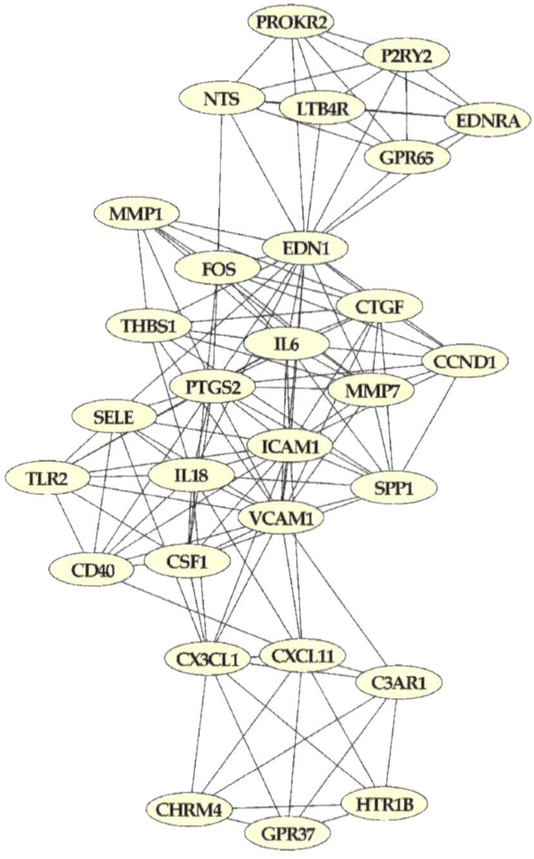

Figure 5. The most significant PPI module obtained with the analysis of the 947 DEGs in common among the three treatment conditions.

2.7. *Inflammatory Response Signaling Pathway*

Since RNA-*seq* analysis and PPI network displayed a strong enrichment of pathways related to inflammation, we decided to focus our attention on the inflammatory response pathways. In our experimental conditions, we observed a dose-dependent significant increase of TLR2 mRNA expression following AFB1 exposure (Figure 6A). Such an up-regulation was confirmed also at the protein level using flow cytometry; in this respect, the percentage of TLR2 positive cells doubled in treated cells at all AFB1 concentrations (Figure 6B). Likewise, the MFI increased in a dose-dependent way in treated cells, showing a statistically significant boost at the highest dose (i.e., 3.6 μM, Figure 6C,D).

In line with TLR2, p38β MAPK mRNA was significantly increased in a dose-dependent manner (Figure 7A). Immunoblotting investigations confirmed this trend, showing a gradual increase of p38β MAPK protein expression (both native and phosphorylated), which was statistically significant ($p < 0.05$) at the highest AFB1 concentration only (Figure 7B–D).

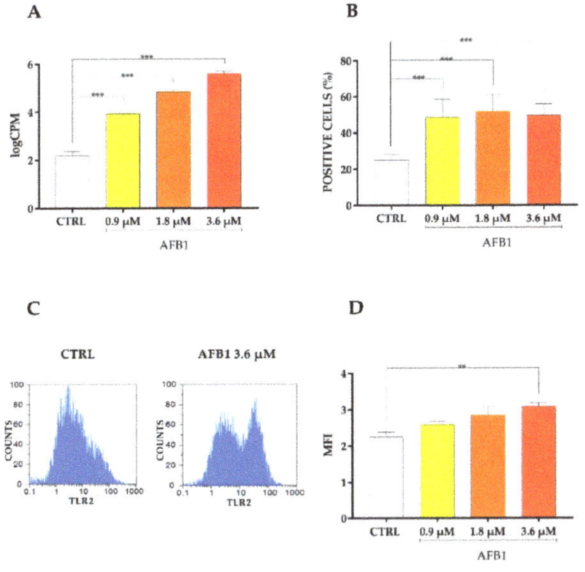

Figure 6. Effect of increasing doses of AFB1 (i.e., 0.9 μM, 1.8 μM and 3.6 μM) on TLR2 mRNA and protein expression. (**A**) Gene expression data are reported as the mean ± SEM of logarithm of counts per million (logCPM) relative to four biological replicates. (**B**) Cells positive (%) for TLR2 protein. Data are expressed as the mean ± SEM of six biological replicates, each one tested in triplicate. (**C**) Mean fluorescence intensity (MFI) of BFH12 cells exposed to 0.1% DMSO only (CTRL) and to 3.6 μM AFB1. The image is representative of six biological replicates. (**D**) Results of MFI analysis are expressed as the mean ± SEM of six biological replicates, each one tested in triplicate. Statistical analysis: one-way ANOVA followed by Bonferroni's multiple comparisons test; **: $p < 0.01$ and ***: $p < 0.001$, treated cells vs. control cells.

Moreover, enrichment analysis, as well as proteins interaction network, revealed the role of FOS in AFB1-mediated inflammatory induction (Figure S5A). Specifically, a dose-dependent mRNA induction ($p < 0.001$) was observed for FOS itself but also for other members of FOS family (i.e., FOSL1 and FOSB, Figure S5B,C), as well as for JUNB (Figure S6). Similarly, the Nuclear Factor Kappa B Subunit 2 (NFKB2) and the RELB Proto-Oncogene (RELB), two members of NF-κB family, exhibited a significant increase of mRNA expression ($p < 0.001$) in accordance with the AFB1 dose (Figure S7A,B).

Among the top ten hub genes highlighted through the PPI network, IL6 displayed a significant increase of mRNA expression after the treatment with all AFB1 concentrations (Figure 8A). Likewise, AFB1-mediated induction of IL6 was confirmed at the protein level by means of ELISA assay, even if the increase in IL6 amount was significant ($p < 0.01$) at 3.6 μM AFB1 only (Figure 8B).

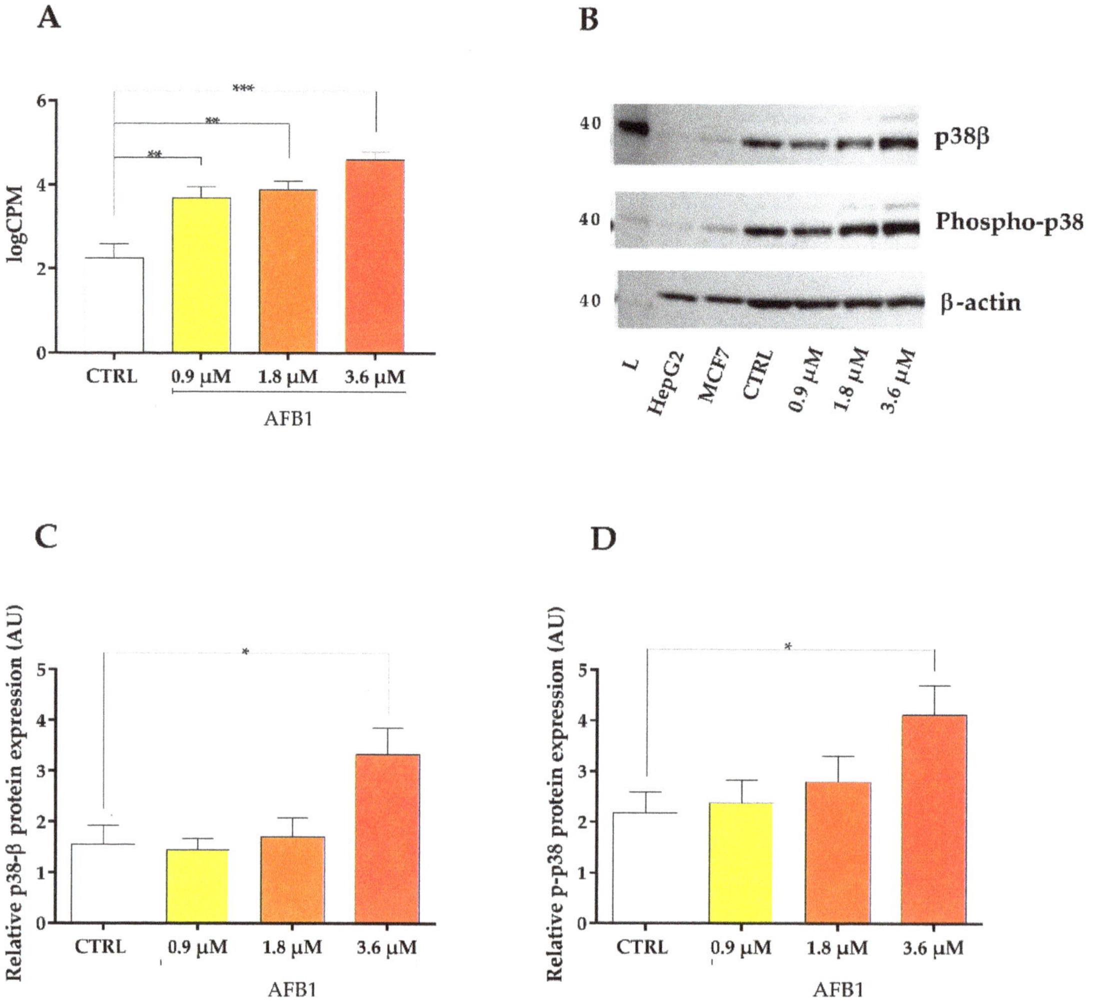

Figure 7. Effect of increasing doses of AFB1 (i.e., 0.9 μM, 1.8 μM and 3.6 μM) on p38β MAPK mRNA and protein expression (native and phosphorylated). (**A**) Gene expression data are reported as the mean ± SEM of logCPM relative to four biological replicates. (**B**) Immunoblotting of p38β and phospho-p38, MAPK using β-actin as loading control. The image is representative of six biological replicates. HepG2 and MCF7 cell lines were used as positive controls. (**C**,**D**) Densitometric analysis of p38β (**C**) and phospho-p38 (**D**) MAPK immunoblottings; data are expressed in arbitrary units (AU) as the mean ± SEM of six biological replicates. Statistical analysis: one-way ANOVA followed by Bonferroni's multiple comparisons test; *: $p < 0.05$, **: $p < 0.01$ and ***: $p < 0.001$, treated cells vs. control cells.

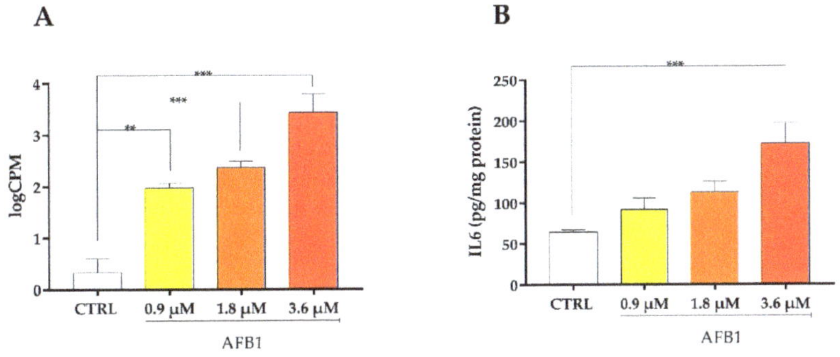

Figure 8. Effect of increasing doses of AFB1 (i.e., 0.9 μM, 1.8 μM and 3.6 μM) on IL6 mRNA and protein expression. (**A**) Gene expression data are reported as the mean ± SEM of logCPM relative to four biological replicates. (**B**) Amount of IL6 quantified by ELISA assay; data are expressed in picogram (pg) per milligram (mg) of total protein of six biological replicates, each one tested in duplicate. Statistical analysis: one-way ANOVA followed by Bonferroni's multiple comparisons test; **: $p < 0.01$ and ***: $p < 0.001$, treated cells vs. control cells.

2.8. Oxidative Stress Signaling Cascade

Looking carefully at the outcome of the transcriptomic analysis, we found out the dysregulation of several genes related to oxidative stress response. Specifically, NQO1 (NAD(P)H Quinone Dehydrogenase 1) was significantly downregulated by AFB1 at all concentrations tested ($p < 0.001$, Figure 9A), and this inhibitory effect was dose-dependent. The same behavior was confirmed at the post-translational level, as AFB1 significantly inhibited NQO1 enzymatic activity starting from the lower concentration ($p < 0.05$, Figure 9B).

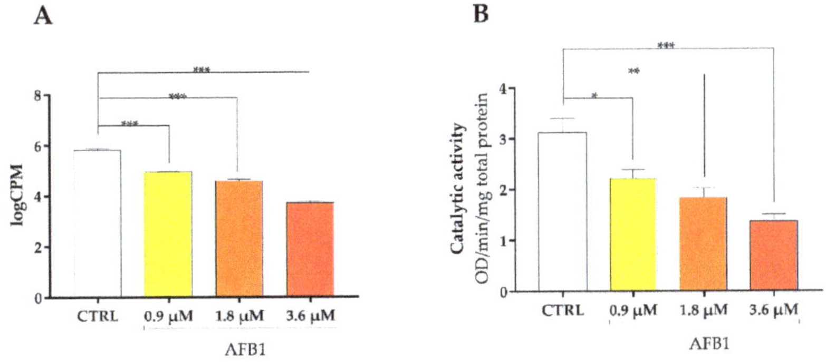

Figure 9. Effect of increasing doses of AFB1 (i.e., 0.9 μM, 1.8 μM and 3.6 μM) on NQO1 mRNA expression and catalytic activity. (**A**) Gene expression data are reported as the mean ± SEM of logCPM relative to four biological replicates. (**B**) Catalytic activity is expressed as median optical density (OD) per minute per mg of total protein ± SEM of four biological replicates, each one tested in duplicate. Statistical analysis: one-way ANOVA followed by Bonferroni's multiple comparisons test; *: $p < 0.05$, **: $p < 0.01$ and ***: $p < 0.001$, treated cells vs. control cells.

The detoxification enzyme UGT1A1 was affected by AFB1, too, but merely at the mRNA level ($p < 0.001$, Figure S8A–C), since immunoblotting data did not show any difference among the different experimental conditions.

With respect to the transcriptional regulators of cytoprotective genes, the master regulator NRF2 (Nuclear Factor Erythroid 2-Related Factor 2) was significantly deregulated by 1.8 and 3.6 μM AFB1 ($p < 0.001$, Figure 10A), while BACH1 (BTB and CNC homology 1), opposed to NRF2, being a repressor of genes involved in the oxidative stress response, was significantly up-regulated by 3.6 μM AFB1 only ($p < 0.001$, Figure 10B). Additionally, the

MAFF (MAF BZIP Transcription Factors F), which is a basic region leucine zipper-type transcription factor involved in the binding of NRF2 and BACH1 to DNA [25], showed an increase of mRNA expression in accordance with the concentration used ($p < 0.001$, Figure 11A), while the protein expression evaluated by immunoblotting analysis highlighted a trend only (Figure 11B,C).

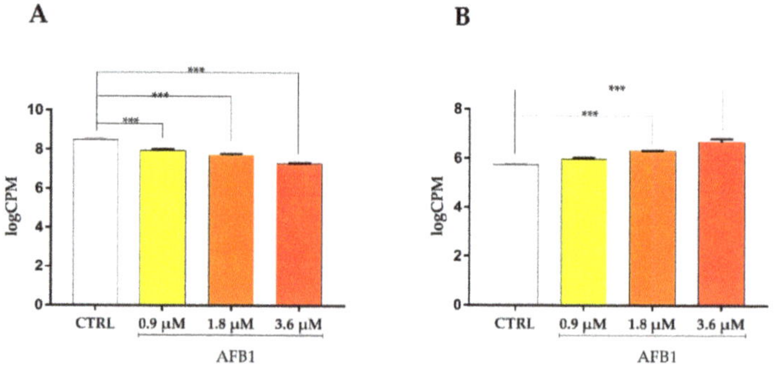

Figure 10. Effect of increasing doses of AFB1 (i.e., 0.9 μM, 1.8 μM and 3.6 μM) on NRF2 (**A**) and BACH1 (**B**) mRNA expression. Data are reported as the mean ± SEM of logCPM relative to four biological replicates. Statistical analysis: one-way ANOVA followed by Bonferroni's multiple comparisons test; ***: $p < 0.001$, treated cells vs. control cells.

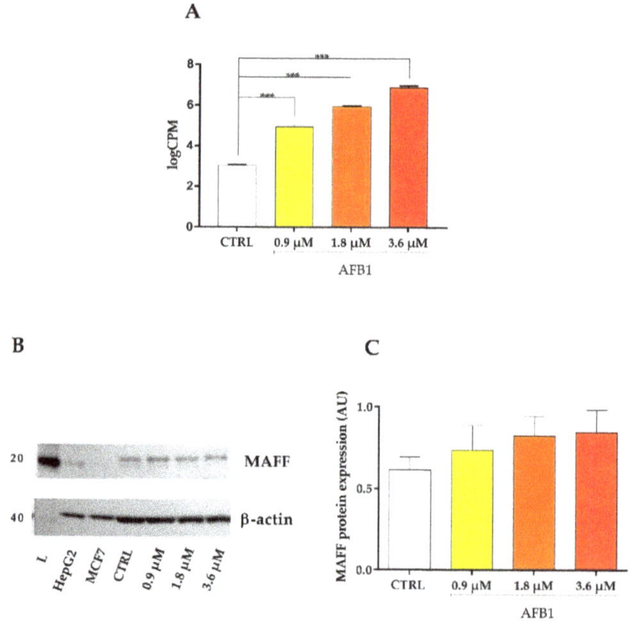

Figure 11. Effect of increasing doses of AFB1 (i.e., 0.9 μM, 1.8 μM and 3.6 μM) on MAFF mRNA and protein expression. (**A**) Gene expression data are reported as the mean ± SEM of logCPM relative to four biological replicates. (**B**) MAFF immunoblotting, using β-actin as loading control. The image is representative of six biological replicates. HepG2 and MCF7 cell lines were used as positive controls. (**C**) Densitometric analysis of MAFF immunoblottings; data are expressed in arbitrary units (AU) as the mean ± SEM of six biological replicates. Statistical analysis: one-way ANOVA followed by Bonferroni's multiple comparisons test; ***: $p < 0.001$, treated cells vs. control cells.

3. Discussion

So far, several studies investigated the molecular mechanism underneath AFB1 induction of hepatocellular damage in human and rat hepatocytes [25–27]. As to food-producing species, a fair amount of information about AFB1-mediated hepatotoxicity is available [22,23]; however, very little is known about AFB1 mechanistic toxicology in cattle liver. In the present study, the molecular mechanisms involved in the exploitation of AFB1 hepatotoxicity have been characterized in BFH12 cells, thereby improving the body of knowledge about the toxicological effects of this mycotoxin in this important farm animal species.

3.1. AFB1 Biotranformation

The metabolite profiling of AFB1, and particularly the production of the two main AFB1 derivatives AFM1 and AFL, was assessed in the cell medium after 48 h of incubation with the mycotoxin (0.9 μM, 1.8 μM and 3.6 μM). The amount of AFB1 and its derivatives increased proportionally to the AFB1 concentration used, thus proving BFH12 cells metabolize the mycotoxin. Present results, at least in part, confirm those previously obtained by [24]. Indeed, the amount of AFB1, AFM1 and AFL detected in BFH12 medium after 48 h of exposure to 3.6 μM AFB1, were higher than those here obtained. Nevertheless, in the former study cells had been pre-treated with a known CYP1A1 inducer (i.e., PCB126) to boost the cellular response to AFB1. Thus, it is conceivable to assume that the observed differences in the detectable concentrations of AFB1, AFM1 and AFL are attributable to the use of the PCB126, which most likely increased the capability of BFH12 to metabolize AFB1.

3.2. AFB1 Cytotoxicity and Mechanism of Cell Death

As already mentioned above, the AFB1 exposure and the resulting hepatotoxicity might substantially differ according to the species susceptibility, dose and time of exposure [20]. Consequently, selecting the optimum mycotoxin concentration to assess its in vitro toxic effects is quite challenging. The highest AFB1 concentration here used (3.6 μM) was chosen in accordance to our previous study [24], and it approximated the IC_{50} estimated in BFH12 cells after 48 h of incubation. Going further, a toxicokinetic study in which purified AFB1 was orally administered to cattle (0.35 mg/kg of body weight), reported a wide array of AF concentrations detected in tissues and plasma, with a maximum pick at 0.18 μM [28]. The lowest concentration tested in this study (i.e., 0.9 μM) is higher than the one mentioned above. Surely, it would be interesting to use realistic AFB1 concentrations; however, treating cells with an amount of AFB1 slightly higher may help to highlight the molecular mechanisms involved in the toxicity. Notably, the increasing AFB1 concentrations chosen in the present study were in accordance to those selected in previous cytotoxic experiments conducted in human and chicken hepatocytes, as well as in similar transcriptomic studies aimed at unveiling AFB1 mechanism of toxicity in chicken hepatocellular carcinoma (LMH) and HepG2 cell lines [29–33]. Overall, these data confirm AFB1 is cytotoxic to bovine hepatocytes, as previously reported [24].

Concerning the mechanism of cell death, apoptosis is a programmed death by which cells cease to grow and are phagocytized before undergoing membrane damage [34]. Conversely, necrosis is an uncontrolled mode of death by which cells reply to danger signals, like pro-inflammatory molecules, releasing cellular constituents after membrane rupture and leading to inflammation processes [35]. Previous studies on poultry species fed on diet contaminated with AFB1 (0.3–0.6 mg AFB1/basal diet) reported that AFB1 is likely to induce the hepatic programmed cell death [36–38]. In the present study, flow cytometry analyses showed that BFH12 cells exposed to 0.9 μM, 1.8 μM and 3.6 μM AFB1 exhibited a higher number of necrotic events compared to apoptotic ones. This experimental evidence was confirmed by transcriptomic results as several pathways linked to inflammation processes and apoptosis were enriched. Specifically, within apoptosis pathway, both pro-apoptotic and anti-apoptotic genes were induced by AFB1. Further details were reported below.

3.3. AFB1 Effects on BFH12 Cell Transcriptome

Overall, the whole-transcriptomic results suggest us that AFB1 deeply affects cattle hepatic transcriptome in a dose-dependent manner. This is clearly demonstrated by the number of DEGs in AFB1-exposed cells compared to control ones. Indeed, the number of DEGs increased in parallel with the AFB1 concentration. Notably, 947 DEGs (22%) were shared by the three different AFB1 concentrations; this leads us to hypothesize these genes represent a sort of core regulation mechanism of response to AFB1, which does not depend on the mycotoxin concentration. However, 13 (0.3%), 65 (1.5%) and 1964 (45.7%) genes were found to be specifically and differentially regulated by 0.9 μM, 1.8 μM and 3.6 μM AFB1, respectively. This would be indicative of an increasing cell responsiveness, with the involvement of specific genes, because of the use of increasing AFB1 concentrations. We decided to focus on DEGs shared among all the AFB1 conditions tested, with the goal of identifying a potential core molecular mechanism at the basis of AFB1 hepatotoxicity promotion.

In such a way, the Functional and Gene Set Enrichment Analysis of DEGs showed that most of the enriched pathways shared by all the AFB1 concentrations tested, and corresponding to the majority of genes modulated by the lowest concentration, encompasses genes associated with inflammation and immune responses. The output of PPI network analysis corroborated this result, as the list of the top ten hub genes included transcripts associated with inflammatory and immune response events. A further support of the main role of inflammation in AFB1-mediated hepatotoxicity was highlighted by the enrichment analysis of the protein interaction module, which resulted in several pathways linked to inflammatory processes. Previous studies conducted on poultry, mice and pigs confirmed the major role played by inflammation in AFB1 hepatotoxicity [39–41]. Thus, the effects of AFB1 on the inflammatory response are extensively discussed in a dedicated subsection.

Genes involved in carcinogenesis were also significantly affected by AFB1, as demonstrated by the KEGG enrichment and PPI network analysis. For instance, the pathway "ECM-receptor interaction" was enriched, which confirms what was previously reported in LMH cell line and turkey liver [32,42]. The extracellular matrix (ECM) consists of several macromolecules and mineral which are essential in maintaining cell architecture and functions; ECM proteins play a central role in this rearrangement by regulating complex cellular processes [43,44]. Therefore, the dysregulation of genes coding for ECM proteins may negatively affect cell network. Among them, laminins are crucial for cellular adhesion, differentiation and proliferation [45]. In our study, the LAMC2, a component of the laminin-332 protein, displayed an increased mRNA expression, which follows the AFB1 concentration used. This agrees with previous studies reporting an increased expression of LAMC2 in several cancer, such as hepatocellular and stomach carcinoma, as well as in lung cancer [46]. Another important ECM-structure glycoprotein whose mRNA level decreased after the treatment with the mycotoxin was the TNXB, an adhesion-modulatory ECM protein playing a crucial role in the cell structure organization [47]. Under pathological conditions, TNXB transcript dramatically decreased in most cancers, including hepatocellular carcinoma, thus corroborating our result [48]. Moreover, the gene expression of two collagenases, the COL4A5 and COL4A6, which are major components of the basement membrane [49] resulted dysregulated in the present study. Specifically, a decreased expression was observed, suggesting the disruption of the cell shape during cancer progression, as reported in previous studies [50,51]. AFB1 exposure also increased the expression of EDN1 in a dose-dependent way. EDN1 protein was also reported in the list of the top-ten hub genes deriving from the PPI network analysis, highlighting its important role in carcinogenesis process. The imbalance expression of EDN1 in cancer development is a well-known event. Indeed, EDN1 is a growth factor playing a critical role in tumorigenesis; it is frequently secreted in several tumors, such as liver, breast and colorectal cancers, in which enhances tumor growth by promoting angiogenesis [52–56].

The "Calcium signaling pathway" was another pathway which resulted significantly enriched in our study. The role of calcium in numerous cell processes, including cell proliferation, differentiation and even apoptosis, is well documented [57–59]. Consequently, the regulation of calcium channels may have a critical role in all the above listed cell functions. In our study, AFB1 reduced in a dose-dependent way the gene expression of the CACNAE1, a voltage-gated calcium channel (VGCC). It has been reported that a high expression of calcium channels is correlated with a reduction in cell proliferation capabilities [60]. Conversely, the downregulation of VGCC family genes, including the CACNAE1, is associated with the development and progression of several cancers [61].

It has been previously described that in human hepatoma cell lines AFB1 caused an imbalance between ROS production and antioxidant defense system, resulting into mitochondrial injury and finally into apoptosis [62,63]. In our study, KEGG GSEA analysis reported the over-representation of the "Apoptosis" term, which embraced both pro-apoptotic and anti-apoptotic genes. In poultry, the mycotoxin leaded to the overexpression of several hepatic death receptors, thus determining the induction of pro-apoptotic genes [36,37]. In our study, the death receptor genes TNFRSF10A and TNFRSF10B resulted in being upregulated in a dose-dependent way, following the exposure to AFB1. The upregulation of death receptor as a consequence of AFB1 exposure was previously reported also in pigs exposed to AFB1 in feed [42]. Moreover, our experiment highlighted an increase of expression of pro-apoptotic genes like GADD45A and CIDEB. The upregulation of such pro-apoptotic genes following the exposure to AFB1 was reported also in HepG2 and HepaRG cell lines, as well as in rat liver [64,65]. At the same time, our transcriptional results highlighted the upregulation of genes associated with anti-apoptotic functions like BCL2L1, a common anti-apoptotic protein that promotes cell survival, and TNFRSF10D, which plays an inhibitory role in TRAIL-induced cell apoptosis [66,67]. This last result would explain the outputs of flow cytometry investigations and the prevalence of necrotic events rather than apoptotic ones. Interestingly, the upregulation of BCL2L1 after AFB1 exposure was also observed in pig liver [42].

DEGs functional analysis found out, as expected, the enrichment of pathways associated with drug metabolism, too. As mentioned above, the CYP family plays a pivotal role in hepatic AFB1 biotransformation, and CYP1A and CYP3A subfamilies seemed to be those mostly involved. Moreover, it has been hypothesized that AFB1 might be a substrate for CYP2B6 in human [68,69]. In our study, CYP1A1 mRNA was significantly modulated by the treatment, as its expression increased in a dose-dependent way. Such a result is partially in contrast with previously published data. Indeed, in primary rabbit hepatocytes the mycotoxin inhibited CYP1A1 mRNA [70], while the inductive effect observed in the present study was reported also in primary human hepatocytes, in a rat hepatoma cell line, as well as in vivo in the liver of ducks fed with an AFB1-contaminated diet [71,72]. Our study also highlighted a strong dose-dependent decrease of CYP2B6 gene expression. By contrast, primary human hepatocytes exposed to AFB1 showed the overexpression of CYP2B6 gene [72]. Albeit contradictory, the observed gene dose-dependent downregulation could suggest that AFB1 is also a substrate of bovine CYP2B6, but such a hypothesis needs to be confirmed. Overall, in the present study a number of detoxifying enzymes was significantly inhibited by AFB1 in a dose-dependent manner. Among them there is ALDH1A1; besides its role in retinoic acid production, an involvement in the detoxification of lipid-derived toxic aldehydes, as well as in the metabolism of oxidative species, was previously described in rat liver [64,73]. Another important enzyme inhibited by AFB1 was MGST1, a membrane-bound enzyme displaying both conjugation and glutathione peroxidase functions, thus participating in the protection of cells against oxidative stress [74]. In mouse, rat and dog, MGST1 gene is expressed to a large extent in the liver; several endogenous and exogenous molecules may be substrate of this enzyme, including carcinogen drugs and products deriving from lipid peroxidation [75,76]. UGT1A1 gene encodes for an enzyme playing a major role in cellular protection. It is involved in the glucuronidation of both endogenous compounds and xenobiotics (e.g., dietary substances, hormones, environmental compounds and drugs) in more hydrophilic derivatives, thus facilitating the excretion with the urine and the bile [77]. Although UGT1A1 gene expression is not affected by AFB1 treatment in HepG2 and bovine mammary epithelial (BME) cell lines and in rat liver [78–80], in our study a strong decrease of UGT1A1 mRNA level was instead observed, suggestive of a peculiar response of cattle liver cells to AFB1 induction of oxidative stress. Interestingly, the expression of the abovementioned cytoprotective genes is regulated by NRF2 [81], a master regulator of the antioxidant response, whose mRNA expression was strongly decreased in the present study. Since oxidative stress has an outstanding role in hepatic initiation and progression of AFB1-induced toxicity, this pathway is discussed in depth below.

3.3.1. AFB1-Mediated Induction of Inflammatory Response

To counteract the broad range of toxic and non-toxic insults occurring during life, mammalian species have evolved a network of complex defense mechanisms in which inflammatory response is a tightly controlled cellular process in response to harmful exogenous and endogenous stimuli [82]. TLRs are a family of pattern recognition receptors (PRRs) playing a pivotal role in innate immune response, as they enhance the inflammatory response by increasing cytokines production [83,84]. TLRs recognize a broad range of exogenous and endogenous ligands, including the pathogen-associated molecular patterns (PAMPs) and damage-associated molecular patterns (DAMPs) [83]. DAMPs are protein or non-protein molecules released by damaged and dying cells, even in the absence of pathogens [85]. Interestingly, danger signals deriving from oxidative stress and lipid peroxidation also seem to activate TLRs, triggering pro-inflammatory cytokines production [86,87]. Previous studies reported the involvement of the TLR family in the modulation of inflammatory and immune responses following AFB1 exposure in human and bovine species, as well as in chicken liver [88–91]. Moreover, in mice fed with a diet supplemented with AFB1, the mRNA levels of TLR2 in the liver increased, thus suggesting a possible role of TLR2 in mediating AFB1-stimulated inflammatory responses [41].

In the present study, a strong upregulation of TLR2 gene was noticed in BFH12 cells exposed to AFB1; hence, we speculate TLR2 could play a central role in AFB1-induced inflammatory response. Furthermore, this upregulation was also confirmed at the protein level by flow cytometry. Thus, it is conceivable to hypothesize that signalling proteins from dying cells, together with danger signals deriving from AFB1-induced oxidative stress, might activate the TLR2 cascade. In particular, the signal transduction through the activation of p38 MAPK family seems to be involved in the development of inflammation. Indeed, once activated, TLRs trigger an intracellular signaling cascade with the involvement of the p38 MAPKs, resulting in the nuclear translocation of certain transcription factors such as AP-1 and NF-κB and the consequent release of pro-inflammatory cytokines and chemokines [92,93]. Notably, an in-vivo experiment made in piglets confirmed the involvement of p38 MAPKs, AP-1 and NF-κB transcription factors in AFB1-induced liver injury [39]. NF-κB gene expression also resulted in being upregulated in the liver of chickens exposed to AFB1 through the feed [90]. In the present study, three members of the p38 MAPK family, as well as AP-1 and NF-κB transcription factors, were upregulated by AFB1. Then, we hypothesize AFB1 might induce the expression of pro-inflammatory cytokines in cattle liver through the involvement of the p38 MAPK pathway. In particular, the p38β gene was strongly induced in a concentration-dependent way. It is worth noting that such evidence was also confirmed at the protein level.

Additionally, in our study we observed a significant and concentration-dependent upregulation of IL6, both at the mRNA and protein level. In this respect, members of the interleukin family of pro-inflammatory cytokines have been proven to be involved in a wide array of cellular processes, including proliferation and differentiation [94]. Our results agreed with those previously published in broilers and pigs fed with a diet containing AFB1, and showing hepatic IL6 mRNA induction [39,40]. Interestingly, inflammatory cytokines are key players in CYP regulation, expression, and ultimately, catalytic activity [95]; specifically, IL6 has been shown to inhibit CYP2B6 [96–98]. As a matter of fact, in our experimental conditions we observed a strong and dose-dependent inhibition of the

CYP2B6 gene (Figure S8A). Intriguingly, this result is in contrast with those observed in human hepatocytes exposed to AFB1, in which CYP2B6 gene was upregulated in a dose-dependent manner [72]. Therefore, it would be conceivable to hypothesize the presence of species-specific mechanisms of gene regulation activated by AFB1. Nevertheless, at protein level neither CYP2B6 expression nor catalytic activity corroborated RNA-*seq* results (Figure S8B–D), maybe suggesting the presence of post-translational (e.g., protein stabilization) mechanisms of gene regulation.

3.3.2. AFB1-Mediated Induction of Oxidative Stress

In liver, AFB1-induction of hepatotoxicity is intimately linked to the ability of this mycotoxin in generating reactive oxygen species (ROS), thereby causing oxidative stress, which in turn triggers oxidative DNA damage and lipid peroxidation of membrane phospholipids [99,100]. ROS production is a common phenomenon in cells, which is normally counterbalanced by physiological defence mechanisms [11,12]. In this regard, NRF2 plays a critical role in the maintenance of cellular redox homeostasis by regulating a plethora of genes playing crucial role in cellular defence [101]. Hence, NRF2 activation by oxidative stress results in transcriptional induction of a battery of cytoprotective genes, such as proteins involved in the cellular antioxidant machinery (e.g., catalase—CAT, glutathione peroxidase—GPX and superoxide dismutase—SOD), as well as proteins involved in xenobiotic detoxification (e.g., NQO1, UGT, GST and ALDH) [12–15,81,102]. Specifically, under unstressed condition, NRF2 is associated to Kelch-like ECH-associated protein 1 (KEAP1), which leads to NRF2 ubiquitination and consequently to its proteasomal degradation. In presence of oxidative stress stimuli, the negative regulator KEAP1 is arrested, and NRF2 translocates into the nucleus where it forms a heterodimer with the sMAF [103].

This heterodimer can bind to the Maf (musculoaponeurotic fibrosarcoma) protein recognition element (MARE), leading to transactivation of cytoprotective genes [84]. Another important transcription factor involved in oxidative stress response is BACH1, which competes with NRF2 for binding to MARE. Notably, once BACH1-sMAF heterodimer is formed, the interaction with MARE inhibits the transcription of many oxidative stress-response genes, thus acting as a transcriptional repressor [104]. Besides heterodimers with NRF2 and BACH1, sMAF can form homodimers, acting as transcriptional repressor of antioxidant genes transcription [105].

In-vivo studies conducted on rat liver reported that AFB1 can impair the fine-tuning defence mechanism by inhibiting the NRF2 protein, enhancing the production of free radicals and lipid peroxides, thereby causing hepatocellular damage [25,106]. In line with these findings, the downregulation of NRF2 and downstream cytoprotective genes was shown also in broilers fed with AFB1-contaminated diet [107]. Moreover, an in-vivo study conducted on mice orally administrated with AFB1 showed a significant decrease of NRF2 as well as of antioxidant-related genes expression in the liver [108]. The same NRF2 downregulation pattern was shown, both at the mRNA and protein level, in the ileum of ducks fed on an AFB1-contaminated diet [109]. Likewise, we previously reported [24] and we confirmed in the present study the inhibition of NRF2 mRNA expression in bovine hepatocytes incubated with AFB1. Consistently with these findings, in the present study NRF2 mRNA was inhibited by AFB1. Interestingly, in the present study, KEAP1 did not show any significant modulation; this allows us to hypothesize an alternative mechanism of NRF2 regulation, possibly KEAP1-independent (e.g., protein kinase phosphorylation cascade, interaction with other proteins), which needs to be further investigated.

Unlike NRF2, in the present study, BACH1 gene expression was upregulated. As with sMAF, we noticed that MAFF was induced by AFB1 both at the mRNA and protein level. Overall, these findings suggest a transcriptional inhibition of cytoprotective genes as a result of AFB1 exposure. In this respect, looking at the RNA-*seq* output, we noticed the downregulation of several cytoprotective genes whose expression is regulated by NRF2. Among them, there is NQO1, a FAD-dependent flavoprotein involved in two-electron reduction of quinone to hydroquinone, thus avoiding the formation of semiquinones and ROS [110]. We measured the antioxidant capacity of this enzyme, and observed a concentration-dependent inhibition of the catalytic activity, confirming the NQO1 inhibition noticed at the protein level. Another important cytoprotective gene whose expression resulted downregulated in a dose-dependent way was the ALDH1A1. Furthermore, p38 MAPKs, in addition to the role in the inflammatory response, seem to also be involved in the antioxidant and cytoprotective responses. In particular, they act as repressors of genes directly involved in the oxidative stress response, whose expression is regulated by NRF2 [111,112]. As an example, it has been reported that UGT1A1 promoter contains one functional ARE that is responsible for the NRF2-dependent induction [112]. In our study, UGT1A1 mRNA expression appeared to be downregulated by AFB1 exposure in a concentration-dependent manner. Similar to CYP2B6, no significant differences in UGT1A1 protein amount were obtained in AFB1-exposed cells; this evidence could be suggestive of a protein stabilization mechanism counteracting UGT1A1 mRNA level decrease. It is also conceivable to suppose a contribution of p38 MAPK pathway in the downregulation of cytoprotective genes, leading to an increase of oxidative stress events.

Figure 12 graphically described the putative connection between inflammation and oxidative stress pathways induced by AFB1 exposure in BFH12 cells, highlighting the central role of TLR2 in hepatotoxicity induction.

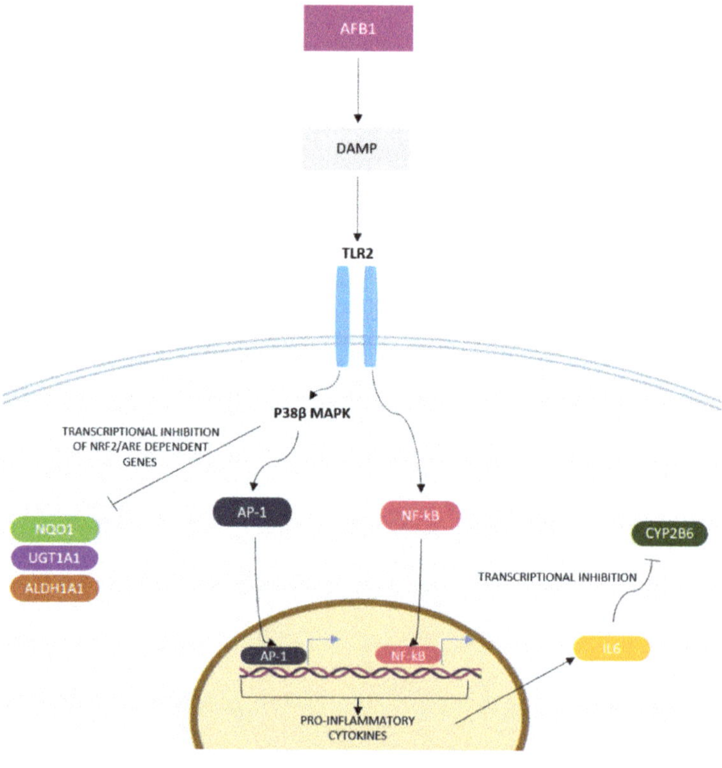

Figure 12. AFB1 induction of hepatotoxicity in BFH12 cells. Danger signals deriving from oxidative stress, as well as the DAMP, lead to TLR2 activation which triggers a downstream signaling cascade with the involvement of p38β MAPK. In turn, p38β MAPK allows the nuclear translocation of AP-1 and NF-kB transcription factors, thus leading the release of IL6 pro-inflammatory cytokine. In addition, p38β MAPK induction negatively regulates the expression of NRF2-dependent cytoprotective genes (e.g., NQO1, UGT1A1 and ALDH1A1), while IL6 inhibited CYP2B6 mRNA expression.

4. Conclusions

Following the need to increase our knowledge on molecular mechanisms underlying AFB1 hepatotoxicity in cattle, the fetal bovine hepatocytes cell line (BFH12) was exposed to increasing AFB1 concentrations (i.e., 0.9 μM, 1.8 μM and 3.6 μM), and then whole-transcriptomic analysis and post-translational confirmatory assays were implemented. AFB1, biotransformed by BFH12 cells, induced cell death via necrosis rather than apoptosis, and deeply affected cattle transcriptome in a dose-dependent manner, as the number of DEGs increased according to AFB1 concentration used. Specifically, the 22% of DEGs was shared among all the three AFB1 concentrations, suggesting a core molecular mechanism, independent from the dose used. Several pathways were affected by AFB1 (e.g., cancer, apoptosis and drug metabolism), including inflammation and oxidative stress. Based on transcriptomic data and post-translation investigations, a putative molecular signaling pathway linking AFB1 to oxidative stress and inflammation response events was proposed. Specifically, TLR2 activation seems to play a central role in AFB1-mediated exploitation of hepatotoxicity by triggering an intracellular signaling cascade, which in turn promotes a pro-inflammatory response. Overall, the present study contributes to gain new insights into the mechanistic toxicology of AFB1 in bovine liver.

5. Materials and Methods

5.1. Chemicals and Reagents

AFB1 (from *A. flavus*; CAS Number 1162-65-8), dimethyl sulfoxide (DMSO), dexamethasone, insulin from bovine pancreas and trypan blue solution were obtained from Sigma-Aldrich (St. Louis, MO, USA). AFB1 (purity 99.9%) and AFM1 (purity 99.1%) was purchased from Lab Service Analytica (Bologna, Italy). AFL (purity 98%) and 13C17-AFB1 (purity 98%) were obtained from DBA Italia (Milano, Italy) and Orsell (Modena, Italy), respectively. The WST-1 Cell Proliferation Reagent was from Roche (Basel, Switzerland), whereas P450-Glo™ CYP2B6 and CellTiter-Glo® Luminescent Cell Viability Assay kits were from Promega Corporation (Madison, WI, USA). The RNeasy Mini kit was from Qiagen (Hilden, Germany). The Bovine IL6 Elisa kit (MBS733925) was purchased from MyBioSource (San Diego, CA, USA). NQO1 Activity Assay kit (ab184867) and rabbit anti-CYP2B6 (ab69652) were from Abcam (Cambridge, UK). The rabbit anti-rat/mouse/human UGT1A1 (AB10339) was from Sigma-Aldrich (St. Louis, MO, USA). The mouse anti-human p38β MAPK11 (F-3; sc-390984) and rabbit anti-human p-p38 MAPK (E-1; sc-166182) were obtained from Santa Cruz Biotechnology (Dallas, TX, USA). The rabbit anti-human MAFF (PA5-85462), Annexin V-FITC Apoptosis Detection Kit (BMS500FI), the Accutase Enzyme Cell Detachment Medium (00-4555-56) and the BCA assay kit were purchased from Invitrogen, Life Technologies (Carlsbad, CA, USA). The rabbit anti-beta actin (ACTB, GTX109639) and the goat anti-mouse IgG (GTX213111-01) were from GeneTex (Irvine, CA, USA). The human anti-bovine CD282:FITC (AbD12542) and HuCAL Fab-dHLX-MH Negative Control antibody (AbD04652) were purchased from Bio-Rad (Hercules, CA, USA). Analytical standards of AFB1, AFL, AFM1 and AFB2 were purchased from DBA Italia (Milano, Italy). All solvents used for analysis were of LC–MS grade: acetonitrile and formic acid were from Sigma Aldrich (Milano, Italy), ultrapure water was freshly produced in-house (Millipore, Milano, Italy).

5.2. Cell Culture

The bovine SV40 large T-antigen-transduced fetal hepatocyte-derived cell line (BFH12) [113] was provided by Dr. Axel Schoeniger (Institute of Biochemistry, University of Leipzig, Leipzig, Germany). Conditions of cell line maintenance have already been reported in [24].

5.3. AFB1 Cytotoxicity

Cells were seeded in 96-well flat-bottom plates at a density of 6×10^3 cells/well, and were exposed for 48 h to three AFB1 concentrations (0.9 μM, 1.8 μM and 3.6 μM) dissolved in 0.1% DMSO as a vehicle. Cells exposed to 0.1% DMSO were used as control. AFB1 concentrations were selected based on the dose-response curve defined in our previous study [24]; specifically, the chosen AFB1 concentrations were sub-cytotoxic and below the IC_{50} (the mycotoxin concentration inhibiting cell viability by 50%). At the end of the incubation time, BFH12 viability was determined using the WST-1 Cell Proliferation Reagent and CellTiter-Glo^{TM} cell viability assay kit, following manufacturers' instructions. Four and six independent cell cultures (i.e., biological replicates) per experimental group were performed, respectively. Each concentration was tested in sextuplicate.

5.4. Cells Incubation for Gene Expression Analysis, Post-Translational and Analytical Investigations

To evaluate the effects of increasing AFB1 concentrations on the whole BFH12 transcriptome, cells were seeded on 6-well culture plates at a density of 5×10^4 cells/well and exposed to the abovementioned AFB1 concentrations, or the vehicle only, for 48 h, as reported in Section 5.3. At the end of the incubation time, the medium was collected and stored at −80 °C for analytical investigations, while cells were washed with 1X PBS/EDTA, re-suspended in 600 µL of RLT buffer (Qiagen) containing 6 µL of β-mercaptoethanol, vortexed and stored at −80 °C until use. Four independent experiments were executed.

For the evaluation of apoptosis and necrosis phenomena, as well as to isolate proteins for downstream analyses (i.e., immunoblotting, ELISA, flow cytometry), cells were seeded in Petri dishes (90 mm diameter) at a density of 3×10^5 cells/Petri dish, and exposed either to AFB1 or DMSO following the same experimental protocol described above. Six independent experiments were executed.

5.5. LC-MS/MS Quantification of AFB1, AFM1 and AFL

The amount of AFB1 and its metabolites (AFM1 and AFL) was measured in the medium of all experimental conditions by LC-MS/MS. Samples were thawed at room temperature, mixed by vortexing for 30 s, and centrifuged at 15,000× *g* for 10 min. Twelve µL of the supernatant was mixed with 1.5 mL of a 0.1% formic acid in water:acetonitrile 85:15 (*v*/*v*) solution, also containing the internal standard AFB2, and then 5 µL was injected in the system. The LC was a Waters Acquity UPLC binary pump, equipped with an Acquity BEH C18 (50 × 2.1 mm, 1.7 µm) reversed-phase column, kept at 40 °C (Waters, Milford, MA, USA). Chromatographic separation was obtained in a 4 min run under programmed conditions, with a variable mixture of 0.1% formic acid in water and acetonitrile flowing at 0.3 mL/min. The detector was a Waters Xevo TQ-S Micro triple quadrupole mass spectrometer (Waters, Milford, MA, USA), operating in positive electrospray ionization (ESI+) with 3.0 kV capillary voltage. Source temperature was 150 °C and desolvation temperature was 600 °C; desolvation and cone gas flow were 900 and 50 L/h, respectively. For each analyte, the following specific transitions (with relative Cone Voltage and Collision Energy values) were monitored: 313.17 > 241.12 m/z (CV 80 V; CE 34 eV) and 313.17 > 284.90 m/z (CV 80 V; CE 20 eV) for AFB1; 329.17 > 273.08 m/z (CV 70 V; CE 24 eV) and 329.17 > 229.11 m/z (CV 70 V; CE 38 eV) for AFM1; 297.15 > 141.04 m/z (CV 78 V; CE 48 eV) and 297.15 > 115.01 m/z (CV 78 V; CE 50 eV) for AFL; 315.13 > 259.03 m/z (CV 70 V; CE 26 eV) and 315.13 > 287.06 m/z (CV 70 V; CE 24 eV) for AFB2. Data acquisition and analysis was carried out with MassLynx 4.2 software (Waters, Milford, MA, USA).

5.6. Total RNA Extraction and RNA-seq Analysis

Total RNA was isolated using RNeasy Mini kit following the manufacturer's instructions. Subsequently, the quali-quantitative evaluation of total RNA was performed as described elsewhere [24]. For each experimental condition (i.e., CTRL, 0.9 µM AFB1, 1.8 µM AFB1 and 3.6 µM AFB1), four independent biological replicates were considered. Library construction and sequencing with NextSeq 500 system were performed at Centro di Ricerca Interdipartimentale per le Biotecnologie Innovative (CRIBI; University of Padova). Reads processing (i.e., quality check, trimming, filtering out and mapping against *Bos taururs* reference genome) was performed, as previously described [24]. Differential expression analysis was handled in EdgeR [114] on samples grouped in accordance with the treatment condition. Pair-wise comparisons between each treatment condition and the control were carried out to highlight transcriptional changes induced by increasing concentration of AFB1 (i.e., 0.9 μM AFB1 vs. CTRL, 1.8 μM AFB1 vs. CTRL, 3.6 μM AFB1 vs. CTRL), setting FDR at <0.05. ClusterProfiler package [115] was then implemented in R environment to functionally interpret significant differentially expressed genes (DEGs) through KEGG over-representation test and KEGG Gene Set Enrichment Analysis (GSEA) [116], as described in [117]. Expression of genes of interest was reported as logarithm of counts per million (logCPM), with a p-value threshold set at $\leq$0.05.

5.7. PPI Network Analysis

PPI network of the 947 DEGs shared among all treatments (i.e., 0.9 μM AFB1, 1.8 μM AFB1, 3.6 μM AFB1) was constructed using the online database Search Tool for the Retrieval of Interacting Genes/Proteins (STRING) with default parameters (https://string-db.org/cgi/input.pl, accessed on 5 June 2021). To better visualize the PPI network, the output data were then imported in Cytoscape software (version 3.8.2; https://cytoscape.org/, accessed on 5 June 2021) setting Degree Filter > 10. The Cytoscape plug-in Molecular Complex Detection (MCODE) was implemented to detect modules with the following parameters: Degree cutoff = 10, Node Score Cutoff = 0.2 and K-Core = 2. KEGG over-representation test was then implemented on genes belonging to the module with the highest score.

5.8. Flow Cytometry

The rate of necrosis and apoptosis induced by AFB1, as well as the protein expression of TLR2, were determined by flow cytometry. Six independent experiments (six independent cell cultures) were executed and each biological replicate (i.e., CTRL, 0.9 μM AFB1, 1.8 μM AFB1 and 3.6 μM AFB1) was analyzed in triplicate. After 48 h of incubation with the mycotoxin, cells were detached using Accutase and centrifuged at 1100 rpm for 10 min at 4 °C. Then, for the assessment of necrosis and apoptosis, the Annexin V-FITC Apoptosis Detection Kit was used following the manufacturer's instruction. Necrotic cells were positive for both Annexin-V FITC and PI, while apoptotic cells were positive for Annexin-V FITC only.

For TLR2 protein quantification, 500,000 cells were harvested and resuspended in 500 μL of RPMI 1640 medium containing sodium azide and FBS, to reach a concentration of 1000 cells/μL. Cells (50,000/tube) were incubated for 1 h at 4 °C with the human anti-bovine TLR2 monoclonal antibody (dilution 1:100) and resuspended in 900 μL of PBS for data acquisition. In both cases, CyFlow Space flow cytometer (Partec-System, Sysmex Europe GmbH, Norderstedt-Amburgo, Germany) was used and data were analyzed with FloMax software (version 2.82). For each replicate, 20,000 events were acquired. The morphology and the complexity of cells were evaluated in forward scatter (FSC) and side scatter (SSC), while TLR2-positive cells were identified on fluorescence channel 5 versus FSC.

5.9. Protein Isolation and Immunoblotting

After collection of the culture medium, monolayers were washed with the lysis buffer (TrisHCl 20 mM, pH 7.4) and scraped off with 500 μL of the same buffer. The cell suspension was then subjected to repeated freeze–thaw cycles in liquid nitrogen and in a water bath (37 °C), respectively. The cell debris was removed by centrifugation at 12,000× g for 10 min (4 °C), and the supernatant was collected. Protein content was quantified using the BCA assay kit following the manufacturer's instructions. Fifteen μg of total proteins were separated on NuPAGE® Novex® 4–12% Bis–Tris Gels by using the XCell SureLock™ Mini-Cell electrophoresis system (Invitrogen, Eugene, OR, USA), and then transferred onto nitrocellulose filters as previously described [118]. Two human positive controls (proteins isolated from HepG2 and MCF7 human cell lines) were loaded on each minigel. Membranes were incubated with anti-ACTB (1:6000 final concentration, 2 h), p38β MAPK11 and p-p38 MAPK (1:1000, overnight), CYP2B6 (1:500, overnight), MAFF (1:1000, overnight) and UGT1A1 (1:2000, overnight) primary antibodies, then with horseradish peroxidase-conjugated goat anti-rabbit (1:6000, 1.5 h, for ACTB, CYP2B6, MAFF, UGT1A1) or anti-mouse XX (1:1000, 1.5 h, for p38β MAPK11 and p-p38 MAPK) IgG. The specific proteins were detected by using SuperSignal® West Pico chemiluminescence substrate (Pierce, Life Technologies, Foster City, CA, USA) according to the manufacturer's instructions. Immunopositive bands were captured by the iBright Imaging Systems (iBright FL1500, Thermo Fisher Scientific, Waltham, MA, USA) and their Integrated Optical Density (IOD) was acquired by means of ImageJ 1.44p image analysis software. For the semi-quantification analysis, IOD of each sample was firstly normalized to the IOD of the loading control (ACTB), and then to the IOD of the positive control (MCF7 proteins for p38 β and p-p38, HepG2 proteins for CYP2B6, MAFF and UGT1A1).

5.10. Interleukin 6 Detection by ELISA

IL6 was quantified using the competitive Bovine IL6 Elisa kit, following manufacturer's instructions. Each experimental condition (i.e., CTRL, 0.9 μM AFB1, 1.8 μM AFB1 and 3.6 μM AFB1) was analyzed in duplicate. The optical density (O.D.) was measured spectrophotometrically at 450 nm in a Multiskan Go multiwell plate reader (Thermo Fisher Scientific, Waltham, MA, USA).

5.11. CYP2B6 and NQO1 Enzymatic Activity

The catalytic activity of two enzymes significantly modulated at the mRNA level by AFB1, i.e., NQO1 and CYP2B6, was evaluated. For each experimental condition (i.e., CTRL, 0.9 μM AFB1, 1.8 μM AFB1 and 3.6 μM AFB1), six independent biological replicates were available and each of them was tested in duplicate (NQO1) or in sextuplicate (CYP2B6). NQO1 catalytic activity was measured as previously reported [117,119], while CYP2B6 enzymatic activity was measured using the P450-Glo™ CYP2B6 luminescent assay kit and VICTOR™X4 Multilabel Plate Reader (Perkin Elmer, Waltham, MA, USA). In both cases, the instructions provided by the manufacturer were followed. The luminescence was expressed in relative luminescence units (RLU) as the luminescence units normalized to the number of metabolically active cells obtained with CellTiter-Glo™ cell viability assay kit (Section 5.3).

5.12. Statistical Analysis

Statistical analysis was performed using GraphPad Prism software (version 8.0.2, San Diego, CA, USA). One-way ANOVA followed by Bonferroni's multiple comparisons test was used for the analysis of gene expression, enzymatic activity, ELISA and immunoblotting output data. For the assessment of apoptotic and necrotic cell rate, one-way ANOVA with Dunnett's multiple comparisons test was implemented. A value of $p < 0.05$ was considered significant.

Supplementary Materials: The following supporting information can be downloaded at: https://www.mdpi.com/article/10.3390/toxins14070504/s1, Table S1: Results of RNA sequencing, filtering and mapping; Table S2: list of DEGs for each treatment; Table S3: Over-representation analysis (KEGG pathways); Table S4: Gene Set Enrichment Analysis (KEGG pathways); Table S5: Log fold change (logFC) for each gene of interest; Figure S1: MDS plot; Figure S2: Number of DEGs between each AFB1 treatment (i.e., 0.9 μM, 1.8 μM, 3.6 μM) vs. control; Figure S3: KEGG enrichment analysis of the top PPI module; Figure S4: Effect of increasing doses of AFB1 (i.e., 0.9 μM, 1.8 μM and 3.6 μM) on FOS (A), FOSL1 (B) and FOSB (C) mRNA expression; Figure S5: Effect of increasing doses of AFB1 (i.e., 0.9 μM, 1.8 μM and 3.6 μM) on JUNB mRNA expression; Figure S6: Effect of increasing doses of AFB1 (i.e., 0.9 μM, 1.8 μM and 3.6 μM) on NFKB2 (A) and RELB (B) mRNA expression; Figure S7: Effect of increasing doses of AFB1 (i.e., 0.9 μM, 1.8 μM and 3.6 μM) on UGT1A1 mRNA and protein expression; Figure S8: Effect of increasing doses of AFB1 (i.e., 0.9 μM, 1.8 μM and 3.6 μM) on CYP2B6 mRNA level, catalytic activity and protein expression.

Author Contributions: Conceptualization, M.D. and M.G.; methodology, M.E.G.; validation, S.I., F.B. and M.G.; formal analysis, S.I. and M.P.; investigation, S.I., I.B., F.B., A.B. (Anisa Bardhi), A.B. (Andrea Barbarossa) and M.G.; resources, A.Z.; data curation, S.I.; writing—original draft preparation, S.I.; writing—review and editing, M.P., M.D. and M.G.; supervision, M.P. and M.G.; project administration, M.G.; and funding acquisition, M.G. All authors have read and agreed to the published version of the manuscript.

Funding: This research was funded by University of Padua, grant number SID 2020-Prot. BIRD 207109. The University of Padua also funded a PhD fellowship to S.I.

Institutional Review Board Statement: Not applicable.

Informed Consent Statement: Not applicable.

Data Availability Statement: Raw Illumina Sequencing Data have been deposited in GenBank (SRA) under the BioProject ID PRJNA847423. Private access to the data is available at the link https://dataview.ncbi.nlm.nih.gov/object/PRJNA847423?reviewer=97vn2eecdlucsqkop4rsru09rl, accessed on 9 June 2022.

Acknowledgments: Authors are grateful to Axel Schoeniger (Institute of Biochemistry, University of Leipzig, Germany) who provided the BFH12 cell line.

Conflicts of Interest: The authors declare no conflict of interest.

References

1. Mahato, D.K.; Lee, K.E.; Kamle, M.; Devi, S.; Dewangan, K.N.; Kumar, P.; Kang, S.G. Aflatoxins in Food and Feed: An Overview on Prevalence, Detection and Control Strategies. *Front. Microbiol.* **2019**, *10*, 2266. [CrossRef] [PubMed]
2. European Food Safety Authority (EFSA). Effects on Public Health of an Increase of the Levels for Aflatoxin Total from 4 Mg/Kg to 10 Mg/Kg for Tree Nuts Other than Almonds, Hazelnuts and Pistachios—Statement of the Panel on Contaminants in the Food Chain. *EFSA J.* **2009**, *7*, 1168. [CrossRef]
3. Ostry, V.; Malir, F.; Toman, J.; Grosse, Y. Mycotoxins as Human Carcinogens—The IARC Monographs Classification. *Mycotoxin Res.* **2017**, *33*, 65–73. [CrossRef] [PubMed]
4. Sweeney, M.J.; Dobson, A.D.W. Mycotoxin Production by Aspergillus, Fusarium and Penicillium Species. *Int. J. Food Microbiol.* **1998**, *43*, 141–158. [CrossRef]
5. Battilani, P.; Toscano, P.; Van Der Fels-Klerx, H.J.; Moretti, A.; Camardo Leggieri, M.; Brera, C.; Rortais, A.; Goumperis, T.; Robinson, T. Aflatoxin B1 Contamination in Maize in Europe Increases Due to Climate Change. *Sci. Rep.* **2016**, *6*, 24328. [CrossRef]
6. Van der Fels-Klerx, H.J.; Vermeulen, L.C.; Gavai, A.K.; Liu, C. Climate Change Impacts on Aflatoxin B1 in Maize and Aflatoxin M1 in Milk: A Case Study of Maize Grown in Eastern Europe and Imported to the Netherlands. *PLoS ONE* **2019**, *14*, e0218956. [CrossRef]
7. Giambrone, J.J.; Diener, U.L.; Davis, N.D.; Panangala, V.S.; Hoerr, F.J. Effects of Aflatoxin on Young Turkeys and Broiler Chickens. *Poult. Sci.* **1985**, *64*, 1678–1684. [CrossRef]
8. Pandey, I.; Chauhan, S.S. Studies on Production Performance and Toxin Residues in Tissues and Eggs of Layer Chickens Fed on Diets with Various Concentrations of Aflatoxin AFB1. *Br. Poult. Sci.* **2007**, *48*, 713–723. [CrossRef]
9. Qureshi, M.A.; Brake, J.; Hamilton, P.B.; Hagler, W.M.; Nesheim, S. Dietary Exposure of Broiler Breeders to Aflatoxin Results in Immune Dysfunction in Progeny Chicks. *Poult. Sci.* **1998**, *77*, 812–819. [CrossRef]
10. Oswald, I.P.; Marin, D.E.; Bouhet, S.; Pinton, P.; Taranu, I.; Accensi, F. Immunotoxicological Risk of Mycotoxins for Domestic Animals. *Food Addit. Contam.* **2005**, *22*, 354–360. [CrossRef]
11. Hussein, H.S.; Brasel, J.M. Toxicity, Metabolism, and Impact of Mycotoxins on Humans and Animals. *Toxicology* **2001**, *167*, 101–134. [CrossRef]
12. Cook, W.O.; Richard, J.L.; Osweiler, G.D.; Trampel, D.W. Clinical and Pathologic Changes in Acute Bovine Aflatoxicosis: Rumen Motility and Tissue and Fluid Concentrations of Aflatoxins B1 and M1. *Am. J. Vet. Res.* **1986**, *47*, 1817–1825. [PubMed]
13. Deng, J.; Zhao, L.; Zhang, N.Y.; Karrow, N.A.; Krumm, C.S.; Qi, D.S.; Sun, L.H. Aflatoxin B1 Metabolism: Regulation by Phase I and II Metabolizing Enzymes and Chemoprotective Agents. *Mutat. Res. Rev. Mutat. Res.* **2018**, *778*, 79–89. [CrossRef] [PubMed]
14. Bedard, L.L.; Massey, T.E. Aflatoxin B1-Induced DNA Damage and Its Repair. *Cancer Lett.* **2006**, *241*, 174–183. [CrossRef]
15. Dohnal, V.; Wu, Q.; Kuča, K. Metabolism of Aflatoxins: Key Enzymes and Interindividual as Well as Interspecies Differences. *Arch. Toxicol.* **2014**, *88*, 1635–1644. [CrossRef]
16. Rawal, S.; Coulombe, R.A. Metabolism of Aflatoxin B1 in Turkey Liver Microsomes: The Relative Roles of Cytochromes P450 1A5 and 3A37. *Toxicol. Appl. Pharmacol.* **2011**, *254*, 349–354. [CrossRef]
17. Reed, K.M.; Mendoza, K.M.; Abrahante, J.E.; Coulombe, R.A. Comparative Response of the Hepatic Transcriptomes of Domesticated and Wild Turkey to Aflatoxin B1. *Toxins* **2018**, *10*, 42. [CrossRef]

18. Elgioushy, M.M.; Elgaml, S.A.; El-Adl, M.M.; Hegazy, A.M.; Hashish, E.A. Aflatoxicosis in Cattle: Clinical Findings and Biochemical Alterations. *Environ. Sci. Pollut. Res.* **2020**, *27*, 35526–35534. [CrossRef]
19. Umar, S.; Tanveer Munir, M.; Ali Shah, M.; Shahzad, M.; Ahmad Khan, R.; Sohoo, M.-R.; Ullah Khan, A.; Ameen, K.; Rafia-Munir, A.; Saleem, F.; et al. Outbreak of Aflatoxicosis on a Local Cattle Farm in Pakistan. *Veterinaria* **2015**, *3*, 13–17.
20. Peles, F.; Sipos, P.; Győri, Z.; Pfliegler, W.P.; Giacometti, F.; Serraino, A.; Pagliuca, G.; Gazzotti, T.; Pócsi, I. Adverse Effects, Transformation and Channeling of Aflatoxins Into Food Raw Materials in Livestock. *Front. Microbiol.* **2019**, *10*, 2861. [CrossRef]
21. Eshelli, M.; Qader, M.M.; Jambi, E.J.; Hursthouse, A.S.; Rateb, M.E. Current Status and Future Opportunities of Omics Tools in Mycotoxin Research. *Toxins* **2018**, *10*, 433. [CrossRef] [PubMed]
22. Zhang, N.Y.; Qi, M.; Gao, X.; Zhao, L.; Liu, J.; Gu, C.Q.; Song, W.J.; Krumm, C.S.; Sun, L.H.; Qi, D.S. Response of the Hepatic Transcriptome to Aflatoxin B1 in Ducklings. *Toxicon* **2016**, *111*, 69–76. [CrossRef] [PubMed]
23. Monson, M.S.; Settlage, R.E.; McMahon, K.W.; Mendoza, K.M.; Rawal, S.; El-Nezami, H.S.; Coulombe, R.A.; Reed, K.M. Response of the Hepatic Transcriptome to Aflatoxin B1 in Domestic Turkey (*Meleagris gallopavo*). *PLoS ONE* **2014**, *9*, e100930. [CrossRef]
24. Pauletto, M.; Tolosi, R.; Giantin, M.; Guerra, G.; Barbarossa, A.; Zaghini, A.; Dacasto, M. Insights into Aflatoxin B1 Toxicity in Cattle: An in Vitro Whole-Transcriptomic Approach. *Toxins* **2020**, *12*, 429. [CrossRef]
25. Ji, Y.; Nyamagoud, S.B.; SreeHarsha, N.; Mishra, A.; Gubbiyappa, S.K.; Singh, Y. Sitagliptin protects liver against aflatoxin B1-induced hepatotoxicity through upregulating Nrf2/ARE/HO-1 pathway. *BioFactors* **2019**, *46*, 76–82. [CrossRef] [PubMed]
26. Zhu, L.; Huang, C.; Yang, X.; Zhang, B.; He, X.; Xu, W.; Huang, K. Proteomics Reveals the Alleviation of Zinc towards Aflatoxin B1-Induced Cytotoxicity in Human Hepatocyes (HepG2 Cells). *Ecotoxicol. Environ. Saf.* **2020**, *198*, 110596. [CrossRef]
27. Zhang, L.Y.; Zhan, D.L.; Chen, Y.Y.; Wang, W.H.; He, C.Y.; Lin, Y.; Lin, Y.C.; Lin, Z.N. Aflatoxin B1 Enhances Pyroptosis of Hepatocytes and Activation of Kupffer Cells to Promote Liver Inflammatory Injury via Dephosphorylation of Cyclooxygenase-2: An in Vitro, Ex Vivo and in Vivo Study. *Arch. Toxicol.* **2019**, *93*, 3305–3320. [CrossRef]
28. Stubblefield, R.D.; Pier, A.C.; Richard, J.L.; Shotwell, O.L. Fate of Aflatoxins in Tissues, Fluids, and Excrements from Cows Dosed Orally with Aflatoxin B1. *Am. J. Vet. Res.* **1983**, *44*, 1750–1752.
29. Yang, X.; Lv, Y.; Huang, K.; Luo, Y.; Xu, W. Zinc Inhibits Aflatoxin B1-Induced Cytotoxicity and Genotoxicity in Human Hepatocytes (HepG2 Cells). *Food Chem. Toxicol.* **2016**, *92*, 17–25. [CrossRef]
30. Chen, X.; Che, C.; Korolchuk, V.I.; Gan, F.; Pan, C.; Huang, K. Selenomethionine Alleviates AFB1-Induced Damage in Primary Chicken Hepatocytes by Inhibiting CYP450 1A5 Expression via Upregulated SelW Expression. *J. Agric. Food Chem.* **2017**, *65*, 2495–2502. [CrossRef]
31. Li, A.P. Metabolism Comparative Cytotoxicity Assay (MCCA) and Cytotoxic Metabolic Pathway Identification Assay (CMPIA) with Cryopreserved Human Hepatocytes for the Evaluation of Metabolism-Based Cytotoxicity in Vitro: Proof-of-Concept Study with Aflatoxin B1. *Chem. Biol. Interact.* **2009**, *179*, 4–8. [CrossRef] [PubMed]
32. Choi, S.Y.; Kim, T.H.; Hong, M.W.; Park, T.S.; Lee, H.; Lee, S.J. Transcriptomic Alterations Induced by Aflatoxin B1 and Ochratoxin A in LMH Cell Line. *Poult. Sci.* **2020**, *99*, 5265–5274. [CrossRef] [PubMed]
33. Ratajewski, M.; Walczak-Drzewiecka, A.; Sałkowska, A.; Dastych, J. Aflatoxins Upregulate CYP3A4 MRNA Expression in a Process That Involves the PXR Transcription Factor. *Toxicol. Lett.* **2011**, *205*, 146–153. [CrossRef] [PubMed]
34. D'arcy, M.S. Cell Death: A Review of the Major Forms of Apoptosis, Necrosis and Autophagy. *Cell Biol. Int.* **2019**, *43*, 582–592. [CrossRef] [PubMed]
35. Wallach, D.; Kang, T.B.; Dillon, C.P.; Green, D.R. Programmed Necrosis in Inflammation: Toward Identification of the Effector Molecules. *Science* **2016**, *352*, aaf2154. [CrossRef] [PubMed]
36. Mughal, M.J.; Peng, X.; Zhou, Y.; Fang, J. Aflatoxin B1 Invokes Apoptosis via Death Receptor Pathway in Hepatocytes. *Oncotarget* **2017**, *8*, 8239–8249. [CrossRef] [PubMed]
37. Wu, B.; Mughal, M.J.; Fang, J.; Peng, X. The Protective Role of Selenium Against AFB1-Induced Liver Apoptosis by Death Receptor Pathway in Broilers. *Biol. Trace Elem. Res.* **2019**, *191*, 453–463. [CrossRef]
38. Liu, X.; Mishra, S.K.; Wang, T.; Xu, Z.; Zhao, X.; Wang, Y.; Yin, H.; Fan, X.; Zeng, B.; Yang, M.; et al. AFB1 Induced Transcriptional Regulation Related to Apoptosis and Lipid Metabolism in Liver of Chicken. *Toxins* **2020**, *12*, 290. [CrossRef]
39. Taranu, I.; Hermenean, A.; Bulgaru, C.; Pistol, G.C.; Ciceu, A.; Grosu, I.A.; Marin, D.E. Diet Containing Grape Seed Meal By-Product Counteracts AFB1 Toxicity in Liver of Pig after Weaning. *Ecotoxicol. Environ. Saf.* **2020**, *203*, 110899. [CrossRef]
40. Ma, Q.; Li, Y.; Fan, Y.; Zhao, L.; Wei, H.; Ji, C.; Zhang, J. Molecular Mechanisms of Lipoic Acid Protection against Aflatoxin B1-Induced Liver Oxidative Damage and Inflammatory Responses in Broilers. *Toxins* **2015**, *7*, 5435–5447. [CrossRef]
41. Huang, L.; Zhao, Z.; Duan, C.; Wang, C.; Zhao, Y.; Yang, G.; Gao, L.; Niu, C.; Xu, J.; Li, S. Lactobacillus Plantarum C88 Protects against Aflatoxin B1-Induced Liver Injury in Mice via Inhibition of NF-KB-Mediated Inflammatory Responses and Excessive Apoptosis. *BMC Microbiol.* **2019**, *19*, 170. [CrossRef] [PubMed]
42. Murarolli, R.A. *Effects of Aflatoxin B1 (AFB 1) on Hepatic Gene Expression in Pigs and Turkeys*; University of Missouri: Columbia, MO, USA, 2013.
43. Rosso, F.; Giordano, A.; Barbarisi, M.; Barbarisi, A. From Cell-ECM Interactions to Tissue Engineering. *J. Cell. Physiol.* **2004**, *199*, 174–180. [CrossRef] [PubMed]
44. Karamanos, N.K.; Theocharis, A.D.; Neill, T.; Iozzo, R.V. Matrix Modeling and Remodeling: A Biological Interplay Regulating Tissue Homeostasis and Diseases. *Matrix Biol.* **2019**, *75–76*, 1–11. [CrossRef] [PubMed]
45. Aumailley, M. The Laminin Family. *Cell Adhes. Migr.* **2013**, *7*, 48–55. [CrossRef]

46. Garg, M.; Braunstein, G.; Koeffler, H.P. LAMC2 as a Therapeutic Target for Cancers. *Expert Opin. Ther. Targets* **2014**, *18*, 979–982. [CrossRef]
47. Valcourt, U.; Alcaraz, L.B.; Exposito, J.-Y.; Lethias, C.; Bartholin, L. Cell Adhesion & Migration Tenascin-X: Beyond the Architectural Function. *Cell Adhes. Migr.* **2015**, *9*, 154–165. [CrossRef]
48. Liot, S.; Aubert, A.; Hervieu, V.; El Kholti, N.; Schalkwijk, J.; Verrier, B.; Valcourt, U.; Lambert, E. Loss of Tenascin-X Expression during Tumor Progression: A New Pan-Cancer Marker. *Matrix Biol. Plus* **2020**, *6–7*, 100021. [CrossRef]
49. Kalluri, R. Basement Membranes: Structure, Assembly and Role in Tumour Angiogenesis. *Nat. Rev. Cancer* **2003**, *3*, 422–433. [CrossRef]
50. Tanjore, H.; Kalluri, R. The Role of Type IV Collagen and Basement Membranes in Cancer Progression and Metastasis. *Am. J. Pathol.* **2006**, *168*, 715–717. [CrossRef]
51. Ikeda, K.; Iyama, K.I.; Ishikawa, N.; Egami, H.; Nakao, M.; Sado, Y.; Ninomiya, Y.; Baba, H. Loss of Expression of Type IV Collagen α5 and α6 Chains in Colorectal Cancer Associated with the Hypermethylation of Their Promoter Region. *Am. J. Pathol.* **2006**, *168*, 856–865. [CrossRef]
52. Komuro, I.; Kurihara, H.; Sugiyama, T.; Takaku, F.; Yazaki, Y. Endothelin Stimulates C-Fos and c-Myc Expression and Proliferation of Vascular Smooth Muscle Cells. *FEBS Lett.* **1988**, *238*, 249–252. [CrossRef]
53. Pedram, A.; Razandi, M.; Hu, R.M.; Levin, E.R. Vasoactive Peptides Modulate Vascular Endothelial Cell Growth Factor Production and Endothelial Cell Proliferation and Invasion. *J. Biol. Chem.* **1997**, *272*, 17097–17103. [CrossRef]
54. Ali, H.; Loizidou, M.; Dashwood, M.; Savage, F.; Sheard, C.; Taylor, I. Stimulation of Colorectal Cancer Cell Line Growth by ET-1 and Its Inhibition by ETA Antagonists. *Gut* **2000**, *47*, 685. [CrossRef]
55. Kojima, K.; Nihei, Z. Expression of Endothelin-1 Immunoreactivity in Breast Cancer. *Surg. Oncol.* **1995**, *4*, 309–315. [CrossRef]
56. Ishibashi, M.; Fujita, M.; Nagai, K.; Kako, M.; Furue, H.; Haku, E.; Osamura, Y.; Yamaji, T. Production and Secretion of Endothelin by Hepatocellular Carcinoma. *J. Clin. Endocrinol. Metab.* **1993**, *76*, 378–383. [CrossRef] [PubMed]
57. Jin, T.; Zhang, Z.; Yang, X.F.; Luo, J.S. S100A4 Expression Is Closely Linked to Genesis and Progression of Glioma by Regulating Proliferation, Apoptosis, Migration and Invasion. *Asian Pac. J. Cancer Prev.* **2015**, *16*, 2883–2887. [CrossRef] [PubMed]
58. Lipkin, M.; Newmark, H. Effect of Added Dietary Calcium on Colonic Epithelial-Cell Proliferation in Subjects at High Risk for Familial Colonic Cancer. *N. Engl. J. Med.* **1985**, *313*, 1381–1384. [CrossRef] [PubMed]
59. Galione, A.; Evans, A.M.; Ma, J.; Parrington, J.; Arredouani, A.; Cheng, X.; Zhu, M.X. The Acid Test: The Discovery of Two-Pore Channels (TPCs) as NAADP-Gated Endolysosomal Ca^{2+} Release Channels. *Pflügers Arch. Eur. J. Physiol.* **2009**, *458*, 869–876. [CrossRef]
60. Bertolesi, G.E.; Shi, C.; Elbaum, L.; Jollimore, C.; Rozenberg, G.; Barnes, S.; Kelly, M.E. The Ca^{2+} Channel Antagonists Mibefradil and Pimozide Inhibit Cell Growth via Different Cytotoxic Mechanisms. *Mol. Pharmacol.* **2002**, *62*, 210–219. [CrossRef]
61. Phan, N.N.; Wang, C.Y.; Chen, C.F.; Sun, Z.; Lai, M.D.; Lin, Y.C. Voltage-Gated Calcium Channels: Novel Targets for Cancer Therapy. *Oncol. Lett.* **2017**, *14*, 2059–2074. [CrossRef]
62. Xu, Q.; Shi, W.; Lv, P.; Meng, W.; Mao, G.; Gong, C.; Chen, Y.; Wei, Y.; He, X.; Zhao, J.; et al. Critical Role of Caveolin-1 in Aflatoxin B1-Induced Hepatotoxicity via the Regulation of Oxidation and Autophagy. *Cell Death Dis.* **2020**, *11*, 6. [CrossRef] [PubMed]
63. Zhang, B.; Dai, Y.; Zhu, L.; He, X.; Huang, K.; Xu, W. Single-Cell Sequencing Reveals Novel Mechanisms of Aflatoxin B1-Induced Hepatotoxicity in S Phase-Arrested L02 Cells. *Cell Biol. Toxicol.* **2020**, *36*, 603–608. [CrossRef] [PubMed]
64. Kuwata, K.; Shibutani, M.; Hayashi, H.; Shimamoto, K.; Hayashi, S.M.; Suzuki, K.; Mitsumori, K. Concomitant Apoptosis and Regeneration of Liver Cells as a Mechanism of Liver-Tumor Promotion by β-Naphthoflavone Involving TNFα-Signaling Due to Oxidative Cellular Stress in Rats. *Toxicology* **2011**, *283*, 8–17. [CrossRef] [PubMed]
65. Josse, R.; Dumont, J.; Fautrel, A.; Robin, M.-A.; Guillouzo, A. Identification of Early Target Genes of Aflatoxin B1 in Human Hepatocytes, Inter-Individual Variability and Comparison with Other Genotoxic Compounds. *Toxicol. Appl. Pharmacol.* **2012**, *258*, 176–187. [CrossRef]
66. Opferman, J.T.; Kothari, A. Anti-Apoptotic BCL-2 Family Members in Development. *Cell Death Differ.* **2017**, *25*, 37–45. [CrossRef]
67. Sheikh, M.S.; Fornace, A.J. Death and Decoy Receptors and P53-Mediated Apoptosis. *Leukemia* **2000**, *14*, 1509–1513. [CrossRef]
68. Macé, K.; Aguilar, F.; Wang, J.-S.; Vautravers, P.; Gómez-Lechón, M.; Gonzalez, F.J.; Groopman, J.; Harris, C.C.; Pfeifer, A.M. Aflatoxin B1-induced DNA adduct formation and p53 mutations in CYP450-expressing human liver cell lines. *Carcinogenesis* **1997**, *18*, 1291–1297. [CrossRef]
69. Code, E.L.; Crespi, C.L.; Penman, B.W.; Gonzalez, F.J.; Chang, T.K.H.; Waxman, D.J. Human cytochrome P4502B6: Interindividual hepatic expression, substrate specificity, and role in procarcinogen activation. *Drug Metab. Dispos.* **1997**, *25*, 985–993.
70. Guerre, P.; Pineau, T.; Costet, P.; Burgat, V.; Galtier, P. Effects of AFB1 on CYP 1A1, 1A2 and 3A6 MRNA, and P450 Expression in Primary Culture of Rabbit Hepatocytes. *Toxicol. Lett.* **2000**, *111*, 243–251. [CrossRef]
71. Mary, V.S.; Valdehita, A.; Navas, J.M.; Rubinstein, H.R.; Fernández-Cruz, M.L. Effects of Aflatoxin B1, Fumonisin B1 and Their Mixture on the Aryl Hydrocarbon Receptor and Cytochrome P450 1A Induction. *Food Chem. Toxicol.* **2015**, *75*, 104–111. [CrossRef]
72. Ayed-Boussema, I.; Pascussi, J.M.; Maurel, P.; Bacha, H.; Hassen, W. Effect of Aflatoxin B1 on Nuclear Receptors PXR, CAR, and AhR and Their Target Cytochromes P450 MRNA Expression in Primary Cultures of Human Hepatocytes. *Int. J. Toxicol.* **2012**, *31*, 86–93. [CrossRef] [PubMed]

73. Xiao, T.; Shoeb, M.; Siddiqui, M.S.; Zhang, M.; Ramana, K.V.; Srivastava, S.K.; Vasiliou, V.; Ansari, N.H. Molecular Cloning and Oxidative Modification of Human Lens ALDH1A1: Implication in Impaired Detoxification of Lipid Aldehydes. *J. Toxicol. Environ. Health Part A* **2009**, *72*, 577–584. [CrossRef] [PubMed]
74. Mosialou, E.; Ekström, G.; Adang, A.E.P.; Morgenstern, R. Evidence That Rat Liver Microsomal Glutathione Transferase Is Responsible for Glutathione-Dependent Protection against Lipid Peroxidation. *Biochem. Pharmacol.* **1993**, *45*, 1645–1651. [CrossRef]
75. Hayes, J.D.; Pulford, D.J. The Glut Athione S-Transferase Supergene Family: Regulation of GST and the Contribution of the Lsoenzymes to Cancer Chemoprotection and Drug Resistance Part I. *Crit. Rev. Biochem. Mol. Biol.* **1995**, *30*, 445–520. [CrossRef]
76. Mattes, W.B.; Daniels, K.K.; Summan, M.; Xu, Z.A.; Mendrick, D.L. Xenobiotica the Fate of Foreign Compounds in Biological Systems Tissue and Species Distribution of the Glutathione Pathway Transcriptome. *Xenobiotica* **2006**, *36*, 1081–1121. [CrossRef]
77. Tukey, R.H.; Strassburg, C.P. Human Udp-Glucuronosyltransferases: Metabolism, Expression, and Disease. *Annu. Rev. Pharmacol. Toxicol* **2000**, *40*, 581–616. [CrossRef]
78. Ghadiri, S.; Spalenza, V.; Dellafiora, L.; Badino, P.; Barbarossa, A.; Dall'Asta, C.; Nebbia, C.; Girolami, F. Modulation of Aflatoxin B1 Cytotoxicity and Aflatoxin M1 Synthesis by Natural Antioxidants in a Bovine Mammary Epithelial Cell Line. *Toxicol. Vitr.* **2019**, *57*, 174–183. [CrossRef] [PubMed]
79. Hanioka, N.; Nonaka, Y.; Saito, K.; Negishi, T.; Okamoto, K.; Kataoka, H.; Narimatsu, S. Effect of Aflatoxin B1 on UDP-Glucuronosyltransferase MRNA Expression in HepG2 Cells. *Chemosphere* **2012**, *89*, 526–529. [CrossRef]
80. Merrick, B.A.; Phadke, D.P.; Auerbach, S.S.; Mav, D.; Stiegelmeyer, S.M.; Shah, R.R.; Tice, R.R. RNA-Seq Profiling Reveals Novel Hepatic Gene Expression Pattern in Aflatoxin B1 Treated Rats. *PLoS ONE* **2013**, *8*, e61768. [CrossRef]
81. Shaw, P.; Chattopadhyay, A. Nrf2-ARE Signaling in Cellular Protection: Mechanism of Action and the Regulatory Mechanisms. *J. Cell. Physiol.* **2020**, *235*, 3119–3130. [CrossRef]
82. Medzhitov, R. Inflammation 2010: New Adventures of an Old Flame. *Cell* **2010**, *140*, 771–776. [CrossRef] [PubMed]
83. El-Zayat, S.R.; Sibaii, H.; Mannaa, F.A. Toll-like Receptors Activation, Signaling, and Targeting: An Overview. *Bull. Natl. Res. Cent.* **2019**, *43*, 187. [CrossRef]
84. Janeway, C.A.; Medzhitov, R. Innate Immune Recognition. *Annu. Rev. Immunol.* **2002**, *20*, 197–216. [CrossRef] [PubMed]
85. Rubartelli, A.; Lotze, M.T. Inside, Outside, Upside down: Damage-Associated Molecular-Pattern Molecules (DAMPs) and Redox. *Trends Immunol.* **2007**, *28*, 429–436. [CrossRef]
86. West, X.Z.; Malinin, N.L.; Merkulova, A.A.; Tischenko, M.; Kerr, B.A.; Borden, E.C.; Podrez, E.A.; Salomon, R.G.; Byzova, T.V. Oxidative Stress Induces Angiogenesis by Activating TLR2 with Novel Endogenous Ligands. *Nature* **2010**, *467*, 972–976. [CrossRef]
87. Kadl, A.; Sharma, P.R.; Chen, W.; Agrawal, R.; Meher, A.K.; Rudraiah, S.; Grubbs, N.; Sharma, R.; Leitinger, N. Oxidized Phospholipid-Induced Inflammation Is Mediated by Toll-like Receptor 2. *Free Radic. Biol. Med.* **2011**, *51*, 1903–1909. [CrossRef]
88. Malvandi, A.M.; Mehrzad, J.; Saleh-Moghaddam, M. Biologically Relevant Doses of Mixed Aflatoxins B and G Up-Regulate MyD88, TLR2, TLR4 and CD14 Transcripts in Human PBMCs. *Immunopharmacol. Immunotoxicol.* **2013**, *35*, 528–532. [CrossRef]
89. Mohammadi, A.; Mehrzad, J.; Mahmoudi, M.; Schneider, M. Environmentally Relevant Level of Aflatoxin B1 Dysregulates Human Dendritic Cells through Signaling on Key Toll-like Receptors. *Int. J. Toxicol.* **2014**, *33*, 175–186. [CrossRef]
90. Li, S.; Liu, R.; Xia, S.; Wei, G.; Ishfaq, M.; Zhang, Y.; Zhang, X. Protective Role of Curcumin on Aflatoxin B1-Induced TLR4/RIPK Pathway Mediated-Necroptosis and Inflammation in Chicken Liver. *Ecotoxicol. Environ. Saf.* **2022**, *233*, 113319. [CrossRef]
91. Mehrzad, J.; Milani, M.; Mahmoudi, M. Naturally Occurring Level of Mixed Aflatoxins B and G Stimulate Toll-like Receptor-4 in Bovine Mononuclear Cells. *Vet. Q.* **2013**, *33*, 186–190. [CrossRef]
92. Chen, L.; Deng, H.; Cui, H.; Fang, J.; Zuo, Z.; Deng, J.; Li, Y.; Wang, X.; Zhao, L. Inflammatory Responses and Inflammation-Associated Diseases in Organs. *Oncotarget* **2018**, *9*, 7204–7218. [CrossRef]
93. Tak, P.P.; Firestein, G.S. NF-KB: A Key Role in Inflammatory Diseases. *J. Clin. Investig.* **2001**, *107*, 7–11. [CrossRef]
94. Mansell, A.; Jenkins, B.J. Dangerous Liaisons between Interleukin-6 Cytokine and Toll-like Receptor Families: A Potent Combination in Inflammation and Cancer. *Cytokine Growth Factor Rev.* **2013**, *24*, 249–256. [CrossRef]
95. Stipp, M.C.; Acco, A. Involvement of Cytochrome P450 Enzymes in Inflammation and Cancer: A Review. *Cancer Chemother. Pharmacol.* **2021**, *87*, 295–309. [CrossRef] [PubMed]
96. Dickmann, L.J.; Patel, S.K.; Rock, D.A.; Wienkers, L.C.; Slatter, J.G. Effects of Interleukin-6 (IL-6) and an Anti-IL-6 Monoclonal Antibody on Drug-Metabolizing Enzymes in Human Hepatocyte Culture. *Drug Metab. Dispos.* **2011**, *39*, 1415–1422. [CrossRef] [PubMed]
97. Aitken, A.E.; Morgan, E.T. Gene-Specific Effects of Inflammatory Cytokines on Cytochrome P450 2C, 2B6 and 3A4 MRNA Levels in Human Hepatocytes. *Drug Metab. Dispos.* **2007**, *35*, 1687–1693. [CrossRef] [PubMed]
98. Rubin, K.; Janefeldt, A.; Andersson, L.; Berke, Z.; Grime, K.; Andersson, T.B. Heparg Cells as Human-Relevant in Vitro Model to Study the Effects of Inflammatory Stimuli on Cytochrome P450 Isoenzymes. *Drug Metab. Dispos.* **2015**, *43*, 119–125. [CrossRef]
99. Yilmaz, S.; Kaya, E.; Kisacam, M.A. The Effect on Oxidative Stress of Aflatoxin and Protective Effect of Lycopene on Aflatoxin Damage. *Aflatoxin-Control Anal. Detect. Health Risks* **2017**, *30*, 67–90.
100. Benkerroum, N. Chronic and Acute Toxicities of Aflatoxins: Mechanisms of Action. *Int. J. Environ. Res. Public Health* **2020**, *17*, 423. [CrossRef]
101. Tonelli, C.; Chio, I.I.C.; Tuveson, D.A. Transcriptional Regulation by Nrf2. *Antioxidants Redox Signal.* **2018**, *29*, 1727–1745. [CrossRef]

102. Duong, H.-Q.; You, K.S.; Oh, S.; Kwak, S.-J.; Seong, Y.-S. Silencing of NRF2 Reduces the Expression of ALDH1A1 and ALDH3A1 and Sensitizes to 5-FU in Pancreatic Cancer Cells. *Antioxidants* **2017**, *6*, 52. [CrossRef] [PubMed]
103. Suzuki, M.; Otsuki, A.; Keleku-Lukwete, N.; Yamamoto, M. Overview of Redox Regulation by Keap1–Nrf2 System in Toxicology and Cancer. *Curr. Opin. Toxicol.* **2016**, *1*, 29–36. [CrossRef]
104. Zhang, X.; Guo, J.; Wei, X.; Niu, C.; Jia, M.; Li, Q.; Meng, D. Bach1: Function, Regulation, and Involvement in Disease. *Oxidative Med. Cell. Longev.* **2018**, *2018*, 1347969. [CrossRef]
105. Katsuoka, F.; Yamamoto, M. Small Maf Proteins (MafF, MafG, MafK): History, Structure and Function. *Gene* **2016**, *586*, 197–205. [CrossRef] [PubMed]
106. Vipin, A.V.; Raksha Rao, K.; Kurrey, N.K.; Anu Appaiah, K.A.; Venkateswaran, G. Protective Effects of Phenolics Rich Extract of Ginger against Aflatoxin B1-Induced Oxidative Stress and Hepatotoxicity. *Biomed. Pharmacother.* **2017**, *91*, 415–424. [CrossRef]
107. Wang, H.; Muhammad, I.; Li, W.; Sun, X.; Cheng, P.; Zhang, X. Sensitivity of Arbor Acres Broilers and Chemoprevention of Aflatoxin B1-Induced Liver Injury by Curcumin, a Natural Potent Inducer of Phase-II Enzymes and Nrf2. *Environ. Toxicol. Pharmacol.* **2018**, *59*, 94–104. [CrossRef] [PubMed]
108. Wang, Y.; Liu, F.; Liu, M.; Zhou, X.; Wang, M.; Cao, K.; Jin, S.; Shan, A.; Feng, X. Curcumin Mitigates Aflatoxin B1-Induced Liver Injury via Regulating the NLRP3 Inflammasome and Nrf2 Signaling Pathway. *Food Chem. Toxicol.* **2022**, *161*, 112823. [CrossRef]
109. Yang, H.; Wang, Y.; Yu, C.; Jiao, Y.; Zhang, R.; Jin, S.; Feng, X. Dietary Resveratrol Alleviates AFB1-Induced Ileum Damage in Ducks via the Nrf2 and NF-KB/NLRP3 Signaling Pathways and CYP1A1/2 Expressions. *Agriculture* **2022**, *12*, 54. [CrossRef]
110. Dinkova-Kostova, A.T.; Talalay, P. NAD(P)H:Quinone Acceptor Oxidoreductase 1 (NQO1), a Multifunctional Antioxidant Enzyme and Exceptionally Versatile Cytoprotector. *Arch. Biochem. Biophys.* **2010**, *501*, 116–123. [CrossRef]
111. Yu, R.; Mandlekar, S.; Lei, W.; Fahl, W.E.; Tan, T.H.; Kong, A.N.T. P38 Mitogen-Activated Protein Kinase Negatively Regulates the Induction of Phase II Drug-Metabolizing Enzymes That Detoxify Carcinogens. *J. Biol. Chem.* **2000**, *275*, 2322–2327. [CrossRef]
112. Yueh, M.-F.; Tukey, R.H. Nrf2-Keap1 Signaling Pathway Regulates Human UGT1A1 Expression in Vitro and in Transgenic UGT1 Mice. *J. Biol. Chem.* **2007**, *282*, 8749–8758. [CrossRef] [PubMed]
113. Gleich, A.; Kaiser, B.; Schumann, J.; Fuhrmann, H. Establishment and Characterisation of a Novel Bovine SV40 Large T-Antigen-Transduced Foetal Hepatocyte-Derived Cell Line. *Vitr. Cell. Dev. Biol.—Anim.* **2016**, *52*, 662–672. [CrossRef] [PubMed]
114. Robinson, M.D.; McCarthy, D.J.; Smyth, G.K. EdgeR: A Bioconductor Package for Differential Expression Analysis of Digital Gene Expression Data. *Bioinformatics* **2010**, *26*, 139–140. [CrossRef] [PubMed]
115. Yu, G.; Wang, L.G.; Han, Y.; He, Q.Y. ClusterProfiler: An R Package for Comparing Biological Themes among Gene Clusters. *OMICS J. Integr. Biol.* **2012**, *16*, 284–287. [CrossRef]
116. Subramanian, A.; Tamayo, P.; Mootha, V.K.; Mukherjee, S.; Ebert, B.L.; Gillette, M.A.; Paulovich, A.; Pomeroy, S.L.; Golub, T.R.; Lander, E.S.; et al. Gene Set Enrichment Analysis: A Knowledge-Based Approach for Interpreting Genome-Wide Expression Profiles. *Proc. Natl. Acad. Sci. USA* **2005**, *102*, 15545–15550. [CrossRef]
117. Pauletto, M.; Giantin, M.; Tolosi, R.; Bassan, I.; Barbarossa, A.; Zaghini, A.; Dacasto, M. Curcumin Mitigates Afb1-Induced Hepatic Toxicity by Triggering Cattle Antioxidant and Anti-Inflammatory Pathways: A Whole Transcriptomic In Vitro Study. *Antioxidants* **2020**, *9*, 1059. [CrossRef]
118. Zancanella, V.; Giantin, M.; Lopparelli, R.M.; Nebbia, C.; Dacasto, M. Constitutive Expression and Phenobarbital Modulation of Drug Metabolizing Enzymes and Related Nuclear Receptors in Cattle Liver and Extra-Hepatic Tissues. *Xenobiotica* **2012**, *42*, 1096–1109. [CrossRef]
119. Pauletto, M.; Giantin, M.; Tolosi, R.; Bassan, I.; Barbarossa, A.; Zaghini, A.; Dacasto, M. Discovering the Protective Effects of Resveratrol on Aflatoxin B1-Induced Toxicity: A Whole Transcriptomic Study in a Bovine Hepatocyte Cell Line. *Antioxidants* **2021**, *10*, 1225. [CrossRef]

toxins

Article

Patulin Alters Insulin Signaling and Metabolic Flexibility in HepG2 and HEK293 Cells

Yashodani Pillay *, Savania Nagiah † and Anil Chuturgoon *

Discipline of Medical Biochemistry and Chemical Pathology, Faculty of Health Sciences, George Campbell Building, Howard College, University of KwaZulu-Natal, Durban 4041, South Africa
* Correspondence: yashodani.pillay@bcchr.ca (Y.P.); chutur@ukzn.ac.za (A.C.)
† Current address: Department of Human Biology, Medical School, Nelson Mandela University, Port Elizabeth 6019, South Africa.

Abstract: Non-communicable diseases (NCDs) have risen rapidly worldwide, sparking interest in causative agents and pathways. Patulin (PAT), a xenobiotic found in fruit products contaminated by molds, is postulated to be diabetogenic in animals, but little is known about these effects in humans. This study examined the effects of PAT on the insulin signaling pathway and the pyruvate dehydrogenase complex (PDH). HEK293 and HepG2 cells were exposed to normal (5 mM) or high (25 mM) glucose levels, insulin (1.7 nM) and PAT (0.2 μM; 2.0 μM) for 24 h. The qPCR determined gene expression of key enzymes involved in carbohydrate metabolism while Western blotting assessed the effects of PAT on the insulin signaling pathway and Pyruvate Dehydrogenase (PDH) axis. Under hyperglycemic conditions, PAT stimulated glucose production pathways, caused defects in the insulin signaling pathway and impaired PDH activity. These trends under hyperglycemic conditions remained consistent in the presence of insulin. These findings are of importance, given that PAT is ingested with fruit and fruit products. Results suggest PAT exposure may be an initiating event in insulin resistance, alluding to an etiological role in the pathogenesis of type 2 diabetes and disorders of metabolism. This highlights the importance of both diet and food quality in addressing the causes of NCDs.

Keywords: insulin resistance; kidney; liver; metabolic flexibility; patulin

Key Contribution: The key finding in this study is that exposure to patulin impairs insulin signaling under hyperglycemic conditions and alters metabolic flexibility after 24 h exposure in an in vitro model. This finding was consistent following insulin stimulation; suggestive of an insulin-resistant mechanism.

Citation: Pillay, Y.; Nagiah, S.; Chuturgoon, A. Patulin Alters Insulin Signaling and Metabolic Flexibility in HepG2 and HEK293 Cells. *Toxins* **2023**, *15*, 244. https://doi.org/10.3390/toxins15040244

Received: 27 February 2023
Revised: 22 March 2023
Accepted: 25 March 2023
Published: 27 March 2023

Copyright: © 2023 by the authors. Licensee MDPI, Basel, Switzerland. This article is an open access article distributed under the terms and conditions of the Creative Commons Attribution (CC BY) license (https://creativecommons.org/licenses/by/4.0/).

1. Introduction

Non-communicable diseases (NCDs), including diabetes and cancer, have quickly become the leading cause of mortality worldwide, with disproportionately higher impacts on public health in developing countries [1]. This has raised interest in causative agents, diet and nutritional guidelines. Mycotoxins are secondary metabolites produced by fungi. These potentially damaging xenobiotics are found in food, with a higher incidence in developing countries [2,3].

Patulin (PAT) is a mycotoxin produced by *Penicillium, Bissochlamys* and *Aspergillus* sp. [4]. These molds are found in overripe or rotting fruit and apple products. A safety level of 50 μg/L PAT in consumables was established following evidence of adverse effects in exposed humans and animals [3,5]. Despite this regulation and the assertion that PAT levels are a measure of poor product quality, there are vast variations in PAT concentrations in apple products worldwide [5]. PAT induces oxidative stress, compromises mitochondrial function and causes cell death via depletion of antioxidant glutathione (GSH) [6–11]. PAT is also associated with changes in lipid metabolism and inflammation [2,12]. These deleterious effects are most commonly observed in organs of high metabolic demand—the kidney, gastrointestinal tract, liver and brain [3,13].

Energy demand and glucose homeostasis are typically maintained through the catabolism of glucose to pyruvate during glycolysis. Pyruvate is subsequently taken up by the mitochondria and broken down in the citric acid (TCA) cycle; subsequently, this facilitates the transfer of electrons to the mitochondrial respiratory chain for ATP production via oxidative phosphorylation. Under hypoxic and hypoglycemic conditions, however, these processes are impaired. These conditions divert metabolic operations to produce glucose through gluconeogenesis and glycogenolysis, resulting in ATP production via anaerobic glycolysis [14]. This metabolic environment is potentially simulated by PAT exposure which has been linked to glycogen breakdown, gluconeogenesis, impaired mitochondrial function and ATP levels in previous studies [15].

Metabolic flexibility is determined by an organism's ability to adapt fuel oxidation processes to nutrient availability and rapidly switch between them. Pyruvate Dehydrogenase (PDH), a multienzyme complex stimulated by insulin catalyzes the conversion of pyruvate to acetyl-CoA, connecting the TCA cycle to glucose and fatty acid metabolism. PDH activity is thus a central component of metabolic flexibility. The pyruvate dehydrogenase kinases (PDKs) inhibit PDH activity through phosphorylation on the PDH E1α subunit under hypoglycemic and hypoxic conditions. PDK elevation is hence associated with gluconeogenesis, metabolic inflexibility and has been implicated in the etiology of type 2 diabetes [16,17].

Glucose uptake in mammalian cells is facilitated by a family of glucose transporters (GLUTs). GLUT2, expressed in the intestine, liver, kidney and pancreatic beta cells, is responsible for glucose sensing and uptake. Loss of GLUT2 expression leads to organ damage, persistent hyperglycemia and diabetes [18]. Insulin promotes the uptake, utilization and storage of glucose by stimulating glycolytic enzyme phosphofructokinase (PFK) and repressing gluconeogenesis and glycogenolysis enzymes, phosphoenolpyruvate carboxykinase (PCK) and glycogen phosphorylase (PYGL) [19]. Insulin activity is initiated by the binding and autophosphorylation of tyrosine residues on the insulin receptor (IR). The activated receptor phosphorylates the insulin receptor substrate 1 (IRS-1). Phosphorylated IRS-1 (pIRS-1) recruits phosphatidylinositide (PI) 3-kinase (PI3K) (p85/p110 subunits) to the plasma membrane which stimulates PI3K-mediated phosphorylation of Akt. Glycogen synthesis kinase (GSK-3) is phosphorylated and inactivated by pAkt; relieving its inhibitory phosphorylation of glycogen synthase (GS), promoting glycogen synthesis. pIRS-1 activates the extracellular signal-regulated kinases (ERK) and mitogen-activated protein kinases (MAPK) pathway in parallel which can be used as a marker for the activation of proximal insulin signaling [20,21]. Underlying insulin resistance in type 2 diabetes has been attributed to defects in one or more components of the insulin signaling cascade [22]. Thus, while previous studies have shown PAT causes changes in PI3K/Akt and ERK/MAPK signaling, little is understood about the metabolic implications thereof [23].

Elevated glucose levels resulting from hepatic glucose production are a feature of type 2 diabetes. Insulin resistance precedes increased PCK and PYGL expression causing glucose efflux and hyperglycemia [19]. Previous studies in animals have referred to PAT as diabetogenic describing elevated glucose levels, depleted glycogen stores and mitochondrial dysfunction following PAT exposure, but mechanistic data in human models are lacking [15,24,25]. PAT toxicity has been described previously in the liver and kidney—both of which have high metabolic demands and are regulators in glucose homeostasis [11,17,26,27]. This study aimed to determine the effects of PAT on insulin signaling and metabolic flexibility in HepG2 and HEK293 cells.

2. Results

2.1. PAT Alters Expression of Enzymes Involved in Glucose Generation

Previous studies have alluded to PAT as a diabetogenic agent [15]. In a preliminary screening, HEK293 and HepG2 cells were exposed to a high glucose (25 mM) medium and PAT for 24 h. qPCR was then used to measure gene expression of enzymes vital to glucose homeostasis.

PFK-2, a key enzyme in glycolysis, was not altered by PAT in HEK293 cells (Figure 1A; $p = 0.7270$) but decreased significantly in HepG2 cells (Figure 1B; $p = 0.0231$; 10-fold). Although overall statistical significance was found for glycogenolysis initiator, *PYGL*, in both cell lines (HEK293 (Figure 1A; $p = 0.0137$)), post-hoc tests found that only the distinct two-fold increase in HepG2 cells was significant when compared to the control (Figure 1B; $p = 0.0218$)]. *PCK-1*, a rate limiting enzyme in gluconeogenesis, was significantly altered by PAT at 2 μM in both cell lines, though trends differed between them (HEK293 cells (Figure 1A; $p = 0.0112$) and HepG2 cells (Figure 1B; $p = 0.0214$)). This outcome provides a preliminary indication that PAT caused a shift from glycolysis to gluconeogenesis and glycogenolysis in HepG2 cells with HG media. Most significant changes were observed at 2 μM PAT–above the established safety concentration for PAT in consumables.

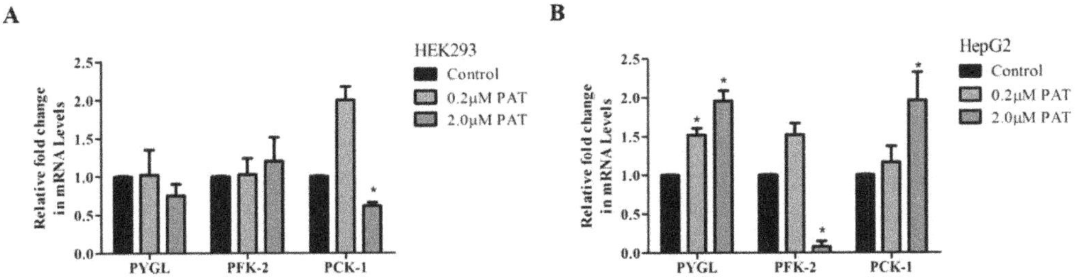

Figure 1. PAT alters expression of enzymes involved in glucose production in hyperglycemic conditions. qPCR was used to assess expression of glucose homeostasis enzymes *PYGL, PFK-1* and *PCK-1* in (**A**) HEK239 and (**B**) HepG2 cells exposed to 25 mM glucose and PAT for 24 h (* $p < 0.05$ relative to control).

2.2. Glucose Availability Determines PAT-Associated Insulin Signaling Defects in HepG2 Cells

IR activation triggers phosphorylation of the receptor and IRS-1. Under NG conditions, PAT stimulated IRS-1 phosphorylation (pIRS-1) (Figure 2A; $p = 0.0388$), though pIR showed no statistical change (Figure 2A; $p = 0.0775$). This trend was potentiated in the presence of insulin shown by a significant two-fold increase in pIR (Figure 2B; $p = 0.0180$) and pIRS-1 (Figure 2B; $p = 0.0145$) in PAT-exposed NG i+ HepG2 cells, exhibiting a similar effect to the positive control, metformin.

This observation was dramatically reversed under hyperglycemic conditions. PAT caused significant two-fold decreases in both pIR (Figure 2C; $p = 0.0178$) and pIRS-1 (Figure 2C; $p = 0.0213$) in HG HepG2 cells. This result was consistent following insulin pathway stimulation (HG i+ HepG2) where both pIR (Figure 2D; $p = 0.0182$) and pIRS-1 (Figure 2D; $p = 0.0188$) were decreased in PAT treatments. These findings are suggestive of defects in insulin signaling under HG conditions following PAT exposure.

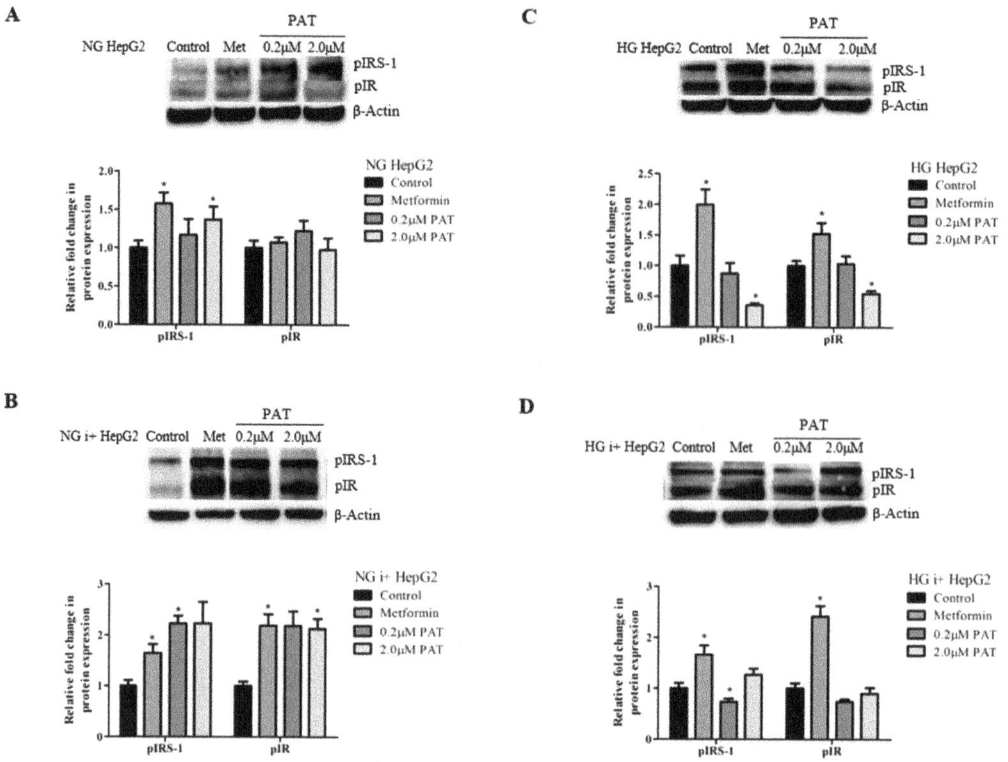

Figure 2. PAT alters insulin signaling in HepG2 cells. Western blotting determined PAT-increased pIR and pIRS-1 in NG media (**A**) and under insulin stimulation (NG i+) (**B**). This effect was reversed in HG media (**C**) decreased pIRS-1 and pIR expression was observed which remained consistent under insulin-stimulated hyperglycemic conditions (HG i+) (**D**) (* $p < 0.05$ relative to control).

2.3. *PAT-Altered ERK/MAPK Signaling Is Influenced by Glucose and Insulin Availability*

Insulin-activated ERK phosphorylation is connected to the downstream regulation of transcription factors involved in metabolic homeostasis. Under NG conditions, PAT increased ERK activation in both HEK293 and HepG2 cells evidenced by increased pERK relative to total ERK (NG HEK293 (Figure 3A; pERK: $p = 0.00610$; ERK; $p = 0.0032$); NG HepG2 (Figure 3C; pERK: $p = 0.0072$; ERK: $p = 0.0788$)).

Conversely, under HG conditions, PAT-exposed HEK293 cells significantly decreased both pERK (Figure 3B; $p = 0.0221$) and total ERK (Figure 3B; $p = 0.0053$) expression. This observation was consistent in HG HepG2 cells (Figure 3D; pERK: $p = 0.0104$ and ERK: $p = 0.0149$). Interestingly, the effects of PAT on ERK signaling in HepG2s under both NG and HG were neutralized in the presence of insulin. No significant effects on ERK activation were observed in PAT-exposed NG i+ HepG2 (Figure 3E; pERK: $p = 0.0630$ and ERK: $p = 0.0399$) or HG i+ HepG2 (Figure 3F; pERK: $p = 0.2027$; and ERK: $p = 0.0920$) treatments.

PAT-altered ERK signaling trends were dissimilar to positive control metformin results, indicating a potential alternate mechanism of action. These data suggest that PAT stimulates ERK/MAPK signaling under NG conditions and inhibits signaling under HG conditions. These effects can, however, be counteracted by stimulating the insulin signaling pathway (Figure 3).

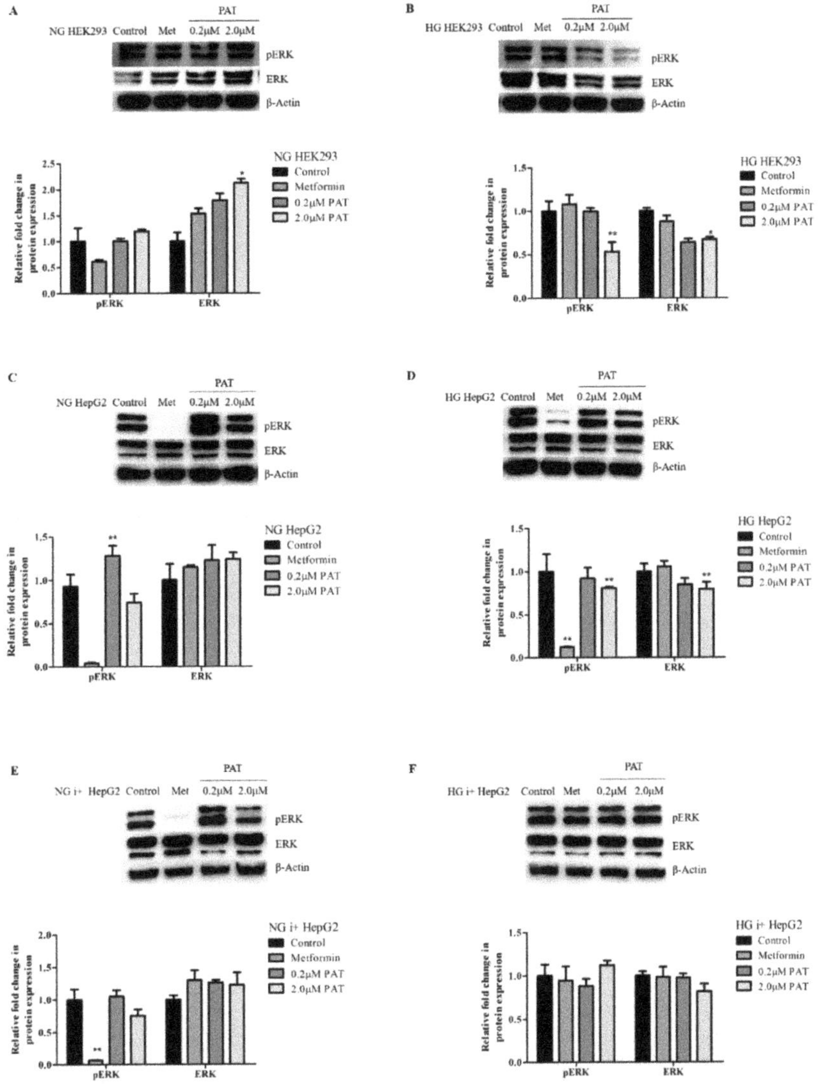

Figure 3. PAT alters ERK/MAPK according to glucose and insulin availability. Western blotting determined PAT caused a significant increase in ERK1/2 phosphorylation activation relative to total ERK expression in NG HEK293 (**A**) and NG HepG2 (**C**) cells. This was reversed in HG HEK293 (**B**) and HG HepG2 (**D**) cells with decreases in pERK and ERK following 24 h PAT exposure. Both trends were neutralized in the presence of insulin (**E**,**F**) (** $p < 0.01$, * $p < 0.05$ relative to control).

2.4. Glucose Availability Affects PAT-Mediated PI3K/Akt Signaling Trends

Insulin-stimulated PI3K activity directs the phosphorylation and activation of downstream target Akt. In PAT-treated NG HEK293 cells, PI3K (Figure 4A; $p = 0.0051$) and pAkt (Figure 4A; $p = 0.0329$) were significantly decreased while total inactive Akt (Figure 4A; $p = 0.0160$) increased. However, in NG HepG2 and NG i+ HepG2 cells PAT significantly increased the PI3K/Akt pathway activation evidenced by increased PI3K and pAkt expression relative to total Akt (NG HepG2 (Figure 4C; PI3K: $p = 0.1431$; pAkt: $p = 0.0043$; Akt: $p = 0.0027$); NG i+ HepG2 (Figure 4E; PI3K: $p = 0.0112$; pAkt: $p = 0.0249$; Akt: $p = 0.0237$)).

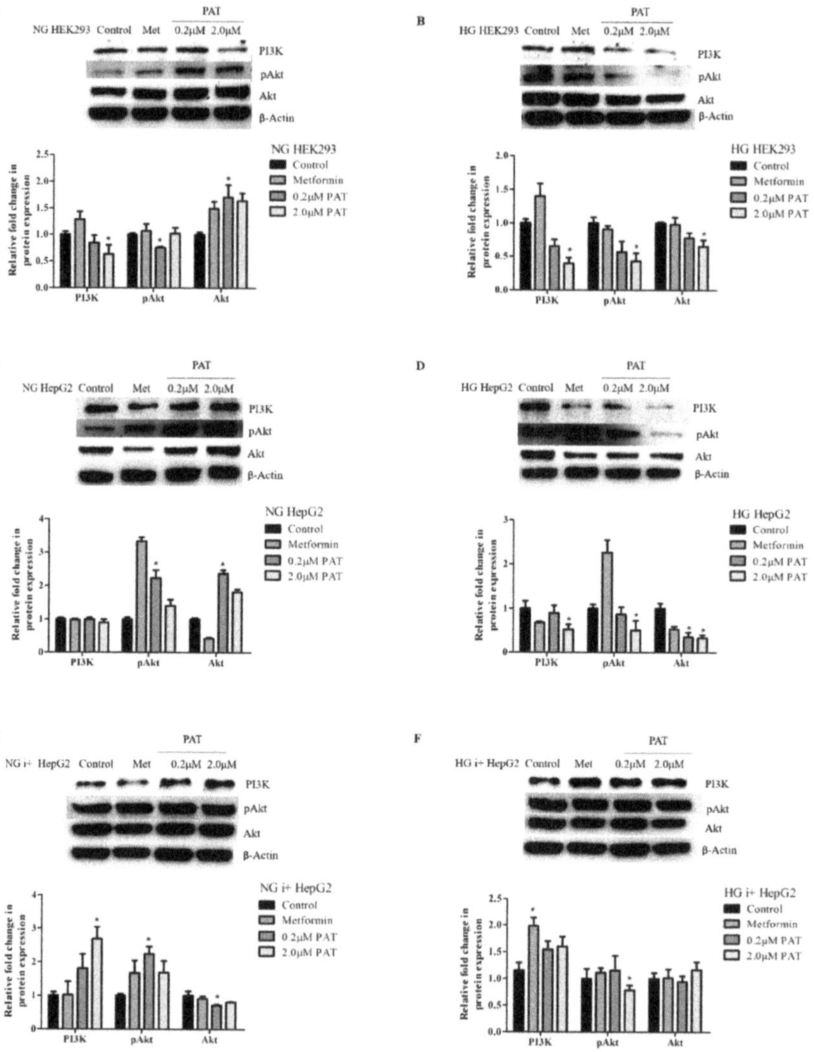

Figure 4. PAT alters PI3K/Akt signaling according to glucose availability irrespective of insulin action. Western blotting established PAT significantly increased PI3K activation and Akt phosphorylation activation relative to total Akt expression in NG HEK293 (**A**) and NG HepG2 (**C**) cells. This was reversed in HG HEK293 (**B**) and HG HepG2 (**D**) cells with decreases in PI3K, pAkt following 24 h PAT exposure. Both trends were maintained in the presence of insulin (**E**,**F**) (* $p < 0.05$ relative to control).

Conversely, PI3K/Akt pathway activation significantly decreased under all PAT HG conditions (HG HEK293 (Figure 4B; PI3K, $p = 0.0043$; pAkt, $p = 0.0072$; Akt, $p = 0.0092$); HG HepG2 (Figure 4D; PI3K, $p = 0.0219$; pAkt, $p = 0.0115$; Akt, $p = 0.0053$); HG i+ HepG2 (Figure 4F; PI3K: $p = 0.0160$; pAkt: $p = 0.0415$; Akt: $p = 0.1681$)).

These results show PAT stimulated PI3K/Akt signaling under NG conditions and inhibited PI3K/Akt activation under HG conditions despite insulin action. This indicated a possible PAT-induced defect in response to elevated glucose levels and insulin-stimulated signaling.

2.5. PAT-Modified GSK-3 Activation Corresponded with PI3K/Akt Signaling Trends While GLUT2 Expression Was Widely Compromised by PAT

Active GSK-3 phosphorylates and inhibits GS, preventing glycogen synthesis. Insulin action phosphorylates and inhibits GSK-3 (pGSK-3) via PI3K/Akt signaling, which enhances glycogen synthesis. Under NG conditions, PAT significantly increased pGSK-3 and decreased total active GSK-3 in both HEK293 and HepG2 cells suggestive of decreased GSK activity (NG HEK293 (Figure 5A; pGSK-3: $p = 0.0073$; GSK-3: $p = 0.0488$); NG HepG2 (Figure 5C; pGSK-3: $p = 0.0415$; GSK-3: $p = 0.0216$)).

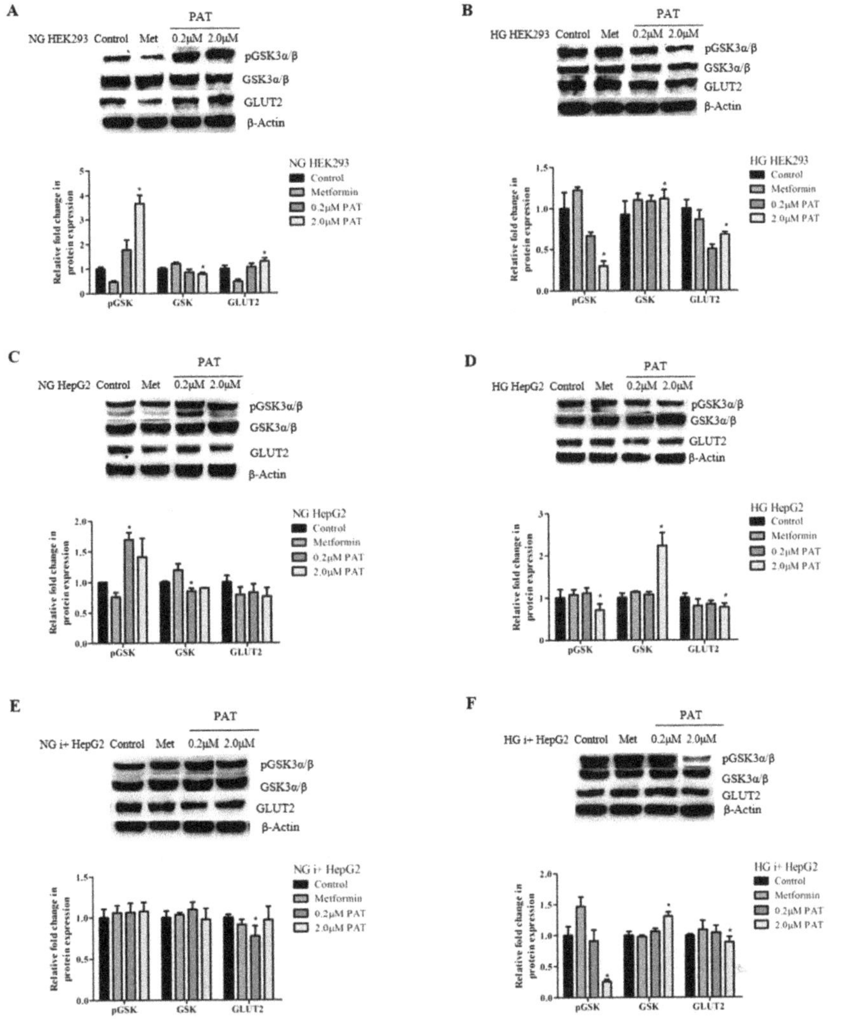

Figure 5. PAT-modified GSK-3 activation corresponded with PI3K/Akt signaling trends while GLUT2 expression was widely compromised by PAT. Western blotting established PAT significantly increased GSK-3 inhibition by phosphorylation relative to total GSK-3 expression with no change GLUT2 in NG HEK293 (**A**) and NG HepG2 (**C**) cells. This was reversed in HG HEK293 (**B**) and HG HepG2 (**D**) cells with decreases in GLUT2, pGSK-3, and increased GSK-3 following 24 h PAT exposure. NG i+ cells (**E**) exposed to PAT showed no significant changes while HG trends were maintained in the presence of insulin (**F**) (* $p < 0.05$ relative to control).

In HG, the media effects were reversed, exhibiting GSK-3 activation in both cell lines. In HG HEK293 cells, inactive pGSK-3 decreased (Figure 5B; $p = 0.0158$) and total active GSK-3 increased (Figure 5C; $p = 0.0216$). This effect was more pronounced in PAT-treated HG HepG2 cells (Figure 5D; pGSK-3: $p = 0.0158$; GSK-3: $p = 0.0149$)—indicative of compromised glycogen synthesis. When stimulated with insulin, inactive pGSK-3 remained decreased in HG i+ HepG2 cells (Figure 5F; $p = 0.0056$). No changes were noted in NG i+ HepG2 cells in either pGSK-3 (Figure 5E; $p = 0.3815$) or GSK-3 (Figure 5E; $p = 0.4969$). These results suggest PAT caused defects in the GSK-3 signaling axis under HG conditions.

GLUT2, involved in glucose sensing and uptake, is linked to impaired glucose-stimulated insulin signaling when suppressed. PAT did not alter GLUT2 expression under NG conditions in both HEK293 and HepG2 cells (NG HEK293 (Figure 5A; $p = 0.0556$); NG HepG2 (Figure 5C; $p = 0.0559$)). In the presence of glucose and insulin, however, PAT significantly decreased GLUT2 expression in both cell lines (NG i+ HepG2 cells (Figure 5E; $p = 0.0496$); HG HEK293 (Figure 5B; $p = 0.0048$); HG HepG2; (Figure 5D; $p = 0.0077$); HG i+ HepG2 cells (Figure 5F; $p = 0.0281$)). This suggests PAT impaired GLUT2-mediated glucose sensing and uptake under conditions of increased glucose and insulin availability.

2.6. *PAT Contributes to Metabolic Inflexibility by PDK-1 Elevation and PDH Inhibition under NG and HG Conditions*

Active PDH, stimulated by insulin, catalyzes the conversion of pyruvate to acetyl-CoA. Inhibition of PDH is catalyzed by PDK-mediated phosphorylation on the E1α subunit.

PAT did not alter PDK-1 and PDH E1α expression in HEK293 cells (NG (Figure 6A; PDK-1: $p = 0.1390$; PDH E1α: $p = 0.2220$); HG (Figure 6B; PDK-1: $p = 0.2223$; PDH E1α: $p = 0.1417$)). pPDH E1α expression remained unchanged in NG HEK293 cells (Figure 6A; $p = 0.3588$) but decreased in HG HEK293 cells (Figure 6B; $p = 0.0221$).

PAT-mediated effects on this pathway were more distinct in HepG2 cells. PDK-1 expression significantly increased in PAT-treated HepG2 cells under all conditions (NG (Figure 6C; $p = 0.0060$), HG (Figure 6D; $p = 0.0186$), NG i+ (Figure 6E; $p = 0.0048$), HG i+ (Figure 6F; $p = 0.0081$)). An associated increase in (inactive) pPDH E1α (NG (Figure 6C; $p = 0.0045$), HG (Figure 6D; $p = 0.0041$), NG i+ (Figure 6E; $p = 0.0041$) and HG i+ (Figure 6F; $p = 0.0179$)) with a concomitant decrease in PDH E1α (NG (Figure 6C; $p = 0.0060$), HG (Figure 6D; $p = 0.0055$), NG i+ (Figure 6E; $p = 0.0032$) and HG i+ (Figure 6F; $p = 0.0124$)) was observed under these conditions in PAT treatments. This was indicative of PDK-1-mediated PDH inhibition in PAT-treated HepG2 cells.

Elevated pPDH E1α suppresses aerobic pyruvate oxidation and can aggravate the Warburg effect. This pathway, characterized predominantly in cancerous cells, favors the anaerobic conversion of pyruvate to lactate. This may explain the distinct differences in PDH E1α results between HepG2 (cancerous) and HEK293 (non-cancerous) cells. Interestingly, lactate levels increased significantly in controls under all HG conditions relative to NG (Table 1; (HEK293; $p = 0.0006$) (HepG2; $p = 0.0013$) (HepG2 i+; $p = 0.0190$)). The highly significant increase observed in HEK293 cells relative to HepG2 cells may be related to metabolic differences in cancerous and non-cancerous cells as well as the pH sensitivity of HEK293 cells. Interestingly, a significant decrease in lactate levels was observed in PAT treatments relative to the controls in NG treatments and HEK293 cells (Table 1; NG HEK293 ($p = 0.0222$); HG HEK293 ($p = 0.0069$); NG HepG2 ($p = 0.0142$); NG i+ ($p = 0.0069$)). HG HepG2 and HG i+ HepG2 treatments however, (Table 1; NG HepG2 ($p = 0.4336$); NG i+ ($p = 0.7427$)) were not statistically changed. This lack of lactate elevation despite changes to the PDH axis in PAT treatments suggests pyruvate was possibly shunted back towards gluconeogenesis or to an alternative fate instead of reduction to lactate by LDH.

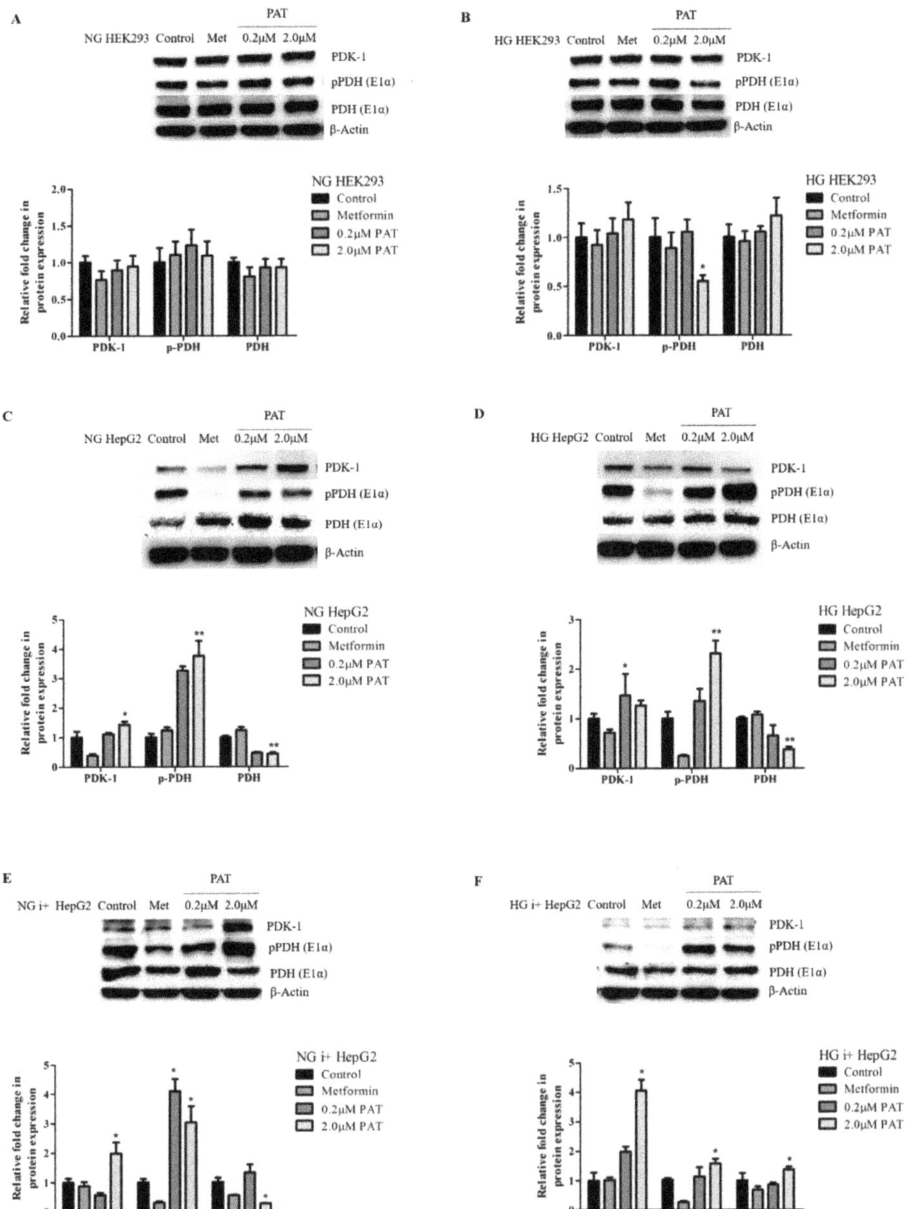

Figure 6. PAT contributes to metabolic inflexibility by PDK-1 elevation and PDH inhibition under NG and HG conditions. Western blotting established PAT caused no significant differences in PDK-1, pPDH E1α and PDH E1α in NG HEK293 (**A**) and HG HEK293 (**B**) cells. PDK-1 was significantly increased with associated changed in pPDH E1α and PDH E1α in NG HepG2 (**C**), HG HepG2 (**D**) NG i+ HepG2 (**E**) and HG i+ HepG2 (**F**) cells following 24 h PAT exposure. (** $p < 0.01$, * $p < 0.05$ relative to control).

Table 1. Extracellular lactate levels (mean ± SD) measured following 24 h PAT exposure in HepG2 and HEK293 cells.

Treatment	Lactate Levels (mmol/L) Normoglycemic Media (NG)	Lactate Levels (mmol/L) Hyperglycemic Media (HG)
HEK293 Control	8.85 ± 0.1306	17.78 ± 0.07348
HEK293 0.2 μM PAT	7.77 ± 0.1796	12.92 ± 0.1878
HEK293 2.0 μM PAT	7.75 ± 0.3103	15.97 ± 0.1470
HepG2 Control	3.36 ± 0.2531	5.58 ± 0.2368
HepG2 0.2 μM PAT	2.64 ± 0.1551	5.35 ± 0.3101
HepG2 2.0 μM PAT	3.12 ± 0.1796	5.17 ± 0.8410
HepG2 (i+) Control	3.15 ± 0.1388	3.63 ± 0.07348
HepG2 (i+) 0.2 μM PAT	2.79 ± 0.1470	3.69 ± 0.4736
HepG2 (i+) 2.0 μM PAT	2.40 ± 0.04899	3.68 ± 0.2286

3. Discussion

The key finding in this study is that exposure to PAT impairs insulin signaling under HG conditions and alters metabolic flexibility after 24 h exposure in an in vitro model. This finding was consistent following insulin stimulation, suggestive of an insulin-resistant mechanism.

A recent study showed PAT was cytotoxic to rat pancreatic β cells but did not affect insulin production [24,25] prompting examination of the insulin signaling cascade in this study. The IR, a tyrosine kinase, undergoes autophosphorylation catalyzing the phosphorylation and activation of downstream cellular proteins including IRS-1, the PI3K/Akt pathway and ERK/MAPK pathway. These pathways act in concert to positively regulate glycolysis, glucose storage as glycogen and lipids and repress glucose synthesis and release by inhibiting glycogenolysis and gluconeogenesis. Insulin action is attenuated by dephosphorylation of the receptor and its substrates. Phospho-tyrosine residues indicated PAT activated the insulin signaling cascade under NG conditions but suppressed activation under HG conditions in HepG2 cells (Figure 2). Defects in this pathway have been linked to impaired glucose tolerance and the pathogenesis of type 2 diabetes [22,28].

The ERK/MAPK pathway is activated in a stepwise sequence within the insulin signaling cascade. pERK catalyzes the activation of transcription factors to initiate cell proliferation. Inhibition of this pathway observed in PAT exposed HG HepG2 and HEK293 cells (Figure 3) prevents insulin-stimulated cell growth. PAT-induced alterations in ERK signaling and proliferation have been reported in the literature, but metabolic consequences have been neglected [4,23]. The findings in this study imply PAT-mediated ERK changes are dependent on insulin and glucose availability (Figure 3). This may be of consequence to glucose metabolism under conditions of PAT-inflicted β-cell death [24].

Proximal insulin signaling activates the PI3K/Akt pathway in parallel to the ERK/MAPK pathway. pIRS-1 recruits PI3K to activate Akt by phosphorylation. pAkt-mediated phosphorylation of GSK-3 (ser 9) relieves GS inhibition and promotes glycogen synthesis. Under NG conditions, PAT-activated pIRS-1 (Figure 2), PI3K/Akt activation (Figure 4) and increased pGSK-3 (Figure 5) in HepG2 cells positively regulating glycogen synthesis. Under HG conditions, however, PAT diminished pIRS-1 (Figure 2) in HepG2 cells and PI3K/Akt (Figure 4) mediated GSK-3 (Figure 5) phosphorylation in both HepG2 and HEK293 cells. These findings reveal distinct differences in PAT-mediated signaling depending on glucose availability. PAT-altered PI3K/Akt signaling has been reported previously in other cell lines [23]; the findings in this study, however, reveal novel metabolic implications. These trends identified in HepG2 cells were observed consistently in both the presence and absence of insulin, implicating a non-insulin dependent mechanism. Elevated total GSK-3 is associated with impaired glucose disposal and glycogen synthesis and type 2 diabetes [20]. Similar dysfunctions in reduced pIRS-1/PI3K/Akt signaling as observed in this study, have been associated with insulin resistance and impaired glucose tolerance [21,29].

Previous studies describe PAT-inflicted glycogen loss and serum glucose elevation in animals, postulating that PAT increases gluconeogenesis and glycogenolysis [15,30]. In this study, active GSK-3, a prominent feature in glycogenolysis, is elevated by PAT under HG conditions in both HEK293 and HepG2 cell lines (Figure 5). PAT also increased the gene expression of glucose-producing enzymes *PYGL* and *PCK-1* and inhibited glycolytic *PFK-1* (Figure 1) under the same conditions in HepG2s. These classic indicators of glycogenolysis and gluconeogenesis, usually inhibited by insulin and HG conditions [28], were increased by PAT under the same conditions. Glucose homeostasis relies on the detection of varying glucose availability. GLUT2, expressed in the kidney and liver, plays a crucial role in glucose sensing and uptake. GLUT2 suppression, induced by PAT in both HEK293 and HepG2 cells (Figure 5), is associated with cellular glucose efflux, impaired activity of glucose sensitive genes, reduced glucose uptake and promoting type 2 diabetes pathogenesis and organ damage [18,31]. While organ damage has been described in previous findings on PAT, this novel finding offers a possible explanation for the increased gluconeogenic markers, impaired glycogen synthesis and glycolysis in PAT treatments in this study.

These findings indicate that PAT compromised cellular metabolic processes in adaption to glucose availability. PDH, the principal enzyme controlling nutrient adaption and metabolic flexibility, is activated by insulin and glucose via dephosphorylation on the E1α subunit. PDH metabolizes pyruvate at the junction between glucose, fatty acid metabolism and the TCA cycle [17]. PDK-1 phosphorylates and inhibits PDH E1α under hypoxic conditions, shifting metabolism to glycolytic processes [32]. PAT significantly increased PDK-1 and pPDH E1α expression in HepG2 cells (Figure 6). In cancerous cells such as HepG2s, this is associated with aggravation of the Warburg effect. The Warburg effect posits cancerous cells inhibit mitochondrial ATP production and increase rates of glycolysis and cytoplasmic conversion of pyruvate to lactate [33,34]. This pathway is linked to glucose scarcity, synthesis and inhibition of pyruvate conversion to acetyl-CoA. Mitochondrial inhibition in cancerous cells is achieved by PDK and pPDH elevation. These actions block excessive mitochondrial ROS production, linked extensively to PAT in previous studies [14]. Thus, PAT-induced pPDH and PDK-1 elevation may be an energy- and mitochondrial conservation strategy [4,10,11]. In addition, lactate levels were decreased in PAT treatments (Table 1), despite PDK-1 and pPDH E1α (Figure 6) elevation. While the lower levels of lactate in i+ treatments suggest insulin may have stimulated mitochondrial function, the high levels in the HEK293 control cells may explain the potent toxicity of PAT in the cell line as PAT loses biological activity in alkaline media [35,36]. Taken together, the results imply pyruvate did not undergo lactic acid fermentation as suggested in the Warburg effect theory. Alternate fates of pyruvate under PDH inhibition include the TCA cycle, fatty acid synthesis and gluconeogenesis [17]. While there is currently insufficient evidence in this study and the literature to conclusively rule out the TCA cycle or fatty acid synthesis these findings, together with gluconeogenic enzyme mRNA levels (Figure 1), GLUT2 and GSK-3 expression (Figure 5), infer pyruvate was shunted back toward gluconeogenesis despite the availability of glucose and insulin. Decreased PDH and metabolic inflexibility are initiating events in impaired glucose oxidation, associated with insulin resistance and type 2 diabetes [14,16,37–39].

Type 2 diabetes is characterized by elevated glucose levels, impaired glucose tolerance, insulin resistance, defects in the insulin signaling pathway, oxidative stress and mitochondrial dysfunction [22]. The literature shows PAT causes oxidative stress and mitochondrial dysfunction which is associated with both causes and consequences of insulin resistance [10,11,26,40]. This study presents novel findings indicating PAT opposes insulin action under HG conditions, suppresses components of the insulin signaling pathway and prevents metabolic flexibility in HepG2 and HEK293 cells. Given that PAT targets the kidney and liver, and the role of these organs in systemic glucose and energy homeostasis, this has strong implications for PAT-induced toxicology and disorders of metabolism.

4. Conclusions

Our findings show PAT caused defects in the insulin signaling pathway under HG conditions, stimulated glucose production pathways and inhibited PDH activity, contributing to metabolic inflexibility. Collectively, these results suggest PAT exposure may be an initiating event in insulin resistance with an etiological role in pathogenesis of type 2 diabetes and other disorders of metabolism. Significant results were observed above and below the safety level of PAT.

5. Materials and Methods

HEK293 and HepG2 cells were obtained from the American Type Culture Collection (Johannesburg, South Africa). Culture reagents were purchased from Lonza BioWhittaker (Basel, Switzerland). Western Blotting and qPCR reagents and consumables were purchased from Bio-Rad (Hercules, CA, USA). Patulin (P1639) was purchased from Sigma-Aldrich (St. Louis, MO, USA). All other reagents were obtained from Merck (Darmstadt, Germany) unless otherwise stated.

5.1. Cell Culture

HepG2 cells were cultured in complete culture medium (CCM) containing Eagle's minimum essential medium (EMEM) supplemented with 2 mM l-glutamine, 1% pen-strep-fungizone (500 units potassium penicillin, 500 µg streptomycin, 1.25 µg amphotericin B/5 mL flask) and 10% fetal bovine serum in 25 cm^3 flasks at 37 °C. HepG2 cells were derived from hepatocellular carcinoma. Despite the limitations of being a cancerous line (relative to primary human hepatocytes), HepG2 cells were used as they exhibit phenotypic stability, genotypic features of normal liver cells and exhibit high glucose consumption and active energy metabolism [41,42]. HEK293 cells were cultured in CCM in comprising Dulbecco's modified essential medium (DMEM) supplemented with 2 mM l-glutamine, 1% pen-strep-fungizone (500 units potassium penicillin, 500 µg streptomycin, 1.25 µg amphotericin B/5 mL flask), 10 mM HEPES and 10% fetal bovine serum in 25 cm^3 flasks at 37 °C; 5% CO_2. HEK293 cells are established from primary human embryonic kidney and have the capacity to switch metabolic processes from glucose consumption and lactate production to glucose–lactate co-consumption [43]. This characteristic makes HEK293 cells an ideal model to study potential changes to glucose metabolism as a result of PAT exposure.

5.2. Dosage Information

PAT (5 mg) was dissolved in 1ml 100% dimethyl sulfoxide (DMSO) (32 mM PAT). Smaller PAT stock solutions were prepared in 0.1 M PBS to a final concentration of 1 mM. When 90% confluent, cells were serum starved for 16 h (h) and then treated with PAT (0.2 µM; 2.0 µM) in 5 mL CCM for 24 h. These PAT concentrations are below and above the safety level of PAT in consumables, respectively, and were determined from incidence and monitor studies [3,5]. These studies indicate that despite a safety level of 50 µg/L (0.3 µM), PAT exposure and levels in consumables can vary from 0.1 µM to 4.0 µM. Cells grown and treated under these conditions (normoglycemic: 5 mM glucose) are referred to as NG in the results.

To determine the effect of PAT on glucose tolerance, once 90% confluent HepG2 and HEK293 cells were serum starved for 16 h and exposed to high glucose CCM (HG) (25 mM) and PAT (0.2 µM; 2.0 µM) for 24 h.

Finally, to determine PAT effect on insulin signaling, 90% confluent HepG2 cells were serum starved for 16 h, exposed to NG (5 mM glucose) or HG (25 mM glucose) media and PAT (0.2 µM; 2.0 µM) for 24 h. Cells were then stimulated with 10 ng/mL (1.7 nM) insulin for 5 min before the media was removed. These are represented by NG i+ and HG i+ in the results. HEK293 cells show a limited insulin signaling profile and were excluded from this parameter in the study.

Metformin (5 mM) was used as a positive control in all protein expression experiments; this concentration was selected from the literature [44].

Results are representative of three independent experiments completed in triplicate.

5.3. Quantitative Polymerase Chain Reaction (PCR)

5.3.1. RNA Isolation and cDNA Preparation

RNA was isolated adding 500 µL QIAzol lysis reagent (Qiagen, Hilden, Germany) to 1×10^6 cells in 500 µL 0.1 M PBS and incubated overnight (−80 °C). Chloroform (100 µL) was then added and samples were centrifuged (15 min, 12,000× *g*, 4 °C). Isopropanol (250 µL) was added to the aqueous phase in a clean tube and incubated overnight (−80 °C). Samples were centrifuged (20 min, 12,000× *g*, 4 °C), supernatant was discarded and pellets were washed with 500 µL cold ethanol and centrifuged (15 min, 7400× *g*, 4 °C) once more. Ethanol was then aspirated; samples were dried and resuspended in 12.5 µL RNase-free water. RNA was quantified using the Nanodrop2000 and standardized to 1000 µg/µL. cDNA was synthesized using Bio-Rad iScript™ reaction mix as per the manufacturer's instructions (iScript™ cDNA Synthesis kit Bio-Rad; 107-8890, Johannesburg, South Africa). Thermocycler conditions were 25 °C for 5 min, 42 °C for 30 min, 85 °C for 5 min and a final hold at 4 °C.

5.3.2. mRNA Quantification

mRNA levels of glucose homeostasis enzymes *PYGL, PFK-1* and *PCK-1* were evaluated using the SsoAdvanced™ Universal SYBR Green Supermix (Bio-Rad; 170-880) according to the manufacturer's instructions. Once synthesized (as above), a reaction volume made up to 25 µL comprising cDNA template, 30 nM sense primer, 30 nM antisense primer, reaction mix and nuclease-free water was made up. Primer sequences and annealing temperatures were as follows *PYGL* [Sense 5′-TGCCCGGCTACATGAATAACA-3′; Antisense 5′-TGTCATTGGGATAGAGACCC-3′; 56.5 °C]; *PCK-1* [Sense 5′-AAAACGGCCTGAACCTCTCG-3′; Antisense 5′-ACACAGCTCAGCGTTATTCTC-3′; 56.5 °C]; *PFK-2* [Sense 5′AGTCCTACG-ACTTCTTTCGGC-3′; Antisense 5′-TCTCCTCAGTGAGATACGCCT-3′; 57 °C] qPCR was completed using CFX Touch™ Real Time PCR Detection System (Bio-Rad, Johannesburg, South Africa). The reaction was subjected to an initial denaturation (95 °C, 4 min), followed by 37 denaturation cycles (95 °C, 15 s), annealing (primer-specific temperature, 40 s), extension (72 °C, 30 s) and a plate read for 37 cycles. *β actin* [Sense 5′-TGACGGGTCACCCACACTGTGCCCAT-3′; Antisense 5′-CTAGAAGCATTTGCGGTGGA-CGATGGAGGG-3′] was run under the same conditions and used as the housekeeping gene. Data were analyzed using the method described by Livak and Schmittgen and represented as fold change ($2^{-\Delta\Delta CT}$) relative to the control [45].

5.4. Western Blotting

5.4.1. Protein Isolation

Cytobuster (Novagen®, Pretoria, South Africa) supplemented with protease inhibitor and phosphatase inhibitor (Roche 05892791001 and 04906837001, respectively) was added to cells (4 °C, 10 min) and centrifuged (5 min, 10,000× *g*, 4 °C). Crude protein extracts were quantified using the bichinchonic acid (BCA) assay (Sigma, St. Louis, MO, USA) and standardized to 1 mg/mL. Samples were denatured by boiling (5 min, 100 °C) in Laemmli buffer (dH_2O, 0.5 M TrisHCl (pH 6.8), glycerol, 10% sodium dodecyl sulphate polyacrylamide (SDS), β-mercaptoethanol, 1% bromophenol blue).

5.4.2. Sodium Dodecyl Sulphate–Polyacrylamide Gel Electrophoresis (SDS-PAGE) and Immunoblotting

Samples were separated by electrophoresis on 7% SDS gels (1 h, 150 V) and transferred to nitrocellulose membranes using the TransBlot Turbo System® (Bio-Rad, Johannesburg, South Africa). Membranes were blocked (1 h, 3%, bovine serum albumin (BSA)) in Tris-buffered saline with Tween20 (TTBS) (20 mM Tris–HCl; pH 7.4), 500 mM NaCl and 0.01% Tween 20). Membranes were then incubated with primary antibody (phosphor-tyrosine (p-tyr100) (CST#9411); phosphor-GSK3β (ser 9) (pGSK3) (ab107166); total GSK3α/β (ab15314);

GLUT2 (ab54460); p44/42 MAPK (ERK1/2) (CST#9102); phosphor-p44/42 MAPK (ERK1/2) (Thr202/Try204) (CST#9106); PI3K (CST#5405); phosphor-ser473 Akt (CST#9271); Akt (CST#9272); PDK-1 (ab110025); phosphor-PDH (E1α) (ab92696); PDH (E1α) (ab110330); (1:1000)) in 1% BSA in TTBS overnight at 4 °C. Membranes were washed four times (10 mL, TTBS, 10 min) and treated with horseradish peroxidase-conjugated secondary antibody (anti-rabbit, CST #7074; anti-mouse, CST #7076, 1:10,000) in 1% BSA for one hour at room temperature. Membranes were then washed four times (10 mL, TTBS, 10 min) and immunoreactivity was detected (Clarity™ Western ECL Blotting Substrate, Bio-Rad, Johannesburg, South Africa) with the Bio-Rad ChemiDoc Imaging System. Protein bands were analyzed with the Bio-Rad Image Lab Analysis 6.0 software and normalized against the corresponding β-actin bands.

5.5. Statistical Analysis

Results are represented as mean fold change ± standard deviation (SD) relative to normalized control. Statistical significance was assessed using one-way ANOVA with appropriate post hoc comparisons on GraphPad Version 5.0 Software. p values less than 0.05 were considered significant.

Author Contributions: Conceptualization, A.C. and Y.P.; methodology, Y.P. and S.N.; lab investigation and formal analysis, Y.P.; funding and resources, A.C. and Y.P.; writing—original draft preparation, Y.P.; writing—review and editing, A.C. and S.N.; supervision, A.C. All authors have read and agreed to the published version of the manuscript.

Funding: This research was funded by the National Research Foundation (Grant UID: 98323), and the University of KwaZulu-Natal College of Health Science.

Institutional Review Board Statement: Ethical review was waived as this study did not involve human or animal subjects.

Informed Consent Statement: Not applicable.

Data Availability Statement: Not applicable.

Conflicts of Interest: The authors declare no conflict of interest.

References

1. Islam, S.M.S.; Purnat, T.D.; Phuong, N.T.A.; Mwingira, U.; Schacht, K.; Fröschl, G. Non-Communicable Diseases (NCDs) in developing countries: A symposium report. *Glob. Health* **2014**, *10*, 81. [CrossRef]
2. Martins, I.J. Overnutrition Determines LPS Regulation of Mycotoxin Induced Neurotoxicity in Neurodegenerative Diseases. *Int. J. Mol. Sci.* **2015**, *16*, 29554–29573. [CrossRef]
3. Pal, S.; Singh, N.; Ansari, K.M. Toxicological effects of patulin mycotoxin on the mammalian system: An overview. *Toxicol. Res.* **2017**, *6*, 764–771. [CrossRef]
4. Liu, B.-H.; Wu, T.-S.; Yu, F.-Y.; Su, C.-C. Induction of Oxidative Stress Response by the Mycotoxin Patulin in Mammalian Cells. *Toxicol. Sci.* **2007**, *95*, 340–347. [CrossRef] [PubMed]
5. Sant'Ana, A.d.S.; Rosenthal, A.; de Massaguer, P.R. The fate of patulin in apple juice processing: A review. *Food Res. Int.* **2008**, *41*, 441–453. [CrossRef]
6. Pfeiffer, E.; Gross, K.; Metzler, M. Aneuploidogenic and clastogenic potential of the mycotoxins citrinin and patulin. *Carcinogenesis* **1998**, *19*, 1313–1318. [CrossRef] [PubMed]
7. Alves, I.; Oliveira, N.; Laires, A.; Rodrigues, A.; Rueff, J. Induction of micronuclei and chromosomal aberrations by the mycotoxin patulin in mammalian cells: Role of ascorbic acid as a modulator of patulin clastogenicity. *Mutagenesis* **2000**, *15*, 229–234. [CrossRef]
8. Schumacher, D.M.; Metzler, M.; Lehmann, L. Mutagenicity of the mycotoxin patulin in cultured Chinese hamster V79 cells, and its modulation by intracellular glutathione. *Arch. Toxicol.* **2005**, *79*, 110–121. [CrossRef] [PubMed]
9. Glaser, N.; Stopper, H. Patulin: Mechanism of genotoxicity. *Food Chem. Toxicol.* **2012**, *50*, 1796–1801. [CrossRef]
10. Zhang, B.; Peng, X.; Li, G.; Xu, Y.; Xia, X.; Wang, Q. Oxidative stress is involved in Patulin induced apoptosis in HEK293 cells. *Toxicon* **2015**, *94* (Suppl. C), 1–7. [CrossRef]
11. Zhong, Y.; Jin, C.; Gan, J.; Wang, X.; Shi, Z.; Xia, X.; Peng, X. Apigenin attenuates patulin-induced apoptosis in HEK293 cells by modulating ROS-mediated mitochondrial dysfunction and caspase signal pathway. *Toxicon* **2017**, *137*, 106–113. [CrossRef]

12. Tsai, W.T.; Lo, Y.C.; Wu, M.S.; Li, C.Y.; Kuo, Y.P.; Lai, Y.H.; Tsai, Y.; Chen, K.C.; Chuang, T.H.; Yao, C.H.; et al. Mycotoxin patulin suppresses innate immune responses by mitochondrial dysfunction and p62/sequestosome-1-dependent mitophagy. *J. Biol. Chem.* **2016**, *291*, 19299–19311. [CrossRef] [PubMed]
13. Puel, O.; Galtier, P.; Oswald, I. Biosynthesis and toxicological effects of patulin. *Toxins* **2010**, *2*, 613–631. [CrossRef] [PubMed]
14. Kim, J.-W.; Tchernyshyov, I.; Semenza, G.L.; Dang, C.V. HIF-1-mediated expression of pyruvate dehydrogenase kinase: A metabolic switch required for cellular adaptation to hypoxia. *Cell Metab.* **2006**, *3*, 177–185. [CrossRef]
15. Sakthisekaran, D.; Shanmugasundaram, K.R.; Shanmugasundaram, E.R. Effect of patulin on some enzymes of carbohydrate metabolism studied in rats. *Biochem. Int.* **1989**, *19*, 37–51. [PubMed]
16. Tao, R.; Xiong, X.; Harris, R.A.; White, M.F.; Dong, X.C. Genetic inactivation of pyruvate dehydrogenase kinases improves hepatic insulin resistance induced diabetes. *PLoS ONE* **2013**, *8*, e71997. [CrossRef]
17. Zhang, S.; Hulver, M.W.; McMillan, R.P.; Cline, M.A.; Gilbert, E.R. The pivotal role of pyruvate dehydrogenase kinases in metabolic flexibility. *Nutr. Metab.* **2014**, *11*, 10. [CrossRef]
18. Cohen, M.; Kitsberg, D.; Tsytkin, S.; Shulman, M.; Aroeti, B.; Nahmias, Y. Live imaging of GLUT2 glucose-dependent trafficking and its inhibition in polarized epithelial cysts. *Open Biol.* **2014**, *4*, 140091. [CrossRef]
19. Haeusler, R.A.; Camastra, S.; Astiarraga, B.; Nannipieri, M.; Anselmino, M.; Ferrannini, E. Decreased expression of hepatic glucokinase in type 2 diabetes. *Mol. Metab.* **2015**, *4*, 222–226. [CrossRef]
20. Henriksen, E.J.; Dokken, B.B. Role of glycogen synthase kinase-3 in insulin resistance and type 2 diabetes. *Curr. Drug Targets* **2006**, *7*, 1435–1441. [CrossRef]
21. Maffucci, T.; Brancaccio, A.; Piccolo, E.; Stein, R.C.; Falasca, M. Insulin induces phosphatidylinositol-3-phosphate formation through TC10 activation. *EMBO J.* **2003**, *22*, 4178–4189. [CrossRef]
22. Fröjdö, S.; Vidal, H.; Pirola, L. Alterations of insulin signaling in type 2 diabetes: A review of the current evidence from humans. *Biochim. Biophys. Acta (BBA) Mol. Basis Dis.* **2009**, *1792*, 83–92. [CrossRef] [PubMed]
23. Alam, S.; Pal, A.; Kumar, R.; Dwivedi, P.D.; Das, M.; Ansari, K.M. EGFR-mediated Akt and MAPKs signal pathways play a crucial role in patulin-induced cell proliferation in primary murine keratinocytes via modulation of Cyclin D1 and COX-2 expression. *Mol. Carcinog.* **2014**, *53*, 988–998. [CrossRef] [PubMed]
24. Nisr, R.B.; Russell, M.A.; Chrachri, A.; Moody, A.J.; Gilpin, M.L. Effects of the microbial secondary metabolites pyrrolnitrin, phenazine and patulin on INS-1 rat pancreatic β-cells. *FEMS Immunol. Med. Microbiol.* **2011**, *63*, 217–227. [CrossRef] [PubMed]
25. Al Seeni, M.N.; Ali, A.M.; Elsawi, N.M.; Abdo, A.S. Assessment of the Secondary Metabolite Patulin and Lycium Barbarum Fruit on INS-1 Rat Pancreatic B-Cells. *Agric. Food Sci. Res.* **2017**, *4*, 24–29.
26. Zhou, S.-M.; Jiang, L.-P.; Geng, C.-Y.; Cao, J.; Zhong, L.-F. Patulin-induced oxidative DNA damage and p53 modulation in HepG2 cells. *Toxicon* **2010**, *55*, 390–395. [CrossRef]
27. Boussabbeh, M.; Ben Salem, I.; Prola, A.; Guilbert, A.; Bacha, H.; Abid-Essefi, S.; Lemaire, C. Patulin induces apoptosis through ROS-mediated endoplasmic reticulum stress pathway. *Toxicol. Sci.* **2015**, *144*, 328–337. [CrossRef]
28. Saltiel, A.R.; Kahn, C.R. Insulin signalling and the regulation of glucose and lipid metabolism. *Nature* **2001**, *414*, 799–806. [CrossRef]
29. Brozinick, J.T.; Roberts, B.R.; Dohm, G.L. Defective Signaling Through Akt-2 and -3 But Not Akt-1 in Insulin-Resistant Human Skeletal Muscle. *Diabetes* **2003**, *52*, 935–941. [CrossRef]
30. Devaraj, H.; Shanmugasundaram, K.; Shanmugasundaram, E. Changes in carbohydrate metabolism during Patulin toxicosis studied in chicks. *J. Indian Inst. Sci.* **2013**, *64*, 65.
31. Thorens, B. GLUT2, glucose sensing and glucose homeostasis. *Diabetologia* **2015**, *58*, 221–232. [CrossRef] [PubMed]
32. Zhang, W.; Zhang, S.L.; Hu, X.; Tam, K.Y. Targeting Tumor Metabolism for Cancer Treatment: Is Pyruvate Dehydrogenase Kinases (PDKs) a Viable Anticancer Target? *Int. J. Biol. Sci.* **2015**, *11*, 1390–1400. [CrossRef] [PubMed]
33. Jha, M.K.; Suk, K. Pyruvate Dehydrogenase Kinase as a Potential Therapeutic Target for Malignant Gliomas. *Brain Tumor Res. Treat.* **2013**, *1*, 57–63. [CrossRef] [PubMed]
34. Golias, T.; Papandreou, I.; Sun, R.; Kumar, B.; Brown, N.V.; Swanson, B.J.; Pai, R.; Jaitin, D.; Le, Q.-T.; Teknos, T.N.; et al. Hypoxic repression of pyruvate dehydrogenase activity is necessary for metabolic reprogramming and growth of model tumours. *Sci. Rep.* **2016**, *6*, 31146. [CrossRef] [PubMed]
35. Chain, E.; Florey, H.W.; Jennings, M.A. An antibacterial substance produced by Penicillium claviforme. *Br. J. Exp. Pathol.* **1942**, *23*, 202.
36. Liste-Calleja, L.; Lecina, M.; Lopez-Repullo, J.; Albiol, J.; Solà, C.; Cairó, J.J. Lactate and glucose concomitant consumption as a self-regulated pH detoxification mechanism in HEK293 cell cultures. *Appl. Microbiol. Biotechnol.* **2015**, *99*, 9951–9960. [CrossRef]
37. Mondon, C.E.; Jones, I.R.; Azhar, S.; Hollenbeck, C.B.; Reaven, G.M. Lactate production and pyruvate dehydrogenase activity in fat and skeletal muscle from diabetic rats. *Diabetes* **1992**, *41*, 1547–1554. [CrossRef]
38. Montes, M.; Chicco, A.; Lombardo, Y.B. The effect of insulin on the uptake and metabolic fate of glucose in isolated perfused hearts of dyslipemic rats. *J. Nutr. Biochem.* **2000**, *11*, 30–37. [CrossRef]
39. Alves, T.C.; Befroy, D.E.; Kibbey, R.G.; Kahn, M.; Codella, R.; Carvalho, R.A.; Petersen, K.F.; Shulman, G.I. Regulation of hepatic fat and glucose oxidation in rats with lipid-induced hepatic insulin resistance. *Hepatology* **2011**, *53*, 1175–1181. [CrossRef]
40. Lim, S.; Ahn, S.Y.; Song, I.-C.; Chung, M.H.; Jang, H.C.; Park, K.S.; Lee, K.-U.; Pak, Y.K.; Lee, H.K. Chronic exposure to the herbicide, atrazine, causes mitochondrial dysfunction and insulin resistance. *PLoS ONE* **2009**, *4*, e5186. [CrossRef]

41. Gerets, H.H.J.; Tilmant, K.; Gerin, B.; Chanteux, H.; Depelchin, B.O.; Dhalluin, S.; Atienzar, F.A. Characterization of primary human hepatocytes, HepG2 cells, and HepaRG cells at the mRNA level and CYP activity in response to inducers and their predictivity for the detection of human hepatotoxins. *Cell Biol. Toxicol.* **2012**, *28*, 69–87. [CrossRef] [PubMed]
42. Chen, Y.; Chen, Z.; Feng, J.H.; Chen, Y.B.; Liao, N.S.; Su, Y.; Zou, C.Y. Metabolic profiling of normal hepatocyte and hepatocellular carcinoma cells via (1) H nuclear magnetic resonance spectroscopy. *Cell Biol. Int.* **2018**, *42*, 425–434. [CrossRef] [PubMed]
43. Martínez-Monge, I.; Albiol, J.; Lecina, M.; Liste-Calleja, L.; Miret, J.; Solà, C.; Cairó, J.J. Metabolic flux balance analysis during lactate and glucose concomitant consumption in HEK293 cell cultures. *Biotechnol. Bioeng.* **2019**, *116*, 388–404. [CrossRef] [PubMed]
44. He, L.; Wondisford, F.E. Metformin action: Concentrations matter. *Cell Metab.* **2015**, *21*, 159–162. [CrossRef] [PubMed]
45. Livak, K.J.; Schmittgen, T.D. Analysis of relative gene expression data using real-time quantitative PCR and the 2(-Delta Delta C(T)) Method. *Methods* **2001**, *25*, 402–408. [CrossRef]

Disclaimer/Publisher's Note: The statements, opinions and data contained in all publications are solely those of the individual author(s) and contributor(s) and not of MDPI and/or the editor(s). MDPI and/or the editor(s) disclaim responsibility for any injury to people or property resulting from any ideas, methods, instructions or products referred to in the content.

toxins

MDPI

Article

Stressful Effects of T-2 Metabolites and Defense Capability of HepG2 Cells

Mercedes Taroncher [1], Fiona Halbig [2], Yelko Rodríguez-Carrasco [1,*] and María-José Ruiz [1]

1 Department of Preventive Medicine and Public Health, Food Science, Toxicology and Forensic Medicine, Faculty of Pharmacy, University of Valencia, 46100 Burjassot, Valencia, Spain
2 Department of Pharmacy, Rheinische Friedrich-Wilhelms-Universität Bonn, 53115 Bonn, Germany
* Correspondence: yelko.rodriguez@uv.es

Abstract: The T-2 toxin (T-2), a mycotoxin produced by several species of *Fusarium* which belongs to group A of trichothecenes, is rapidly metabolized, and its main metabolites are HT-2, Neosolaniol (Neo), T2-triol and T2-tetraol. In this work, the antioxidant defense system of HepG2 cells against oxidative stress induced by T-2 and its metabolites was evaluated. The results obtained demonstrated that there is an overall decrease in glutathione (GSH) levels after all mycotoxins exposure. Moreover, the GSH levels and the enzymatic activities related to GSH (GPx and GST) increased with NAC pre-treatment (glutathione precursor) and decreased with BSO pre-treatment (glutathione inhibitor). The GPx activity is increased by T2-tetraol. The GST activity increased after T-2 and T2-triol exposure; however, T2-tetraol decreased its activity. Furthermore, CAT activity increased after T-2 and T2-triol; nevertheless, Neo decreased its activity. Finally, SOD activity is increased by all mycotoxins, except after T-2 exposure. So, the damage associated with oxidative stress by T-2 and its metabolites is relieved by the antioxidant enzymes system on HepG2 cells.

Keywords: T-2; metabolites; glutathione; antioxidant enzymes

Key Contribution: The role played by oxidative stress of T-2 and mainly its metabolites Neo, T2-triol, and T2-tetraol on HepG2 cells.

Citation: Taroncher, M.; Halbig, F.; Rodríguez-Carrasco, Y.; Ruiz, M.-J. Stressful Effects of T-2 Metabolites and Defense Capability of HepG2 Cells. *Toxins* **2022**, *14*, 841. https://doi.org/10.3390/toxins14120841

Received: 24 October 2022
Accepted: 24 November 2022
Published: 1 December 2022

Publisher's Note: MDPI stays neutral with regard to jurisdictional claims in published maps and institutional affiliations.

Copyright: © 2022 by the authors. Licensee MDPI, Basel, Switzerland. This article is an open access article distributed under the terms and conditions of the Creative Commons Attribution (CC BY) license (https://creativecommons.org/licenses/by/4.0/).

1. Introduction

Mycotoxins are toxic secondary metabolites produced under favorable conditions by fungi that grow on food and feed worldwide [1]. Mycotoxin contamination is an aggravated problem in developing countries. The most detected mycotoxins in food are aflatoxins (AFs), fumonisins (FBs), ochratoxin A (OTA), patulin (PAT), zearalenone (ZEA), and trichothecenes. Among the trichothecenes, the T-2 toxin (T-2) is the most cytotoxic compound [1,2]. The T-2 is produced by various species of *Fusarium*, such as *F. sporotichioides*, *F. poae*, *F. equiseti*, and *F. acuminatum*. It is detected in field crops such as wheat, maize, barley, and oats and processed grains such as malt, beer, and bread [3]. Furthermore, T-2 has also been detected in drinking water in endemic areas of China, such as Qinghai, and in traditional Chinese medicines [4–6]. The most common route of T-2 exposure is through ingestion, although there are other routes. Farmers may be exposed via direct contact with contaminated grain dust and hay, and it could cause organ damage to the kidney, liver, skin, brain, and gut [7,8].

The T-2 is rapidly metabolized to HT-2, Neosolaniol (Neo), T2-triol, and T2-tetraol [1,9]. Consequently, the cytotoxic mechanism of T-2 in vivo and in vitro might be partly attributable to its metabolites [10]. Previously, the cytotoxicity of T-2 metabolites was studied individually and in combinations in various in vitro studies, and it was observed that T-2 showed the highest cytotoxic effect [9]. The T-2 is a predisposing factor in the Kashin–Beck disease (KBD), a chronic endemic type of osteochondropathy that was mainly distributed from northeastern to southwestern China [4,11]. Moreover, the T-2 induces neurotoxicity, emesis, cardiovascular alterations, immunodepression, muscular weakness, ataxia anorectic responses, hepatotoxicity, and bone marrow damage [1,12,13]. The T-2 decreases the levels of antibodies, immunoglobulins as well as diverse cytokines [12]. An important toxicity of T-2 is mitochondrial toxicity. The T-2 may induce the collapse of mitochondrial membrane potential and promote the production of mitochondrial reactive oxygen species (ROS). Mu et al. suggested that alterations in mitochondrial gene expression could alter the coupling of mitochondrial oxidative phosphorylation leading to ROS generation [14].

Oxidative metabolites can be scavenged by GSH. The function of GSH may be evaluated in vitro by adding exogenous compounds capable of decreasing or increasing the GSH synthesis. For instance, N-Acetyl-cysteine (NAC) is an effective source of sulfhydryl groups in cells and a scavenger of free radicals because it may interact with ROS, so it is a precursor of GSH. On the other hand, D-L-buthionine-(S, R)-sulfoximine (BSO) is an inhibitor of γ-glutamylcysteine synthetase, the enzyme that triggers the cytosol's production of GSH.

There are a variety of studies reporting the effects of oxidative stress damage generated by mycotoxins in different cell lines [15–17]. However, there are few reports in the literature concerning the effects of T-2 and its metabolites on enzymatic and non-enzymatic antioxidant defense. We selected the HepG2 cells due to it is a hepatic cell line, and the liver is the main drug-metabolizing organ and the main T-2 target organ [12,18]. In this study, the GSH levels and the antioxidant defense system (GPx, GST, SOD, and CAT activities) of HepG2 cells exposed to T-2, Neo, T2-triol, and T2-tetraol were evaluated. Moreover, to demonstrate the role of GSH in T-2 and its metabolite detoxification, the effect of pre-treatments with NAC or BSO was also assayed.

2. Results

2.1. In Vitro Cytotoxicity

The MTT assay was used to assess the cytotoxicity of T-2 and its modified forms on HepG2 cells after 24 h, and the results were published in our previous work [9]. The results demonstrated that T-2, Neo, T2-triol, and T2-tetraol exposure reduced cell viability in a manner that was dependent on both time and concentration. This decrease in cell viability is of the same order for all mycotoxins, being between 10 and 70% for T-2 and T2-tetraol and between 23 and 77% for Neo and T2-triol.

In this study, we have selected the $IC_{50}/2$, $IC_{50}/4$, and $IC_{50}/8$ values for each mycotoxin obtained from the MTT assay. The values were 7.5, 15, and 30 nM for T-2, 12.5, 25, and 50 nM for Neo, 0.12, 0.25, and 0.45 μM for T2-triol and 0.45, 0.9, and 1.7 μM for T2-tetraol.

2.2. Intracellular ROS Generation

Due to the cytotoxic effect observed of T-2 and its metabolites on HepG2 cells, we studied the generation of ROS to test if oxidative stress is one of the mechanisms by which these mycotoxins exert their cytotoxicity. As can be observed in Figure 1, there was an increase in ROS production after T-2, T2-triol, and T2-tetraol exposure. The increases in ROS production were from 17.79 to 27.92% for T-2, 21.75 for T2-triol, and 23.09 to 40.81% for T2-tetraol.

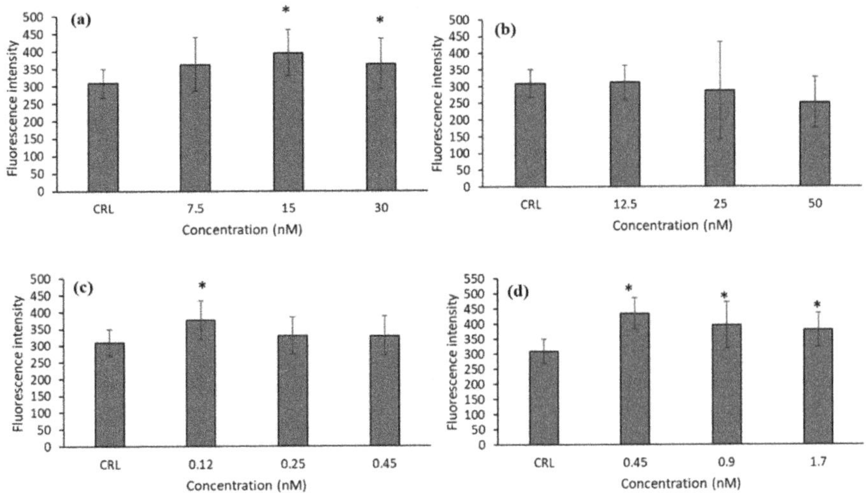

Figure 1. The ROS-induced fluorescence in HepG2 cells exposed to T-2 (7.5, 15, and 30 nM) (**a**), Neo (12.5, 25, and 50 nM) (**b**), T2-triol (0.12, 0.25, and 0.45 µM) (**c**) and T2-tetraol (0.45, 0.9, and 1.7 µM) (**d**) during 24 h. Results are expressed as mean ± SEM (n = 2). Student's t-test for paired samples was used for statistical analysis of results. * $p \leq 0.05$ indicates a significant difference from the control (CRL; fresh medium).

2.3. GSH Determination

As far as the GSH determination after T-2 exposure is concerned, the GSH levels and GSH/GSSG ratio significantly decreased ($p \leq 0.05$) at 7.5 nM by 50.6 and 52.7% and at 15 nM by 25.8 and 18.7%, respectively, compared to control (Figure 2a,c). The GSSG levels also decreased at all the assayed concentrations with respect to control cells (Figure 2b).

Regarding NAC pre-treatment, a significant increase in GSH levels (30.1%) in control cells was observed with respect to control without NAC pre-treatment (Figure 2). Also, at 7.5 and 30 nM T-2, NAC pre-treatment led to an increase in GSH level (68.3 and 19.3%, respectively), GSSG content (45.7 and 14.7%, respectively), and the GSH/GSSG ratio (112.4 and 13.9%, respectively) with respect to cells in fresh medium, showing a NAC protective effect in HepG2 cells. Furthermore, an increase of GSH (13.3%) and GSH/GSSG (10%) at 30 nM was observed, corresponding with a decrease (12.3%) of GSSG levels, compared to control with NAC pre-treatment.

Concerning the BSO pre-treatment, a decrease in GSH level (57.6%), GSSG level (37.4%), and GSH/GSSG ratio (35.2%) was obtained in their own controls compared to controls without pre-treatment. In this respect, a significant decrease in GSH levels was determined at 47.2 and 88.5% at 15 and 30 nM, respectively. The GSH/GSSG ratio also decreased from 17.6 to 81.4% (Figure 2c). On the other hand, a significant decrease in GSH levels (66.6%) was also obtained after the exposure of T-2 at the highest concentration (30 nM) compared to their own control, whereas a significant decrease in GSH/GSSG ratio (from 26.9 to 70.5%) was obtained with respect to control with BSO pre-treatment.

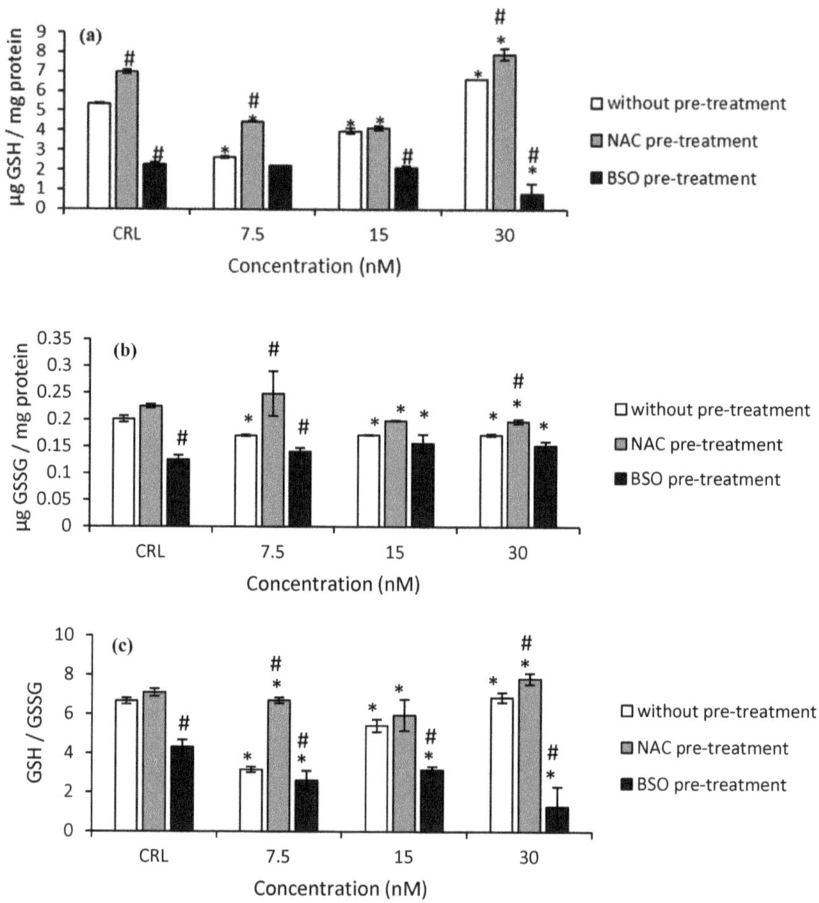

Figure 2. Effect of T-2 (7.5, 15, and 30 nM) with and without NAC or BSO pre-treatment on GSH levels (**a**), GSSG levels (**b**), and on the GSH/GSSG ratio (**c**) after 24 h of exposure. Data are expressed as mean values ± SEM of three independent experiments. One-way ANOVA followed by the Turkey HDS *post-hoc* test for multiple comparisons was used to analyze the difference between groups. * $p \leq 0.05$ indicates a significant difference from the respective control (CRL; fresh medium or without pre-treatment, NAC pre-treatment, and BSO pre-treatment); # $p \leq 0.05$ indicates a significant difference with respect to the fresh medium.

Regarding Neo (Figure 3), in fresh medium, the GSH levels and GSH/GSSG ratio significantly decreased from 56.2 to 73.3%, and from 48.6 to 70.4%, after Neo exposure, respectively, compared to control cells (Figure 3a,c). On the contrary, the GSSG content significantly increased from 13.2 to 37.1% (Figure 3b).

Concerning the NAC pre-treatment, an increase of GSH and GSSG levels and GSH/GSSG ratio (by 16.9, 22.4, and 26.8%, respectively) in control cells was observed with respect to control in fresh medium (Figure 3). Similarly, the pre-treatment with NAC produced an increase in GSH levels (from 137.4 to 197.6%) and GSH/GSSG ratio (from 102.3 to 159.9%) in cells exposed to Neo, with respect to cells in fresh medium, which showed a cytoprotecting effect of NAC in these cells.

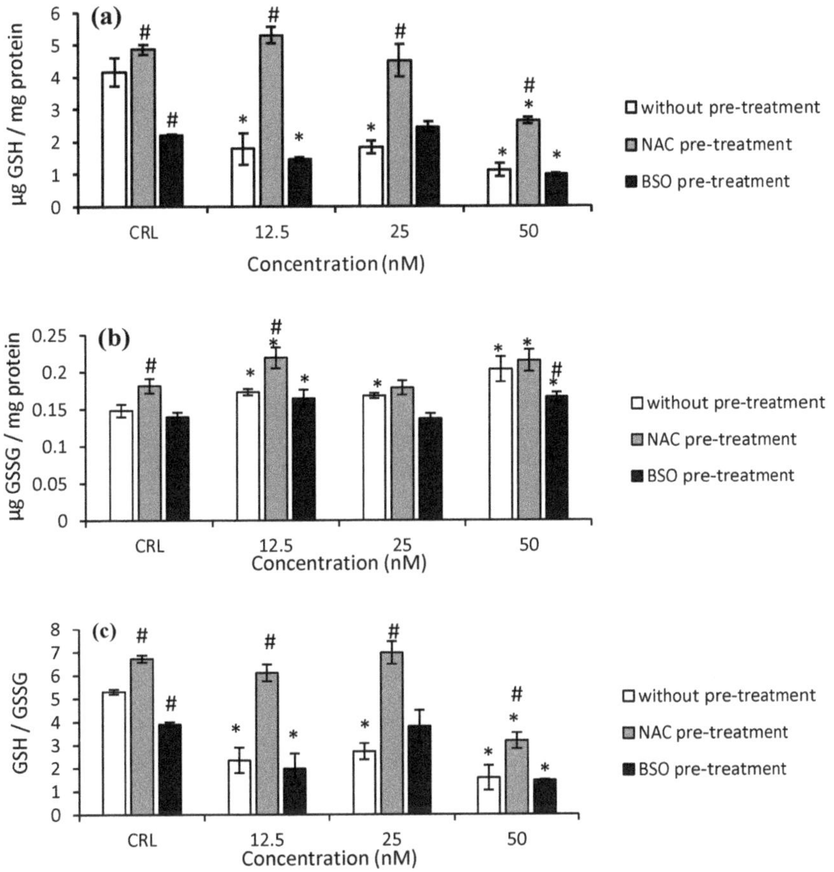

Figure 3. Effect of Neo (12.5, 25, and 50 nM) with and without NAC or BSO pre-treatment on GSH levels (**a**), GSSG levels (**b**), and on the GSH/GSSG ratio (**c**) after 24 h of exposure. Data are expressed as mean values ± SEM of three independent experiments. One-way ANOVA followed by the Turkey HDS *posthoc* test for multiple comparisons was used to analyze the difference between groups. * $p \leq 0.05$ indicates a significant difference from the respective control (CRL; fresh medium or without pre-treatment, NAC pre-treatment, and BSO pre-treatment); # $p \leq 0.05$ indicates a significant difference with respect to the fresh medium.

With respect to BSO pre-treatment, a decrease in the GSH levels and GSH/GSSG ratio (47.4 and 26.4%, respectively) was obtained in controls compared to controls without BSO. The GSSG levels significantly decreased by 18.3% at the highest Neo concentration. Additionally, a significant decrease of GSH levels (by 32.9 and 55.9%) and GSH/GSSG ratio (by 49.4 and 62.6%) and a significant increase in GSSG levels (by 17.6 and 18.6%) were also obtained after 12.5 and 50 nM of Neo exposure, respectively, compared to BSO pre-treated controls.

Regarding T2-triol (Figure 4), the GSH levels and GSH/GSSG ratio decreased after 0.45 µM T2-triol exposure (57.9 and 53.3%, respectively) in fresh medium, whereas the GSSG content significantly increased (34.2%) compared to control cells.

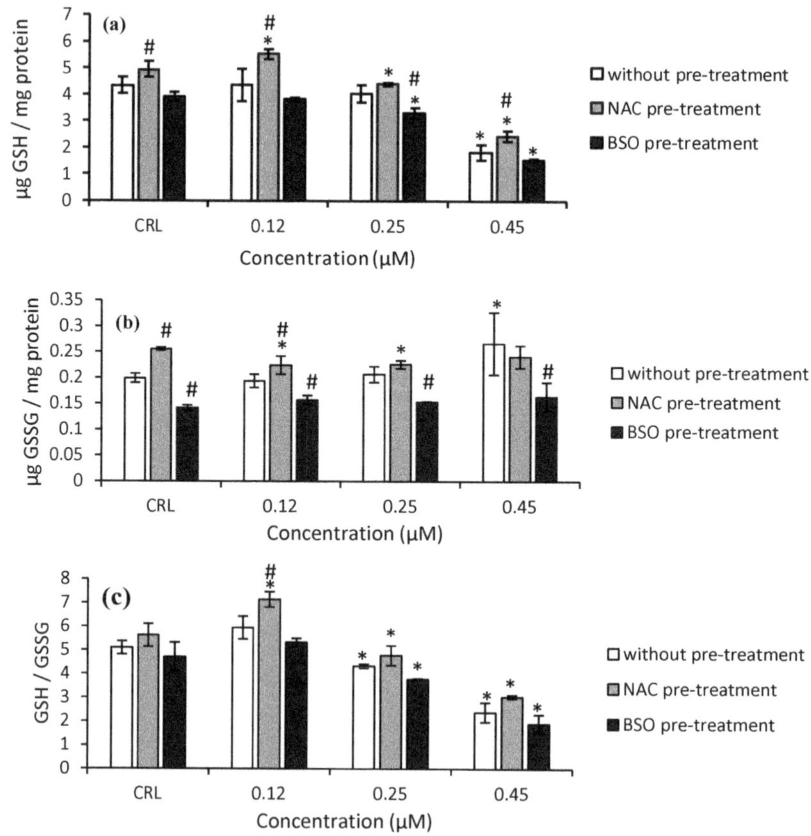

Figure 4. Effect of T2-triol (0.12, 0.25, and 0.45 µM) with and without NAC or BSO pre-treatment on GSH levels (**a**), GSSG levels (**b**), and on the GSH/GSSG ratio (**c**) after 24 h of exposure. Data are expressed as mean values ± SEM of three independent experiments. One-way ANOVA followed by the Turkey HDS *posthoc* test for multiple comparisons was used to analyze the difference between groups. * $p \leq 0.05$ indicates a significant difference from the respective control (CRL; fresh medium or without pre-treatment, NAC pre-treatment, and BSO pre-treatment); # $p \leq 0.05$ indicates a significant difference with respect to the fresh medium.

Concerning the effects of NAC pre-treatment, a significant increase of GSH and GSSG levels was determined in control cells by 14 and 28.8%, respectively, compared to controls without NAC pre-treatment (Figure 4). Additionally, NAC pre-treatment induced an increase in GSH level (from 26.8 to 34.7%), GSSG levels (15.7%), and GSH/GSSG ratio (20.1%) in cells exposed to T2-triol compared to cells in fresh medium (Figure 4).

Regarding the BSO pre-treatment, a significant decrease in the GSH levels (up to 60.3%) and GSH/GSSG ratio (up to 59.9%) were also obtained after T2-triol exposure compared to its own control. Finally, a decrease up to 38.7% of GSSG content of T2-triol was observed compared to cells without BSO pre-treatment (Figure 4b).

Similar results were obtained with T2-tetraol. The GSH levels and GSH/GSSG ratio increased (15.3 and 13.7%, respectively) at 0.45 µM and decreased (31.1 and 22.4%, respectively) at 1.7 µM with respect to the control cells. (Figure 5).

The NAC pre-treatment produced an increase in GSH levels (30.9%), GSSG levels (15.6%), and GSH/GSSG ratio (15.2%) in controls compared to controls without NAC. When cells exposed to NAC pre-treatment were compared to cells exposed to fresh medium, an increase of GSH, GSSG levels, and the GSH/GSSG ratio was observed. These values increased to 66.8%, 32.9%, and 26.4%, respectively.

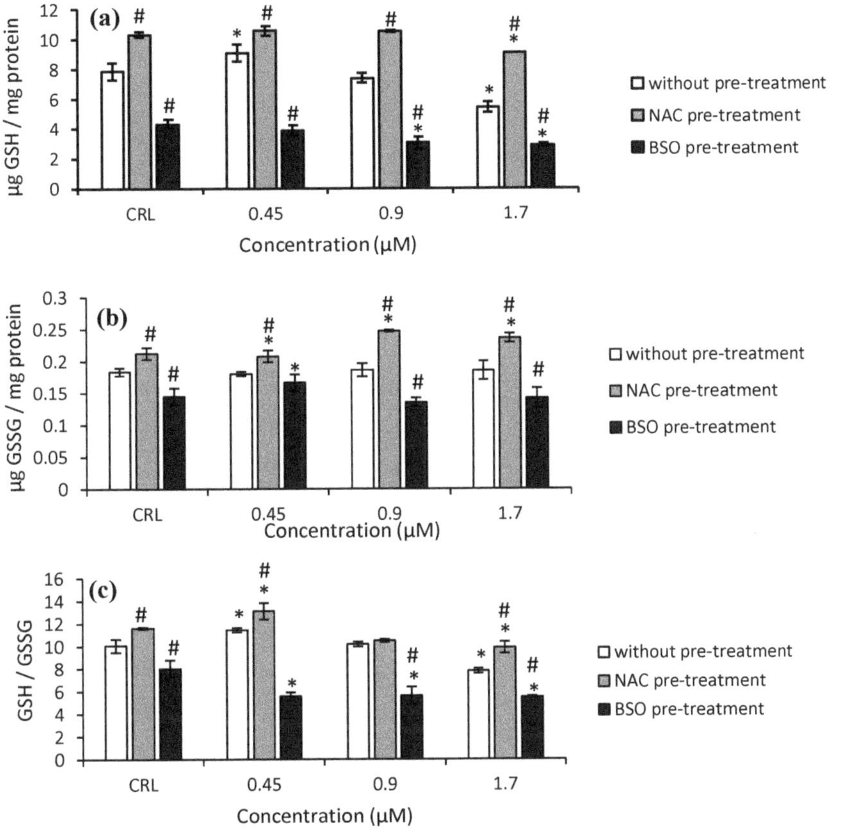

Figure 5. Effect of T2-tetraol (0.45, 0.9, and 1.7 µM) with and without NAC or BSO pre-treatment on GSH levels (**a**), GSSG levels (**b**), and on the GSH/GSSG ratio (**c**) after 24 h of exposure. Data are expressed as mean values ± SEM of three independent experiments. One-way ANOVA followed by the Turkey HDS *posthoc* test for multiple comparisons was used to analyze the difference between groups. * $p \leq 0.05$ indicates a significant difference from the respective control (CRL; fresh medium or without pre-treatment, NAC pre-treatment, and BSO pre-treatment); # $p \leq 0.05$ indicates a significant difference with respect to the fresh medium.

Concerning the BSO pre-treatment, a significant decrease in the GSH, GSSG levels, and GSH/GSSG ratio was observed (Figure 5). The BSO pre-treatment showed a decrease from 47.1 to 58.9% of GSH levels, from 23.3 to 27.1% of GSSG levels and from 30 to 51.6% of GSH/GSSG ratio, with respect to cells without BSO. Finally, with BSO pre-treatment, a significant decrease of up to 31.9% was observed in the GSH/GSSG ratio with respect to its own control.

2.4. Enzymatic Activity

The GPx, GST, CAT, and SOD activities were assessed in HepG2 cells after 24 h of incubation with T-2 (7.5, 15, and 30 nM) and its modified forms, Neo (12.5, 25, and 50 nM), T2-triol (0.12, 0.25 and 0.45 µM) and T2-tetraol (0.45, 0.9 and 1.7 µM).

Concerning GPx activity, a significant decrease (67.9%) was obtained after 30 nM T-2 exposure in the fresh medium, as shown in Figure 6a, whereas a significant increase was observed at this concentration in cells pre-treated with NAC (326.4%) pre-treatment, compared to cells in fresh medium. Regarding to Neo exposure, an increase by 249.2% at 25 nM was determined in NAC pre-treated cells compared to its own control, while an increase (from 242.9 to 548.3%) was observed at all concentrations tested compared to cells in fresh medium (Figure 6b).

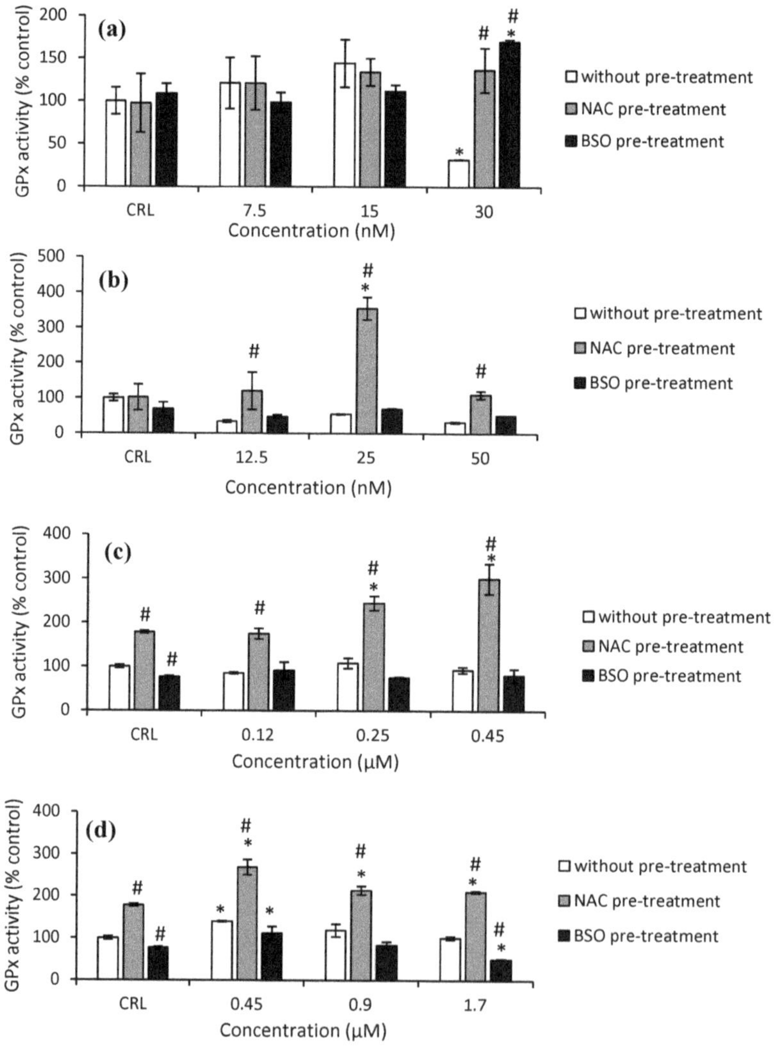

Figure 6. Effect of T-2 (7.5, 15 and 30 nM) (**a**), Neo (12.5, 25, 50 nM) (**b**), T2-triol (0.12, 0.25, 0.45 μM) (**c**) and T2-tetraol (0.45, 0.9, 1.7 μM) (**d**) with and without NAC or BSO pre-treatment on glutathione peroxidase activity after 24 h of exposure. Data are expressed in % of the unexposed control (CRL). In terms of μmol of NADPH oxidized/min/mg of protein, the GPx activity is expressed; mean ± SEM ($n = 3$). One-way ANOVA followed by the Turkey HDS *posthoc* test for multiple comparisons was used to analyze the difference between groups. * $p \leq 0.05$ indicates a significant difference from the respective control (fresh medium or without pre-treatment, NAC pre-treatment, and BSO pre-treatment); # $p \leq 0.05$ indicates a significant difference with respect to the fresh medium.

Respect to T2-triol, control NAC pre-treated increased (77.9%) the GPx activity while control BSO pre-treated significantly decreased (23.2%) with respect to fresh medium control cells. Cells NAC pre-treated increased the GPx activity from 37.2 to 68.8% compared to its own control, and it increased (from 104.5 to 223.7%) with respect to fresh medium cells (Figure 6c). Finally, regarding T2-tetraol, an increase in GPx activity of 77.9% was shown in control NAC pre-treatment compared to control without pre-treatment, while a decrease of 23.2% in control BSO pre-treatment was observed. The GPx activity increased by 39.2%, from 18 to 51.2% and by 79.8 to 108.4%, in HepG2 cells exposed to T2-tetraol in fresh medium and in cells pre-treated with NAC with respect to its own control and respect to the corresponding cells in fresh medium, respectively. On the other hand, with BSO pre-treatment, the GPx activity decreased with respect to cells in fresh medium (50.7%) and with respect to its own control (35.3%) after the highest concentration of T2-tetraol (Figure 6d).

As shown in Figure 7a, the GST activity increased significantly in HepG2 cells from 49.8% to 115.7% after T-2 exposure in the fresh medium. After NAC pre-treatment, the highest increase in GST activity was observed at 15 nM T-2 compared to its own control cells (by 102.3%) and with respect to cells in fresh medium (by 38.2%). Regarding Neo exposure, it only was observed in cells with NAC pre-treatment and led to an increase to 221.8% (50 nM) and 39.8% (25 nM) compared to cells without pre-treatment and to control with NAC pre-treatment, respectively (Figure 7b).

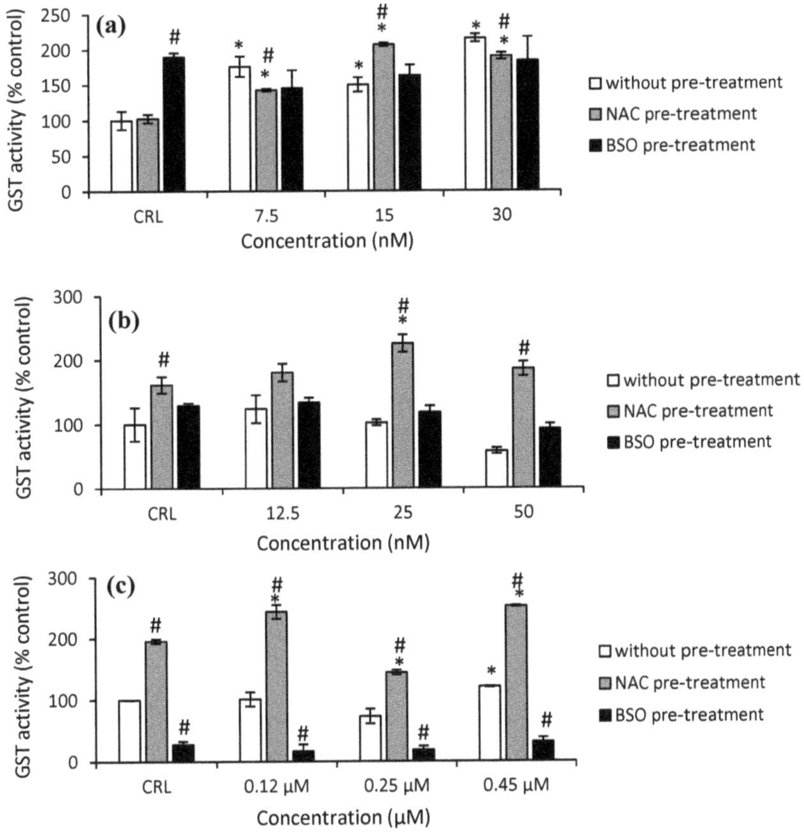

Figure 7. *Cont.*

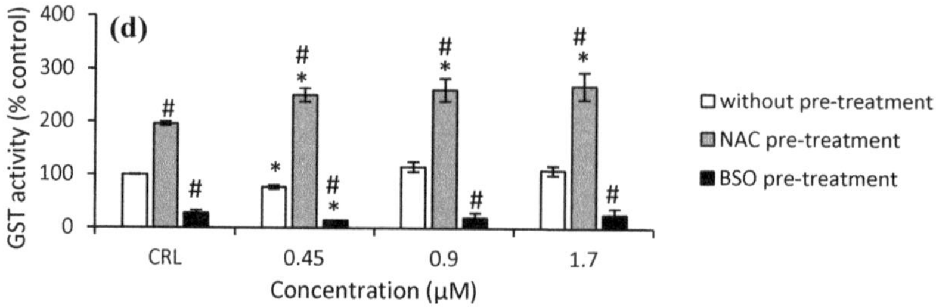

Figure 7. Effect of T-2 (7.5, 15 and 30 nM) (**a**), Neo (12.5, 25, 50 nM) (**b**), T2-triol (0.12, 0.25, 0.45 μM) (**c**) and T2-tetraol (0.45, 0.9, 1.7 μM) (**d**) with and without NAC or BSO pre-treatment on glutathione S-transferase activity after 24 h of exposure. Data are expressed in % of the unexposed control (CRL). In terms of mol of product formed/min/mg of protein, the GST activity is expressed; mean ± SEM (n = 3). One-way ANOVA followed by the Turkey HDS *posthoc* test for multiple comparisons was used to analyze the difference between groups. * $p \leq 0.05$ indicates a significant difference from the respective control (fresh medium or without pre-treatment, NAC pre-treatment, and BSO pre-treatment); # $p \leq 0.05$ indicates a significant difference with respect to the fresh medium.

With regard to T2-triol exposure, a significant increase (20.9%) at 0.45 μM was shown in cells with fresh medium compared to the control. Regarding the effects of NAC pre-treatment, an increase of 95.6% in control NAC pre-treated cells compared to control no pre-treated was obtained. Moreover, an increase of GST activity after T2-triol exposure up to 29% compared to control cells with NAC pre-treatment was observed as well as an increase from 97.5 to 140.6% compared to cells without pre-treatment. Conversely, a decrease of 72.6% in control BSO pre-treated compared to control cells not pre-treated was shown. Similarly, a decrease of up to 83.7% after T2-triol exposure was observed compared to cells without BSO pre-treatment (Figure 7c). Finally, regarding T2-tetraol exposure, an increase of 95.6% was shown in control cells NAC pre-treated compared to control no pre-treated, whereas BSO pre-treated showed a decrease of 72.6%. Furthermore, a significant increase was obtained at all concentrations tested in cells pre-treated with NAC with respect to its own control cells (from 28.3 to 37.2%) and also to the corresponding cells in fresh medium (from 146.5 to 227.7%). On the other hand, the GST activity decreased in cells BSO pre-treated exposed to 0.45 μM by 45.8% compared to control BSO pre-treated, whilst a decrease at all concentrations tested in cells pre-treated with BSO with respect to cells without pre-treatment from 80.6 to 76.9% was also observed (Figure 7d).

Figure 8a shows that the CAT activity increased at the lower T-2 and T2-triol concentrations from 68.9 to 102.2% and from 41.9 to 48.6%, respectively, with respect to control cells (Figure 8c). Conversely, at the lower Neo concentrations, the CAT activity significantly decreased from 38.9 to 49.3%, while an increase of 58.2% was detected at the highest Neo concentration with respect to control (Figure 8b,d).

Finally, with respect to SOD activity in HepG2 cells, an increase at all concentrations tested of Neo (from 325.3 to 445.9%), T2-triol (from 250.65 to 441.2%), and T2-tetraol (from 70.4 to 140.2%) compared to their respective controls was obtained (Figure 9b–d).

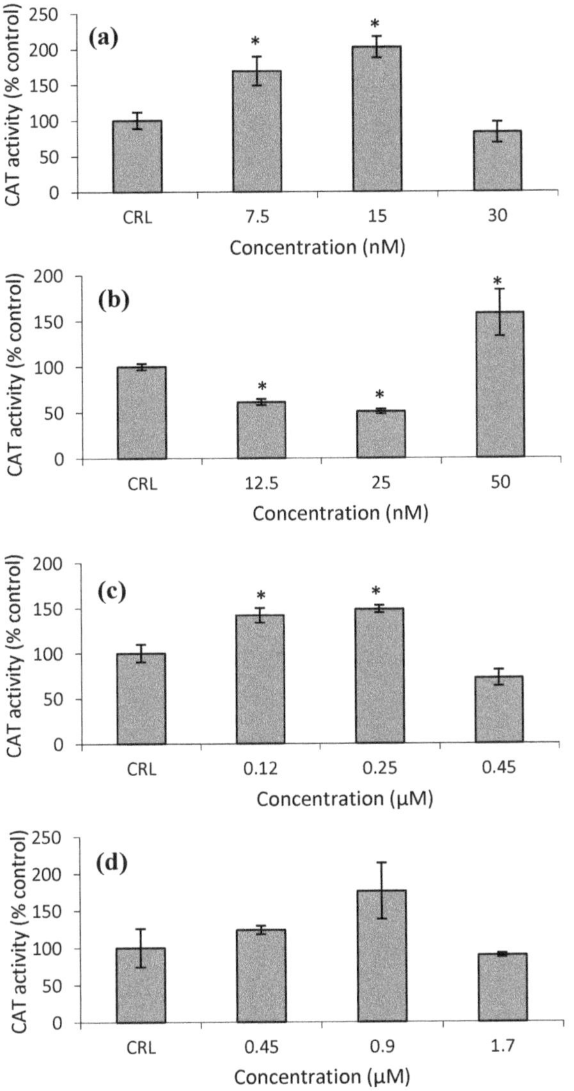

Figure 8. Effect of T-2 (7.5, 15 and 30 nM) (**a**), Neo (12.5, 25, 50 nM) (**b**), T2-triol (0.12, 0.25, 0.45 μM) (**c**) and T2-tetraol (0.45, 0.9, 1.7 μM) (**d**) on catalase activity after 24 h of exposure. Data are expressed in % of the unexposed control (CRL). The CAT activity is expressed as μmol of H_2O_2/min/mg of protein; mean ± SEM (n = 3). Student's t-test for paired samples was used for statistical analysis of results. * $p \leq 0.05$ indicates a significant difference with respect to the control.

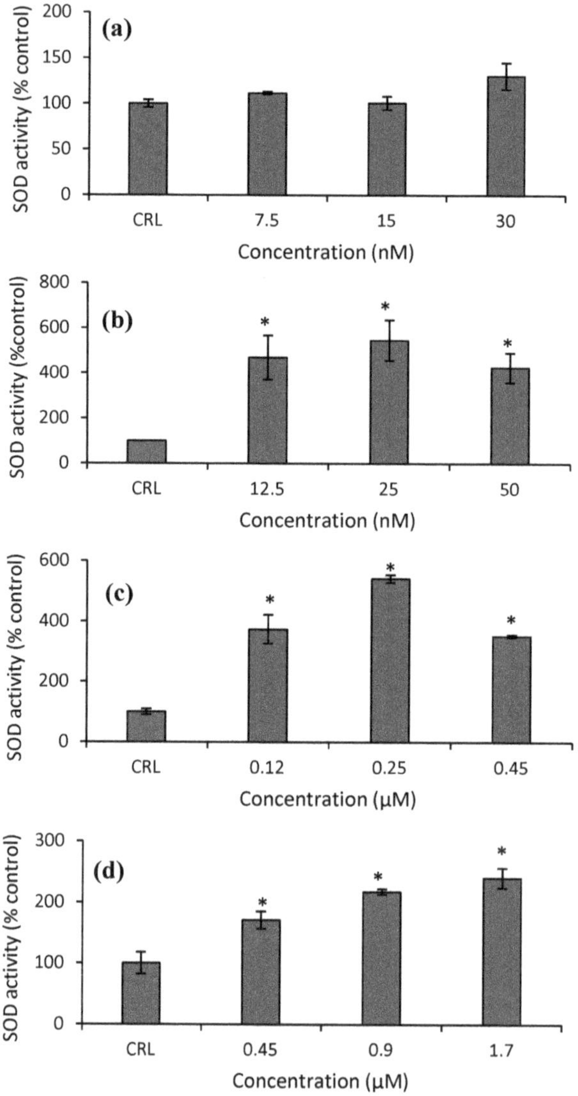

Figure 9. Effect of T-2 (7.5, 15 and 30 nM) (**a**), Neo (12.5, 25, 50 nM) (**b**), T2-triol (0.12, 0.25, 0.45 μM) (**c**) and T2-tetraol (0.45, 0.9, 1.7 μM) (**d**) on superoxide dismutase activity after 24 h of exposure. Data are expressed in % of the unexposed control (CRL); mean ± SEM (n = 3). Student's t-test for paired samples was used for statistical analysis of results. * $p \leq 0.05$ indicates a significant difference with respect to the control.

3. Discussion

This study showed that in HepG2 cells, T-2, T2-triol, and T2-tetraol triggered an increase of intracellular ROS levels after 24 h of exposure (Figure 1), demonstrating that oxidative stress plays a role in cellular toxicity.

The induction of oxidative stress by T-2 has been previously demonstrated by other authors [3,10,11,19–27]. The results obtained for T-2 in HepG2 cells grown in medium showed a decrease in GSH levels at 7.5 and 15 nM compared to control cells. However, intracellular GSH increases at 30 nM and it may be to keep the cells in a reduced state and to avoid oxidative stress to high doses (Figures 1a and 2a).

In our results, correlated with the decrease in GSH levels, an increase in GST activity was observed. However, GPx activity remained unchanged (Figures 6a and 7a). According to Gouze et al. [28], the 12,13-epoxide group, which characterizes all epoxy-trichothecene toxins, could be a substrate for GST conjugation, as GST is active in the epoxides group. These authors demonstrated that deoxynivalenol (DON) trichothecene has the ability to conjugate GST, as DON is a substrate of GST [28]. Therefore, in this work, the increase of GST enzymatic activity after T-2 exposure leads us to speculate that this mycotoxin could be a substrate for GST due to the structural similarity of both trichothecenes. So, the GST-mediated conjugation could represent a detoxification pathway in protecting cells from injury by T-2. On the other hand, GPx activity in HepG2 is not the main implicated enzyme in detoxifying against the cytotoxicity of T-2. Also, it is possible that GPx inactivates itself by the decrease of its own substrate. Similarly, SOD activity remains unchanged after T-2 exposure. Nevertheless, CAT activity increased (Figures 8a and 9a). This can be due to the fact that T-2 increases ROS (Figure 1a), including H_2O_2, and it is transformed rapidly to H_2O and O_2 by CAT. Thus, our results suggest that the main enzymes involved in detoxifying T-2 in HepG2 cells are GST and CAT. Therefore, GST activity remains high since this enzyme is important to protect against oxidative stress to which the HepG2 cells are submitted, as shown by the equally high CAT activity and GSH content at the highest T-2 concentration tested.

Similar results to our SOD activity were reported in HepG2 cells exposed to OTA, in Hek-293 cells after DON exposure, and in SH-SY5Y cells after beauvericin (BEA), α-ZEL, and β-ZEL exposure. These authors demonstrated that the activity of SOD tended to the control levels, that are, cells exposed to culture medium only, without mycotoxins [29–31]. Regarding enzymatic activities in cell culture exposed to mycotoxin, there is a variety of results. According to our results, Yang et al. [10,18] observed an increase in CAT, SOD, and GPx activity after T-2 exposure (from 10 to 100 nM) in broiler hepatocytes. Conversely, Li et al. [19] and Zhang et al. [3] reported a decrease in CAT activity after T-2 exposure to porcine kidney and Leydig cells. The increase of CAT, GPx, and SOD activities suggests that in hepatocytes, the T-2 increases ROS which is catalyzed by SOD, and the high levels of H_2O_2 as a result of the reaction are detoxified by CAT and GPx, decreasing the GSH levels. On the contrary, the SOD, GPx, and CAT enzymes do not seem to be the main enzymes responsible for the detoxification in Leydig or porcine kidney cells in response to T-2 exposure. So, these enzymes are not activated by the cells to prevent oxidative stress.

The mechanism of toxicity of T-2 modified forms is the innovative part of this study, as little or no information is available. Regarding Neo, a significant decrease in GSH and an increase in GSSG levels at all concentrations tested were determined in HepG2 cells (Figure 3a,b). These results are similar to the effects of T-2 in this work. Nevertheless, the depletion in GSH levels seems unrelated to its involvement through the activity of GPx, due to its activity remaining unchanged (Figures 6b and 7b). So, the enzymatic cell system not related to GSH is the most important enzymatic defense system in HepG2 cells exposed to Neo. The increase in SOD activity has also been observed in GH3 and broiler hepatocytes [10,11,18] after T-2 exposure. Other authors reported differences in the results of the CAT activity depending on the range of concentration tested in different cell types exposed to T-2 [3,19,29,32]. However, no data were reported about Neo.

A significant decrease in GSH and an increase in GSSG levels at the highest concentration of T2-triol (0.45 μM) tested were found in HepG2 cells (Figure 4a,b). Similar effects were observed with ZEA and its metabolites on CHO-K1, HepG2, and SH-SY5Y cells [29,32,33], BEA in CHO-K1 cells [34], and OTA in HepG2 cells [31]. The decrease in GSH levels is accompanied by an increase in GST at 0.45μM. However, GPx activity remains unchanged (Figures 6c and 7c). This indicates that the T2-triol detoxifying enzyme is mainly GST.

On the other hand, the increase of ROS after T2-triol exposure (Figure 1c) provide the stimulus for increasing the CAT and SOD activities at all concentrations (Figures 8c and 9c). Much research has been carried out to assess the impact of different mycotoxins on the antioxidant defense system by different authors. However, as far as the author´s knowledge extends, no data about the effects of T-2 metabolites on the enzymatic antioxidant protective system has yet to be found in the literature.

On HepG2 cells, intracellular GSH increased immediately at the lowest concentration (0.45 μM) of T2-tetraol, acting as a major thiol-disulfide redox buffer as a response to oxidative damage to keep the cells in a reduced state (Figure 5a). However, with the higher concentration (1.7 μM), GSH is depleted due to high levels of oxidative stress. On the other hand, the increase in GSH content was accompanied by an increase in GPx activity at 0.45 μM. Nonetheless, the GST activity decreased (Figures 6d and 7d). So, regarding T2-tetraol, the main enzymatic detoxifying mechanism involved was GPx. On the other hand, the oxidizing compounds produced by T2-tetraol were catalyzed by the SOD enzyme because their activity increased at all concentrations tested. The increase of ROS after T2-tretraol exposure (Figure 1d) leads to GSH action (Figure 5a) and also to SOD activity (Figure 9d). On the contrary, the T2-tetraol did not stimulate CAT activity, so we can conclude that T2-tetraol was detoxified by GPx and SOD enzymatic activities (Figures 8d and 9d). In previous work, we demonstrated that the T-2 is converted by hydrolysis to Neo and T2-triol, but all three metabolites end up as T2-tetraol [9], and it is known that the GPx is the most important enzyme for the extra peroxisomal inactivation of H_2O_2, particularly in liver cells. This is related to our results, which show that in HepG2 cells, the final metabolite T2-tetraol is detoxified mainly by the GPx enzyme. According to our results, the increase in GPx and SOD enzymatic activities was previously reported for T-2 [10,11,18] and for DON [35]. Furthermore, similar to our results, the tendency of CAT to basal rate levels [17,30,31] and the decrease in GST [17,29] was previously observed after OTA, DON, and STE exposure in cell cultures.

The role of the pre-treatment with NAC and BSO, a GSH promoter, and a GSH depletory, respectively, was also evaluated. For T-2, Neo, T2-triol, and T2-tetraol, the GSH levels were increased in cells with NAC pre-treatment because the GSH is involved in the cellular defense mechanism when HepG2 cells were exposed to these mycotoxins. It was evidenced by the increase in GPx and GST activities after NAC pre-treatment in all mycotoxin treatments. We determined that pre-treatment with the antioxidant NAC suppressed oxidative damage induced by T-2 and its metabolites on HepG2 cells. On the contrary, overall, BSO showed a significant decrease in GSH levels in cells exposed to T-2, T2-triol, and T2-tetraol. However, no effects were observed after Neo exposure compared to cells without BSO pre-treatment. This can be because the GPx and GST enzymes (depending on GSH levels) are not involved in the Neo detoxification mechanism. In fact, no changes were observed in GPx (after Neo and T2-triol exposure) and GST (after Neo exposure) activities after BSO pre-treatment compared to cells without pre-treatment. On the contrary, GST activity after BSO exposure is highly suppressed because the GSH levels decrease, and GST is a GSH-related enzyme. Similar results related to the antioxidant defense system after NAC and BSO pre-treatment have already been described on HepG2 [36,37] and SH-SY5Y cells [17]. According to these authors, the NAC blocks ROS induced by STE on SH-SY5Y cells because the GSH levels and GPx and GST activity increased compared to cells without NAC pre-treatment [17]. The same cells have more predisposition to oxidative damage induced by STE after BSO pre-treatment because GSH levels decreased in cells treated with STE at all concentrations tested further than in cells without pre-treatment, leading to a decrease in GPx and GST activities. The results obtained in this work are similar to those obtained by these authors and confirm that GSH was involved in the defense mechanism of the HepG2 cells after trichothecenes exposure.

4. Conclusions

In conclusion, our results suggest that T-2, Neo, T2-triol and T2-tetraol lead to the actuation of the GSH redox system in HepG2 cells because the decrease of intracellular GSH levels was observed after all mycotoxins and the stimulation of GSH dependent enzyme system. So, the GSH depletion confirmed oxidative stress as an underlying mechanism involved in the cytotoxicity of T-2 and its metabolites. Nevertheless, regarding antioxidant enzymes, each mycotoxin acts in a different way. Furthermore, the GSH levels and the enzymatic activities related to GSH increased in NAC pre-treated and decreased in BSO pre-treated cells. So, the damage associated with oxidative stress by T-2 and its metabolites is relieved by the antioxidant enzymes system on HepG2 cells.

5. Materials and Methods

5.1. Reagents

The substances utilized for cell culture and reagent-grade chemicals include Dulbecco's Modified Eagle's Medium (DMEM), streptomycin, penicillin, phosphate-buffered saline (PBS), trypsin/EDTA solutions, newborn calf serum (NBCS), sodium azide (NaN_3), β–nicotinamide adenine dinucleotide phosphate (β-NADPH), GSH, GSSG, NAC, BSO, N-ethylmaleimide (NEM), o-phtaldialdehyde (OPT), t-octylphenoxypolyethoxyethanol (Triton-X100), 1-chloro-2,4-dinitrobenzene (CDNB), tris hydroxymethyl aminomethane (Tris), ethylenediaminetetraacetic acid (EDTA), and H_2O_2 were purchased from Sigma-Aldrich (St. Louis, MO, USA). Methanol (MeOH) was acquired from Merck Life Science S.L. (Madrid, Spain). A Milli-Q water purification system (Millipore, Bedford, Burlington, MA, USA) was used to produce deionized water (resistivity < 18 MΩ cm). Standards of T-2 (MW: 466.52 g/mol), Neosolaniol (Neo; MW: 382.40 g/mol), T-2 triol (MW: 382.45 g/mol), and T-2 tetraol (MW: 298.33 g/mol) were purchased from Sigma-Aldrich (St. Louis, MO, USA). Stock solutions of the mycotoxins were made in MeOH at the proper working concentrations and kept at a constant temperature of −20 °C in the dark.

5.2. Cell Culture and Treatment

Human hepatocarcinoma (HepG2) cells (ATCC: HB-8065) were grown in DMEM medium with 10% NBCS, 100 U/mL penicillin, and 100 mg/mL streptomycin. A pH of 7.4, 5% CO_2 at 37 °C, and 95% air atmosphere at constant humidity were the conditions used for incubation. In order to maintain genetic homogeneity, cells were sub-cultivated twice a week with only a small number of sub-passages. After trypsinization, the HepG2 cells were sub-cultivated in a 1:3 split ratio. The final mycotoxin concentrations assayed were obtained by adding each mycotoxin to the culture medium with a final MeOH concentration ≤1% (*v*/*v*). Controls containing the same amount of solvents were used in each experiment.

5.3. In Vitro Cytotoxicity

Cytotoxic effects were determined in HepG2 cells by the MTT assay. In this test, only metabolically active cells are used to evaluate whether a cell is viable by reducing soluble yellow tetrazolium salt to an insoluble purple formazan crystal through a mitochondrial-dependent reaction. The MTT viability assay was carried out as described by Ruiz et al. [38]. Briefly, at a density of 2×10^4 cells/well, the HepG2 cells were seeded in 96-well tissue culture plates. After cells reached 80% of confluence, serial dilutions of T-2, Neo, T2-triol, and T2-tetraol in a fresh medium were added. Mycotoxin concentrations ranged from: 12.5 to 100 nM for T-2, 11 to 164 nM for Neo, 164 to 2620 nM for T2-triol, and 209 to 3350 nM for T2-tetraol. The range of concentrations tested for each mycotoxin was chosen based on assays done previously with all the mycotoxins. The mycotoxin was exposed for 24 h. After that, 200 μL of fresh medium was added to each well after the medium was removed. Then, 50 μL/well of MTT was added, and the plates were put back into the incubator in the dark. The MTT solution was removed after 3 h of incubation, and 200 μL of DMSO and 25 μL Sorensen's glycine buffer were added. In order to achieve complete dissolution, plates were shaken for 5 min. Using an automatic ELISA plate reader (MultiSkanEX, Thermo Scientific, Walthman, MA, USA), the absorbance was determined at 540 nm.

Cell viability was expressed as a percentage relative to the control solvent (1% MeOH). Determinations were performed in three independent experiments. The mean inhibition concentration (IC_{50}) values were obtained using SigmaPlot version 11 (Systat Software Inc., GmbH, Düsseldorf, Germany).

5.4. Intracellular ROS Generation

Adding H_2-DCFDA allowed researchers to observe intracellular ROS generation in HepG2 cells after 24 h. Intracellular esterases deacetylate the H_2-DCFDA after it has been absorbed by cells, and the non-fluorescent 2′, 7′-dichlorodihydrofluorescein (H_2-DCF) that results is converted by ROS into the highly fluorescent dichlorofluorescein (DCF). Briefly, a 96-well black culture microplate was seeded with 2×10^4 cells/well. After cells reached 80% of confluence, the culture medium was changed to fresh medium containing different concentrations of T-2 (7.5, 15, 30 nM), Neo (12.5, 25, 50 nM), T2-triol (0.12, 0.25, 0.45 μM) and T2-tetraol (0.45, 0.9, 1.70 μM) for 24 h of incubation. The range of concentrations tested for each mycotoxin was chosen in accordance with earlier research done in our laboratory based on the IC_{50} values and considering the best concentration range for each mycotoxin $IC_{50}/2$, $IC_{50}/4$, and $IC_{50}/8$ values [9]. Afterward, the culture medium was removed, and cells were incubated with 50 μM H_2-DCFDA/well in the fresh medium for 30 min. Later, the H_2-DCFDA was removed, and cells were washed with PBS after adding 200 μL PBS/well. Increases in fluorescence were measured on a Wallace Victor2, model 1420 multilabel counter (PerkinElmer, Turku, Finland) at excitation/emission wavelengths of 485/535 nm. Results are expressed as increases in fluorescence with respect to solvent control. Determinations were performed in two independent experiments with 12 replicates each.

5.5. GSH Determination

The determination of GSH and GSSG was evaluated according to Maran et al. [39]. Briefly, a six-well plate was seeded with 2.25×10^5 cells/well. The cells were exposed to different treatments: (a) fresh medium (DMEM medium), (b) NAC pre-treatment (DMEM medium plus 1 mM NAC), and (c) BSO pre-treatment (DMEM medium plus 60 μM BSO). Once the cells reached 80% confluence, the culture medium was replaced with fresh medium containing different concentrations of T-2 (7.5, 15, 30 nM), Neo (12.5, 25, 50 nM), T2-triol (0.12, 0.25, 0.45 μM) and T2-tetraol (0.45, 0.9, 1.70 μM) incubated for 24 h. Following removal of the medium, the cells were washed with PBS and homogenized in 0.25 mL of 20 mM Tris and 0.1% Triton. The microplate reader Wallace Victor^2, model 1420 multilabel counter (PerkinElmer, Turku, Finland) with excitation and emission wavelength of 345 and 424 nm, was used to measure the concentrations of GSH and GSSG, respectively. The level of GSH and GSSG was represented in μg/mg proteins. Determinations were performed in three independent experiments.

5.6. Determination of Enzymatic Activities

In order to determine the scavenging procedures in HepG2 cells exposed to T-2 and its metabolites, the GST, GPx, CAT, and SOD activities were determined at the concentrations previously selected. For these assays, 2.25×10^5 cells/well were seeded in six-well plates. The cells were exposed to different treatments: (a) fresh medium (DMEM medium), (b) NAC pre-treatment (DMEM medium plus 1 mM NAC), and (c) BSO pre-treatment (DMEM medium plus 60 μM BSO). After cells achieved the 80% confluence, cells were treated with the T-2, Neo, T2-triol, and T2-tetraol for 24 h. Then, cells were homogenized in 0.1 M phosphate buffer pH 7.5 with 2 mM EDTA after the medium was eliminated. All the enzyme determinations were performed in triplicate.

Following the conjugation of GSH with CDNB for 5 min, the GST activity was assessed using the method of [39]. The reaction mixture contained in a final volume of 1 mL: 100 μL of 20 mM GSH, 825 μL of 0.1 M Na/K phosphate buffer at pH 6.5, 25 μL of 50 mM

CDNB dissolved in ethanol, and 50 μL of homogenized cell sample. The GST activity was expressed as mol of product formed/min/mg of protein using a molar absorptivity of CDNB (9.6 mM^{-1} cm^{-1}). Enzymatic activity was evaluated in a thermocirculator of PerkinElmer UV/vis spectrometer Lambda 2 version 5.1 (PerkinElmer, Turku, Finland). The absorbance was measured at 340 nm.

According to Maran et al. [39], the GPx activity was measured spectrophotometrically utilizing H_2O_2 as a substrate for Se-dependent peroxidase activity of GPx by following oxidation of NADPH during the first five min in a coupled reaction with GR, as described by Maran et al. [39]. In 1 mL final volume, the reaction mixture contained 4 mM NaN_3 with 500 μL of 0.1 M phosphate buffer, pH 7.5 and 2 mM EDTA, 250 μL of ultrapure water, 100 μL of 20 mM GSH, 2U GR, 50 μL of 5 mM H_2O_2 and 20 μL of 10 mM NADPH. Fifty microliters of homogenized cell samples were added to the reaction mixture. At pH 7.5, one unit of GPx will reduce 1 μmol of GSSG per min. Using a molar absorptivity of NADPH (6.22 mM^{-1} cm^{-1}), the enzymatic activity of the GPx enzyme was determined and expressed as μmol of NADPH oxidized/min/mg of protein. Assays were carried out at 25 °C in a thermocirculator of PerkinElmer UV/vis spectrometer Lambda 2 version 5.1 (PerkinElmer, Turku, Finland). At 340 nM, the absorbance was measured.

According to Zingales et al. [17], the CAT activity was measured. Briefly, 100 μL of homogenized cell sample were combined with 400 μL of 40 mM H_2O_2 and 500 μL of 0.5 M potassium phosphate buffer at pH 7.2. Using a spectrophotometer (Super Aquarius CECIL 9500 CE, Milton Technical Center, Cambridge, United Kingdom), the rate of enzymatic decomposition of H_2O_2 was measured as decreases in absorbance at 240 nm for 3 min at 30 °C. The molar absorptivity of H_2O_2 (43.6 mM^{-1} cm^{-1}) was used to calculate the CAT activity, which is represented as μmol H_2O_2/min/mg of protein.

The Ransod kit (Randox Laboratories Ltd., Antrim, United Kingdom) modified for 1.5 mL cuvettes was used to measure the SOD activity. The free radical superoxide is destroyed by the SOD by being changed into peroxide. A spectrophotometer of PerkinElmer UV/Vis Lambda 2 version 5.1 (PerkinElmer, Turku, Finland) was used to measure the SOD activity at 505 nm for 3 min at 37 °C. Units of SOD/mg protein were used to express the SOD values.

5.7. Determination of Protein Content

The Bio-Rad DC Protein Assay (Bio-Rad, Hercules, CA, USA), with the catalog number 5000116, was used to determine the protein content. Using an automatic ELISA plate reader (MultiSkanEX, Labsystem, Helsinki, Finland), the concentration of protein (μg/mL) was measured at 690 nm.

5.8. Pre-Treatment with BSO or NAC

Cells were pre-treated with NAC or BSO prior to the mycotoxin exposure to examine the effects of these compounds on the modulation of intracellular GSH content and on the enzymatic activities related to GSH content. Approximately 2.25 × 10^5 cells/well were exposed to 1 mM NAC or 60 μM BSO dissolved in the medium for 24 h. After that, the cells were treated with fresh medium containing T-2, Neo, T2-triol, and T2-tetraol at the designated concentrations for 24 h. As a control, cells with 1% MeOH in the medium were used. The GSH content and enzymatic activities were evaluated after 24 h of exposure, as previously mentioned. Comparisons between cells exposed to different concentrations of each mycotoxin in fresh medium and NAC or BSO pre-treatment were performed.

5.9. Statistical Analysis

Data from various independent experiments were expressed as mean ± SEM. Student's *t*-test for paired samples was used for statistical analysis of results. One-way ANOVA followed by the Turkey HDS *posthoc* test for multiple comparisons was used to analyze the difference between groups. A difference level of $p \leq 0.05$ was considered statistically significant.

Author Contributions: Conceptualization, M.T. and M.-J.R.; methodology, M.T. and F.H.; software, M.T.; validation, M.T. and M.-J.R.; formal analysis, M.T. and F.H.; investigation, M.T.; resources, M.T.; data curation, M.T. and F.H.; writing—original draft preparation, M.T.; writing—review and editing, M.-J.R. and Y.R.-C.; visualization, M.-J.R. and Y.R.-C.; supervision, M.-J.R. and Y.R.-C.; project administration, M.-J.R.; funding acquisition, M.-J.R. All authors have read and agreed to the published version of the manuscript.

Funding: This research was funded by Ministerio de Ciencia e Innovación (Spain) (PID2020-115871RB-I00). M. Taroncher thanks the Ministerio de Ciencia e Innovación for providing the Ph.D. grant (PRE2021-096941).

Institutional Review Board Statement: Not applicable.

Informed Consent Statement: Not applicable.

Data Availability Statement: Data is contained within the article.

Conflicts of Interest: The authors declare no conflict of interest.

References

1. Dai, C.; Xiao, X.; Sun, F.; Zhang, Y.; Hoyer, D.; Shen, J.; Tang, S.; Velkov, T. T-2 Toxin Neurotoxicity: Role of Oxidative Stress and Mitochondrial Dysfunction. *Arch. Toxicol.* **2019**, *93*, 3041–3056. [CrossRef] [PubMed]
2. Wild, C.P.; Gong, Y.Y. Mycotoxins and Human Disease: A Largely Ignored Global Health Issue. *Carcinogenesis* **2009**, *31*, 71–82. [CrossRef] [PubMed]
3. Zhang, Y.F.; Yang, J.Y.; Li, Y.K.; Zhou, W. Toxicity and Oxidative Stress Induced by T-2 Toxin in Cultured Mouse Leydig Cells. *Toxicol. Mech. Methods* **2017**, *27*, 100–106. [CrossRef] [PubMed]
4. Lei, R.; Jiang, N.; Zhang, Q.; Hu, S.; Dennis, B.S.; He, S.; Guo, X. Prevalence of Selenium, T-2 Toxin, and Deoxynivalenol in Kashin–Beck Disease Areas in Qinghai Province, Northwest China. *Biol. Trace Elem. Res.* **2016**, *171*, 34–40. [CrossRef]
5. Sun, L.-Y.; Li, Q.; Meng, F.-G.; Fu, Y.; Zhao, Z.-J.; Wang, L.-H. T-2 Toxin Contamination in Grains and Selenium Concentration in Drinking Water and Grains in Kaschin-Beck Disease Endemic Areas of Qinghai Province. *Biol. Trace Elem. Res.* **2012**, *150*, 371–375. [CrossRef]
6. Wang, X.; Liu, X.; Liu, J.; Wang, G.; Wan, K. Contamination Level of T-2 and HT-2 Toxin in Cereal Crops from Aba Area in Sichuan Province, China. *Bull. Environ. Contam. Toxicol.* **2012**, *88*, 396–400. [CrossRef]
7. Arcella, D.; Gergelova, P.; Innocenti, M.L.; Steinkellner, H. Human and Animal Dietary Exposure to T-2 and HT-2 Toxin. *EFSA J.* **2017**, *15*, e05043. [CrossRef]
8. Máté, M.; Gábor, M.; Zsuzsanna, N. The Effects of T-2 Toxin on Animal Health, Focusing Especially on Poultry: Literature review. *Magy. Allatorv. Lapja* **2018**, *140*, 475–483.
9. Taroncher, M.; Rodríguez-Carrasco, Y.; Ruiz, M.J. Interactions between T-2 Toxin and Its Metabolites in HepG2 Cells and in Silico Approach. *Food Chem. Toxicol.* **2021**, *148*, 111942. [CrossRef]
10. Yang, L.; Yu, Z.; Hou, J.; Deng, Y.; Zhou, Z.; Zhao, Z.; Cui, J. Toxicity and Oxidative Stress Induced by T-2 Toxin and HT-2 Toxin in Broilers and Broiler Hepatocytes. *Food Chem. Toxicol.* **2016**, *87*, 128–137. [CrossRef]
11. Deyu, H.; Luqing, C.; Xianglian, L.; Pu, G.; Qirong, L.; Xu, W.; Zonghui, Y. Protective Mechanisms Involving Enhanced Mitochondrial Functions and Mitophagy against T-2 Toxin-Induced Toxicities in GH3 Cells. *Toxicol. Lett.* **2018**, *295*, 41–53. [CrossRef] [PubMed]
12. Petruška, P.; Latacz, A.; Kolesárová, A.; Capcarová, M. Effect of quercetin and T-2 toxin on antioxidant parameters of porcine blood in vitro. *J. Microbiol. Biotechnol. Food Sci.* **2012**, *2*, 510–516.
13. Sheng, K.; Lu, X.; Yue, J.; Gu, W.; Gu, C.; Zhang, H.; Wu, W. Role of Neurotransmitters 5-Hydroxytryptamine and Substance P in Anorexia Induction Following Oral Exposure to the Trichothecene T-2 Toxin. *Food Chem. Toxicol.* **2019**, *123*, 1–8. [CrossRef] [PubMed]
14. Mu, P.; Xu, M.; Zhang, L.; Wu, K.; Wu, J.; Jiang, J.; Chen, Q.; Wang, L.; Tang, X.; Deng, Y. Proteomic Changes in Chicken Primary Hepatocytes Exposed to T-2 Toxin Are Associated with Oxidative Stress and Mitochondrial Enhancement. *Proteomics* **2013**, *13*, 3175–3188. [CrossRef]
15. Juan-García, A.; Bind, M.-A.; Engert, F. Larval Zebrafish as an in Vitro Model for Evaluating Toxicological Effects of Mycotoxins. *Ecotoxicol. Environ. Saf.* **2020**, *202*, 110909. [CrossRef]
16. Taroncher, M.; Rodríguez-Carrasco, Y.; Ruiz, M.J. T-2 Toxin and Its Metabolites: Characterization, Cytotoxic Mechanisms and Adaptive Cellular Response in Human Hepatocarcinoma (HepG2) Cells. *Food Chem. Toxicol.* **2020**, *145*, 111654. [CrossRef]
17. Zingales, V.; Fernández-Franzón, M.; Ruiz, M.J. Sterigmatocystin-Induced Cytotoxicity via Oxidative Stress Induction in Human Neuroblastoma Cells. *Food Chem. Toxicol.* **2020**, *136*, 110956. [CrossRef]
18. Yang, L.; Tu, D.; Wang, N.; Deng, Z.; Zhan, Y.; Liu, W.; Hu, Y.; Liu, T.; Tan, L.; Li, Y.; et al. The Protective Effects of DL-Selenomethionine against T-2/HT-2 Toxins-Induced Cytotoxicity and Oxidative Stress in Broiler Hepatocytes. *Toxicol. In Vitro* **2019**, *54*, 137–146. [CrossRef]

19. Li, X.; Wang, X.; Liu, S.; Wang, J.; Liu, X.Y.; Zhu, Y.; Zhang, L.; Li, R. Betulinic Acid Attenuates T-2 Toxin-Induced Cytotoxicity in Porcine Kidney Cells by Blocking Oxidative Stress and Endoplasmic Reticulum Stress. *Comp. Biochem. Physiol. Part C Toxicol. Pharmacol.* **2021**, *249*, 109124. [CrossRef]
20. Gaigé, S.; Djelloul, M.; Tardivel, C.; Airault, C.; Félix, B.; Jean, A.; Lebrun, B.; Troadec, J.D.; Dallaporta, M. Modification of Energy Balance Induced by the Food Contaminant T-2 Toxin: A Multimodal Gut-to-Brain Connection. *Brain Behav. Immun.* **2014**, *37*, 54–72. [CrossRef]
21. Wu, J.; Jing, L.; Yuan, H.; Peng, S.Q. T-2 Toxin Induces Apoptosis in Ovarian Granulosa Cells of Rats through Reactive Oxygen Species-Mediated Mitochondrial Pathway. *Toxicol. Lett.* **2011**, *202*, 168–177. [CrossRef] [PubMed]
22. He, S.J.; Hou, J.F.; Dai, Y.Y.; Zhou, Z.L.; Deng, Y.F. N-Acetyl-Cysteine Protects Chicken Growth Plate Chondrocytes from T-2 Toxin-Induced Oxidative Stress. *J. Appl. Toxicol.* **2012**, *32*, 980–985. [CrossRef] [PubMed]
23. Agrawal, M.; Bhaskar, A.S.B.; Rao, P.V.L. Involvement of Mitogen-Activated Protein Kinase Pathway in T-2 Toxin-Induced Cell Cycle Alteration and Apoptosis in Human Neuroblastoma Cells. *Mol. Neurobiol.* **2015**, *51*, 1379–1394. [CrossRef]
24. Zhang, X.; Wang, Y.; Velkov, T.; Tang, S.; Dai, C. T-2 Toxin-Induced Toxicity in Neuroblastoma-2a Cells Involves the Generation of Reactive Oxygen, Mitochondrial Dysfunction and Inhibition of Nrf2/HO-1 Pathway. *Food Chem. Toxicol.* **2018**, *114*, 88–97. [CrossRef]
25. Fatima, Z.; Guo, P.; Huang, D.; Lu, Q.; Wu, Q.; Dai, M.; Cheng, G.; Peng, D.; Tao, Y.; Ayub, M.; et al. The Critical Role of P16/Rb Pathway in the Inhibition of GH3 Cell Cycle Induced by T-2 Toxin. *Toxicology* **2018**, *400–401*, 28–39. [CrossRef] [PubMed]
26. Guo, P.; Liu, A.; Huang, D.; Wu, Q.; Fatima, Z.; Tao, Y.; Cheng, G.; Wang, X.; Yuan, Z. Brain Damage and Neurological Symptoms Induced by T-2 Toxin in Rat Brain. *Toxicol. Lett.* **2018**, *286*, 96–107. [CrossRef]
27. Liu, X.; Guo, P.; Liu, A.; Wu, Q.; Xue, X.; Dai, M.; Hao, H.; Qu, W.; Xie, S.; Wang, X.; et al. Nitric Oxide (NO)-Mediated Mitochondrial Damage Plays a Critical Role in T-2 Toxin-Induced Apoptosis and Growth Hormone Deficiency in Rat Anterior Pituitary GH3 Cells. *Food Chem. Toxicol.* **2017**, *102*, 11–23. [CrossRef]
28. Gouze, M.E.; Laffitte, J.; Rouimi, P.; Loiseau, N.; Oswald, I.P.; Galtier, P. Effect of Various Doses of Deoxynivalenol on Liver Xenobiotic Metabolizing Enzymes in Mice. *Food Chem. Toxicol.* **2006**, *44*, 476–483. [CrossRef]
29. Agahi, F.; Juan-García, A.; Font, G.; Juan, C. Study of Enzymatic Activity in Human Neuroblastoma Cells SH-SY5Y Exposed to Zearalenone's Derivates and Beauvericin. *Food Chem. Toxicol.* **2021**, *152*, 112227. [CrossRef]
30. Dinu, D.; Bodea, G.O.; Ceapa, C.D.; Munteanu, M.C.; Roming, F.I.; Serban, A.I.; Hermenean, A.; Costache, M.; Zarnescu, O.; Dinischiotu, A. Adapted Response of the Antioxidant Defense System to Oxidative Stress Induced by Deoxynivalenol in Hek-293 Cells. *Toxicon* **2011**, *57*, 1023–1032. [CrossRef]
31. Ramyaa, P.; Krishnaswamy, R.; Padma, V.V. Quercetin Modulates OTA-Induced Oxidative Stress and Redox Signalling in HepG2 Cells-Up Regulation of Nrf2 Expression and down Regulation of NF-KB and COX-2. *Biochim. Biophys. Acta Gen. Subj.* **2014**, *1840*, 681–692. [CrossRef] [PubMed]
32. Tatay, E.; Espín, S.; García-Fernández, A.J.; Ruiz, M.J. Oxidative Damage and Disturbance of Antioxidant Capacity by Zearalenone and Its Metabolites in Human Cells. *Toxicol. In Vitro* **2017**, *45*, 334–339. [CrossRef] [PubMed]
33. Tatay, E.; Font, G.; Ruiz, M.J. Cytotoxic Effects of Zearalenone and Its Metabolites and Antioxidant Cell Defense in CHO-K1 Cells. *Food Chem. Toxicol.* **2016**, *96*, 43–49. [CrossRef] [PubMed]
34. Mallebrera, B.; Font, G.; Ruiz, M.J. Disturbance of Antioxidant Capacity Produced by Beauvericin in CHO-K1 Cells. *Toxicol. Lett.* **2014**, *226*, 337–342. [CrossRef]
35. Krishnaswamy, R.; Devaraj, S.N.; Padma, V.V. Lutein Protects HT-29 Cells against Deoxynivalenol-Induced Oxidative Stress and Apoptosis: Prevention of NF-KB Nuclear Localization and down Regulation of NF-KB and Cyclo-Oxygenase-2 Expression. *Free Radic. Biol. Med.* **2010**, *49*, 50–60. [CrossRef]
36. Gao, W.; Jiang, L.; Ge, L.; Chen, M.; Geng, C.; Yang, G.; Li, Q.; Ji, F.; Yan, Q.; Zou, Y.; et al. Sterigmatocystin-Induced Oxidative DNA Damage in Human Liver-Derived Cell Line through Lysosomal Damage. *Toxicol. In Vitro* **2015**, *29*, 1–7. [CrossRef]
37. Woelflingseder, L.; del Favero, G.; Blažević, T.; Heiss, E.H.; Haider, M.; Warth, B.; Adam, G.; Marko, D. Impact of Glutathione Modulation on the Toxicity of the Fusarium Mycotoxins Deoxynivalenol (DON), NX-3 and Butenolide in Human Liver Cells. *Toxicol. Lett.* **2018**, *299*, 104–117. [CrossRef]
38. Ruiz, M.J.; Festila, L.E.; Fernández, M. Comparison of Basal Cytotoxicity of Seven Carbamates in CHO-K1 Cells. *Toxicol. Environ. Chem.* **2006**, *88*, 345–354. [CrossRef]
39. Maran, E.; Fernández, M.; Barbieri, P.; Font, G.; Ruiz, M.J. Effects of Four Carbamate Compounds on Antioxidant Parameters. *Ecotoxicol. Environ. Saf.* **2009**, *72*, 922–930. [CrossRef]

toxins MDPI

Article

Targeted Sphingolipid Analysis in Heart, Gizzard, and Breast Muscle in Chickens Reveals Possible New Target Organs of Fumonisins

Philippe Guerre [1,*], Caroline Gilleron [2], Maria Matard-Mann [2] and Pi Nyvall Collén [2]

[1] National Veterinary School of Toulouse, ENVT, Université de Toulouse, F-31076 Toulouse, France
[2] Olmix S.A., ZA du Haut du Bois, F-56580 Bréhan, France
* Correspondence: philippe.guerre@envt.fr

Abstract: Alteration of sphingolipid synthesis is a key event in fumonisins toxicity, but only limited data have been reported regarding the effects of fumonisins on the sphingolipidome. Recent studies in chickens found that the changes in sphingolipids in liver, kidney, lung, and brain differed greatly. This study aimed to determine the effects of fumonisins on sphingolipids in heart, gizzard, and breast muscle in chickens fed 20.8 mg FB1 + FB2/kg for 9 days. A significant increase in the sphinganine:sphingosine ratio due to an increase in sphinganine was observed in heart and gizzard. Dihydroceramides and ceramides increased in the hearts of chickens fed fumonisins, but decreased in the gizzard. The dihydrosphingomyelin, sphingomyelin, and glycosylceramide concentrations paralleled those of ceramides, although the effects were less pronounced. In the heart, sphingolipids with fatty acid chain lengths of 20 to 26 carbons were more affected than those with 14–16 carbons; this difference was not observed in the gizzard. Partial least squares-discriminant analysis on sphingolipids in the heart allowed chickens to be divided into two distinct groups according to their diet. The same was the case for the gizzard. Pearson coefficients of correlation among all the sphingolipids assayed revealed strong positive correlations in the hearts of chickens fed fumonisins compared to chickens fed a control diet, as well as compared to gizzard, irrespective of the diet fed. By contrast, no effect of fumonisins was observed on sphingolipids in breast muscle.

Keywords: fumonisin; sphingolipid; ceramide; sphingomyelin; muscle; heart

Key Contribution: This study reports for the first time major changes in the sphingolipidome of the heart, gizzard, and pectoral muscle in chickens exposed to a fumonisins dose considered non-toxic.

Citation: Guerre, P.; Gilleron, C.; Matard-Mann, M.; Nyvall Collén, P. Targeted Sphingolipid Analysis in Heart, Gizzard, and Breast Muscle in Chickens Reveals Possible New Target Organs of Fumonisins. *Toxins* **2022**, *14*, 828. https://doi.org/10.3390/toxins14120828

Received: 7 November 2022
Accepted: 23 November 2022
Published: 24 November 2022

Publisher's Note: MDPI stays neutral with regard to jurisdictional claims in published maps and institutional affiliations.

Copyright: © 2022 by the authors. Licensee MDPI, Basel, Switzerland. This article is an open access article distributed under the terms and conditions of the Creative Commons Attribution (CC BY) license (https://creativecommons.org/licenses/by/4.0/).

1. Introduction

Fumonisins are mycotoxins produced by molds of the genus *Fusarium*, mainly *F. verticillioides*. This class of contaminant is found worldwide in maize, maize byproducts, and animal feed [1,2]. Among the various types of fumonisins, fumonisins B are the most abundant, and among fumonisins B, fumonisin B1 (FB1), and to a lesser extent fumonisin B2 (FB2), are the most abundant, toxic, and regulated. Fumonisins B are considered carcinogenic in rodents and probable carcinogens for humans, with a provisional maximum tolerable daily intake (PMTDI) being set for humans and maximum levels of fumonisins being recommended in animal feed [2–5]. Fumonisins have multiple toxic effects in animals as they are neurotoxic in horses, pneumotoxic in pigs, hepatotoxic, and nephrotoxic in most animal species, including poultry. At the cellular level, oxidative stress, mitochondrial dysfunction, and induction of apoptosis or cell proliferation have been reported [6–9]. Although the effects of FB1 vary greatly, inhibition of ceramide synthase (CerS) due to structural analogy between FB1 and sphingoid bases is considered one of the main mechanisms underlying the toxicity [10,11]. This inhibition leads to reduction of the de novo synthesis of ceramides and an increase in sphinganine (Sa) concentrations, while sphingosine (So) concentrations tend to decrease (Figure 1). Ceramide levels in cells can be maintained by hydrolysis of sphingomyelins, which is known as the salvage pathway for ceramide synthesis.

The pronounced effect of fumonisins on sphingolipids has led to widespread use of the Sa:So ratio as an exposure biomarker in humans and animals, including poultry [9,11,12]. There have, however, only been a limited number of studies to date aimed at characterization of the effects of fumonisins on the sphingolipidome. Sphingolipidomic studies in chickens and turkeys conducted at the maximum recommended concentration for fumonisins in feed in the European Union have revealed numerous alterations in sphingolipids in the liver [13,14]. These changes are consistent with the previously reported inhibition of CerS in vitro and in vivo [10,11]. Because the effects of fumonisins on sphingolipids in chicken and turkey livers were characterized by a decrease in C14–C16 ceramides and an increase in C20–C24 ceramides, it has been hypothesized that inhibition of CerS may be more pronounced for CerS5 than that for CerS2 (Figure 1), whereas the opposite has been suggested to occur in mammals [11,15]. This difference is important because a decrease in C20–C24 ceramides, and a compensatory increase in C16 ceramide, have been reported in CerS2 knockout mice, and these changes were responsible for hepatotoxicity that resembles fumonisins toxicity [16,17]. Sphingolipid analysis in chickens and turkeys also revealed that sphingomyelin levels did not decrease in liver [13,14]. This latter result is also important for explaining the lack of toxicity of fumonisins in these studies, with a decrease in sphingomyelins being reported in tissue and cell cultures at toxic concentrations of fumonisins, and this event was found to precede ceramide-induced apoptosis [11,15,18–21].

Targeted sphingolipids analysis conducted in the kidneys, lungs, and brains of chickens revealed that the effects of fumonisins were not limited to changes in sphingolipids in the liver, and that the effects varied according to the organ studied [22]. The effects in kidneys were similar to those observed in livers (Figure 1). By contrast, sphinganine and ceramides were unchanged in the lungs and brain, suggesting no inhibition of CerS in these organs. However, major changes in sphingolipid concentrations in lung and brain were found in chickens fed with a diet that was contaminated with fumonisins compared to controls [22]. In the lungs, these changes corresponded to an increase in sphinganine-1-phosphate and a decrease in glycosylceramides. In the brain, these changes corresponded to an increase in deoxysphinganine, sphingosine, ceramides, and sphingomyelins (Figure 1). Interestingly, the most affected sphingolipids in lung and in brain in chickens fed fumonisins corresponded to compounds that have key functions in these organs in human diseases [23–26]. As only traces of FB1 were found in lungs and no FB1 was detected in brains, it has been hypothesized that the effects of fumonisins are due to indirect effects, metabolites, or cell-mediated (Figure 1) [22].

Although fumonisins have been reported to exert cardiac toxicity in mammals and avian species [27–30], the effect of fumonisins on sphingolipids in the heart remain unknown. These effects seem interesting to study as sphingolipids have a key role in the onset of cardiovascular diseases in humans [31]. Additionally, diets containing fumonisins have been reported to damage the gizzard and to alter its weight [32–36], although there have been no studies to date regarding the effect of fumonisins on sphingolipids in the gizzard. Moreover, sphingolipids are known to play a key role in skeletal muscle [37–40], but the effects of fumonisins on this tissue remain unknown. The purpose of this study was to investigate the effects of 20.8 mg FB1 + FB2/kg fed to chicken for 9 days on sphingolipids in the heart, gizzard, and breast muscle. Targeted sphingolipid analysis was conducted by UHPLC-MSMS of samples obtained from animals by a method that has already been used to characterize the effects of fumonisins on sphingolipids in liver, kidney, lung, and brain [13,22].

MDPI

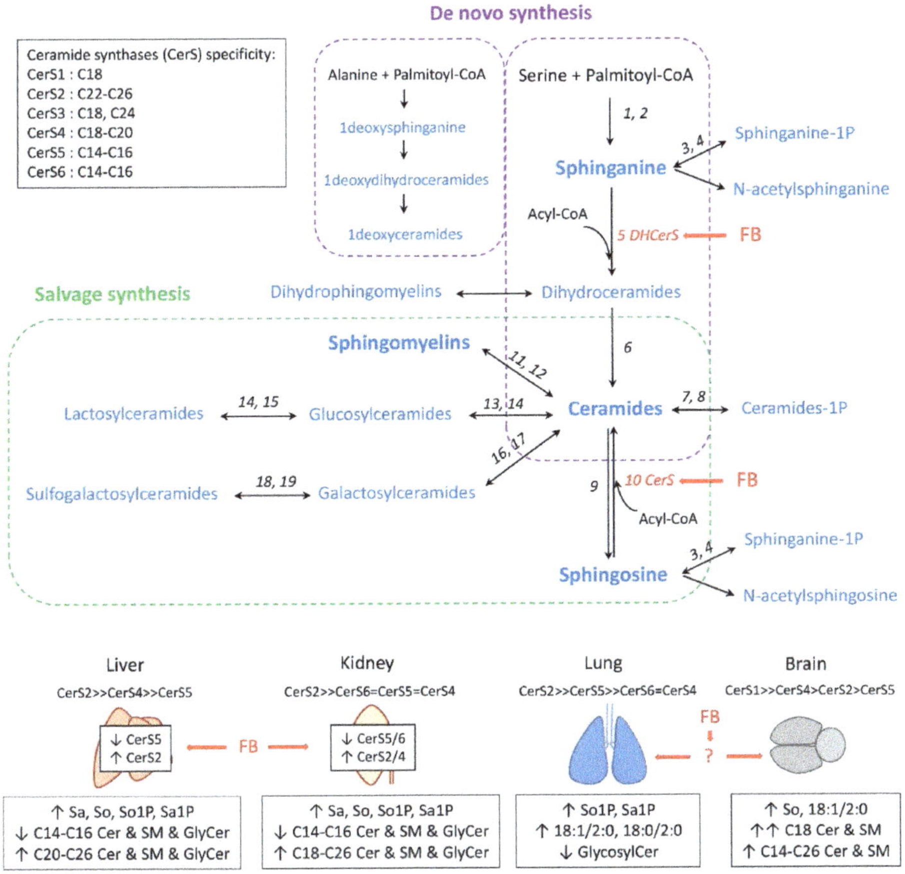

Figure 1. Schematic outline of the sphingolipid synthesis pathways [11]. Ceramide synthases (CerS) specificity and tissue expression were obtained from [41]. Fumonisins B (FB) are inhibitors of CerS (red arrow). 1-Deoxysphinganine is generated from alanine [42]. The effects of fumonisins on sphingolipids in liver, kidney, lung, and brain are summarized from [13,22]. 1: Serine palmitoyl-transferase; 2: Reductase; 3: Sphinganine kinase; 4: Phosphatase; 5: (Dihydro)ceramide synthase; 6: Dihydroceramide desaturase; 7: Ceramide kinase; 8: Phosphatase; 9: Ceramidase; 10: Ceramide synthase; 11: Sphingomyelin synthase; 12: Sphingomyelinase; 13: Glucosylceramide synthase; 14: β-glucosidase; 15: Lactosylceramide synthase; 16: Ceramide galactosyltransferase; 17: Galactosylase; 18: Cerebroside sulfotransferase; and 19: Arylsulfatase.

2. Results

2.1. Effect of Fumonisins According to the Class of Sphingolipids

The concentrations of sphingolipids measured in heart, gizzard, and breast muscle of chickens fed the control diet and chickens fed a diet containing 20.8 mg FB1 + FB2/kg over a period of 9 days are presented in Table 1. The effects of fumonisins according to the class of sphingolipids are presented in Figure 2.

Table 1. Concentrations of sphingolipids measured in the heart, gizzard, and breast muscle of broilers fed a control diet free of mycotoxins and broilers fed 20.8 mg FB1 + FB2 for 9 days [1].

	Heart		Gizzard		Breast Muscle	
	Control	FB	Control	FB	Control	FB
			Sphingoid bases and derivates			
d18:1 (So)	5326 ± 596	5891 ± 1086	1367 ± 387	1157 ± 183	2585 ± 636	3107 ± 388 *
d18:0 (Sa)	868 ± 133	1204 ± 222 ***	233 ± 19	213 ± 24 *	127 ± 89	110 ± 72
18:1/2:0	62 ± 13	69 ± 11	27 ± 8	29 ± 9	24 ± 8	23 ± 7
18:0/2:0	ND	ND	16 ± 5	25 ± 7 **	ND	ND
LysoSM	249 ± 41	234 ± 33	353 ± 34	327 ± 47	130 ± 29	122 ± 23
			Ceramides and dehydroceramides			
18:1/14:0	5277 ± 894	6217 ± 1588	307 ± 85	204 ± 45 **	75 ± 22	70 ± 9
18:1/16:0	77,884 ± 12,241	90,785 ± 21,079	110,479 ± 33,863	81,436 ± 18,040 *	14,530 ± 5548	13,523 ± 2337
18:0/16:0	1125 ± 193	1353 ± 298	1753 ± 533	1089 ± 104 **	375 ± 79	355 ± 40
18:1/18:1	1036 ± 225	1232 ± 243	343 ± 132	329 ± 102	ND	ND
18:1/18:0	25,264 ± 4163	30,555 ± 10,317	22,600 ± 8108	17,194 ± 4879	15,646 ± 4753	15,281 ± 3584
18:0/18:0	62 ± 20	82 ± 35	328 ± 81	225 ± 44 **	33 ± 12	40 ± 13
18:1/20:0	17,052 ± 2930	22,897 ± 5917 *	17,282 ± 5215	13,521 ± 3411	3574 ± 1426	3510 ± 889
18:1/22:2	1556 ± 344	1882 ± 549	49,559 ± 14,820	36,320 ± 11,872 *	308 ± 97	290 ± 67
18:1/22:1	681 ± 130	972 ± 292 *	12,002 ± 5214	8371 ± 2826	314 ± 103	299 ± 59
18:1/22:0	31,576 ± 4466	39,187 ± 8219 *	26,966 ± 7331	21,517 ± 4277	6466 ± 2186	6222 ± 1146
18:0/22:0	60 ± 19	76 ± 33	223 ± 53	168 ± 46 **	ND	ND
18:1/23:1	ND	ND	991 ± 344	772 ± 215	ND	ND
18:1/23:0	2140 ± 341	2553 ± 496 *	7308 ± 1401	6070 ± 988 *	445 ± 128	402 ± 80
18:0/23:0	23 ± 9	31 ± 12	ND	ND	ND	ND
18:1/24:2	3273 ± 714	4170 ± 1002 *	272,361 ± 78,578	177,125 ± 43,980 **	1380 ± 435	1377 ± 292
18:1/24:1	25,974 ± 4601	32,914 ± 9442	71,565 ± 28,535	54,187 ± 16,562	11,736 ± 3967	11,664 ± 1939
18:1/24:0	20,773 ± 2780	23,963 ± 4458	14,143 ± 2922	12,127 ± 1801	4563 ± 1261	4399 ± 1009
18:0/24:0	683 ± 139	852 ± 349	ND	ND	ND	ND
18:1/26:2	539 ± 125	713 ± 197 *	2815 ± 985	1862 ± 383 *	26 ± 20	21 ± 8
18:1/26:1	444 ± 91	526 ± 135	1270 ± 333	881 ± 243 *	32 ± 16	33 ± 17
18:1/26:0	ND	ND	230 ± 82	151 ± 38 *	ND	ND
			Sphingomyelins and dehydrosphingomyelins			
SM18:1/14:0	13,516 ± 2450	12,256 ± 1114	1962 ± 424	1749 ± 274	1024 ± 250	1054 ± 150
SM18:1/16:0	279,024 ± 24,417	269,938 ± 20,805	296,493 ± 37,427	283,011 ± 26,615	142,118 ± 28,443	134,059 ± 11,388
SM18:0/16:0	29,224 ± 3338	30,528 ± 3387	287,685 ± 53,159	268,997 ± 52,074	14,195 ± 4236	14,297 ± 2589
SM18:1/18:1	7638 ± 1747	8262 ± 1449	ND	ND	ND	ND
SM18:1/18:0	718,620 ± 70,463	774,687 ± 75,203	455,089 ± 63,676	452,100 ± 59,844	557,238 ± 118,116	576,469 ± 66,525
SM18:0/18:0	9964 ± 2069	10,078 ± 2031	20,824 ± 4824	19,992 ± 6472	2876 ± 682	2993 ± 579
SM18:1/20:0	131,232 ± 8241	147,199 ± 14,149 **	46,568 ± 5174	46,010 ± 6491	33,633 ± 10,149	34,817 ± 5905
SM18:0/20:0	1780 ± 228	1898 ± 404	2872 ± 646	2899 ± 819	413 ± 94	414 ± 131
SM18:1/22:2	7332 ± 1102	8022 ± 786	70,886 ± 24,225	61,750 ± 14,671	2554 ± 559	2480 ± 283
SM18:1/22:1	8259 ± 747	9199 ± 1216	32,164 ± 5132	31,112 ± 8408	4092 ± 1083	3931 ± 687
SM18:1/22:0	509,447 ± 41,889	546,720 ± 65,977	221,170 ± 32,543	201,609 ± 30,819	69,740 ± 18,958	69,604 ± 15,227
SM18:0/22:0	5074 ± 925	5230 ± 1059	3921 ± 1054	3492 ± 1033	520 ± 140	498 ± 102
SM18:1/23:1	7885 ± 1113	8422 ± 1406	5465 ± 1250	4849 ± 958	2787 ± 927	2969 ± 955
SM18:1/23:0	27,322 ± 3979	29,310 ± 5566	33,713 ± 7600	34,101 ± 12,246	2134 ± 578	2062 ± 468
SM18:0/23:0	3984 ± 755	4392 ± 1105	466 ± 90	423 ± 115	765 ± 269	705 ± 180
SM18:1/24:3	ND	ND	19 ± 2	18 ± 1	23 ± 1	24 ± 2

Table 1. *Cont.*

	Heart		Gizzard		Breast Muscle	
	Control	FB	Control	FB	Control	FB
SM18:1/24:2	30,205 ± 4115	31,127 ± 3310	1078,170 ± 385,433	876,991 ± 143,010	14,609 ± 4097	14,083 ± 2864
SM18:1/24:1	254,344 ± 30,783	271,904 ± 47,893	381,839 ± 65,468	378,196 ± 73,027	57,450 ± 14,974	58,443 ± 14,194
SM18:1/24:0	152,411 ± 18,191	166,908 ± 26,357	76,543 ± 12,027	70,781 ± 14,501	15,543 ± 4195	15,449 ± 3559
SM18:0/24:1	2453 ± 434	2665 ± 483	6900 ± 2267	6308 ± 2712	319 ± 75	333 ± 89
SM18:0/24:0	605 ± 104	564 ± 124	954 ± 281	794 ± 179	ND	ND
SM18:1/25:2	3295 ± 2038	3682 ± 2300	7145 ± 1906	6169 ± 1291	239 ± 60	211 ± 38
SM18:1/25:1	1527 ± 243	1771 ± 348	5960 ± 1313	5792 ± 1904	303 ± 77	285 ± 93
SM18:1/25:0	817 ± 215	948 ± 221	1493 ± 422	1478 ± 453	108 ± 26	106 ± 28
SM18:1/26:3	5393 ± 677	5784 ± 586	6079 ± 2098	5165 ± 1483	324 ± 119	381 ± 123
SM18:1/26:2	8088 ± 1363	9027 ± 1297	35,708 ± 11,840	30,033 ± 8733	529 ± 182	565 ± 217
SM18:1/26:1	2500 ± 483	2742 ± 485	12,702 ± 3216	11,639 ± 3881	307 ± 101	329 ± 97
SM18:1/26:0	ND	ND	857 ± 245	841 ± 291	ND	ND
Hexosyl-, and lactosylceramides, and ceramides sulfatides						
Hex18:1/16:0	2695 ± 477	2789 ± 640	730 ± 250	601 ± 143	832 ± 257	781 ± 203
Hex18:1/18:0	20,024 ± 4349	23,297 ± 5049	2913 ± 1341	2329 ± 1234	ND	ND
Hex18:1/20:0	3123 ± 538	3534 ± 1098	676 ± 283	688 ± 419	ND	ND
Hex18:1/22:0	2199 ± 370	2811 ± 752 *	1179 ± 716	874 ± 385	ND	ND
Hex18:1/24:1	14,817 ± 1554	17,483 ± 3396 *	1160 ± 660	637 ± 261 *	1340 ± 976	976 ± 518
Hex18:1/24:0	4285 ± 875	5119 ± 1405	934 ± 351	905 ± 466	ND	ND
Lac18:1/16:0	3280 ± 663	3293 ± 1192	5131 ± 1257	3919 ± 1036 *	1172 ± 441	1091 ± 162
Lac18:1/18:0	164,969 ± 26,340	195,360 ± 49,026	24,678 ± 6030	23,262 ± 8744	16,877 ± 6682	18,010 ± 4202
Lac18:1/20:0	16,080 ± 3467	20,537 ± 7002	971 ± 667	764 ± 424	7237 ± 3232	7812 ± 2430
Lac18:1/22:0	11,757 ± 4980	17,071 ± 10,225	2325 ± 821	2803 ± 1798	1432 ± 654	1783 ± 610
Lac18:1/24:1	17,053 ± 2330	22,119 ± 5235 *	3281 ± 970	3747 ± 2019	18,040 ± 8822	23,515 ± 7061
Lac18:1/24:0	18,598 ± 3893	24,212 ± 8371	ND	ND	2308 ± 1149	2441 ± 785
ST18:1/24:1	39,266 ± 6545	46,422 ± 11,226	ND	ND	ND	ND
ST18:1/24:0	56,643 ± 9617	71,979 ± 13,510 **	ND	ND	ND	ND

[1] Results are expressed in pmol sphingolipids/g as means ± SD, n = 10. ANOVA was used to assess the difference between groups. Significant differences among groups were reported as follow: (*) 0.05 < p <0.01; (**) p < 0.01; (***) p < 0.001. ND: not detected.

Feeding fumonisins contaminated diet significantly increased the Sa:So ratio in heart, whereas no significant differences between the controls and the treated samples were observed in gizzard and breast muscle. The effects of fumonisins on dihydroceramides and on ceramides also varied according to the tissue studied. A significant increase in dihydroceramides and ceramides was observed in the heart, whereas a significant decrease was observed in the gizzard. No effect was observed on breast muscle (Figure 2). No significant effect of fumonisins was observed on the concentrations of dihydrosphingomyelins, sphingomyelins, or glycosylceramides, irrespective of the tissue.

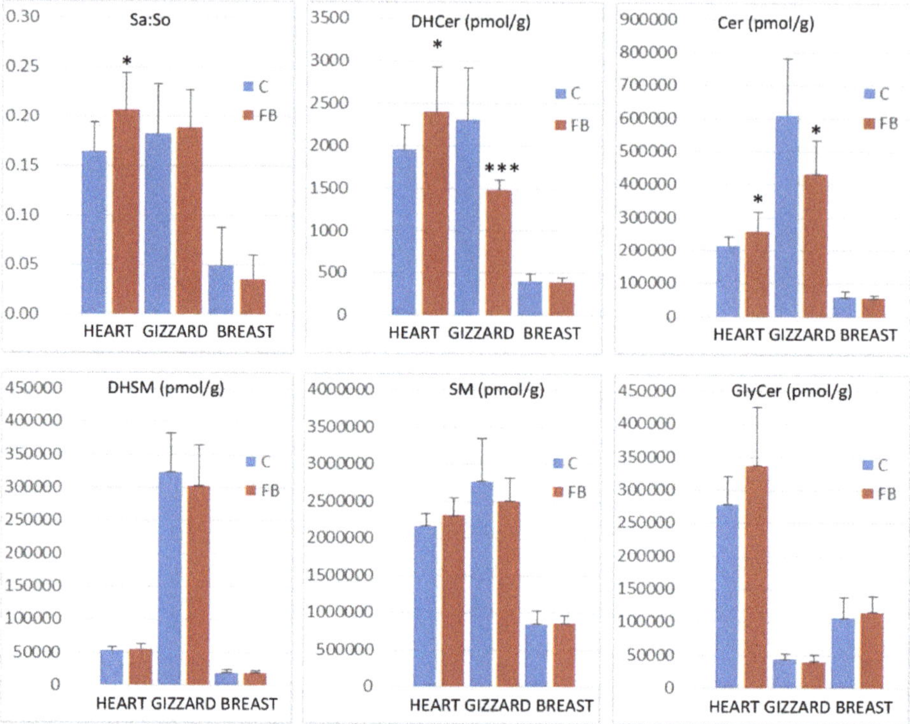

Figure 2. Sphinganine:sphingosine (Sa:So) ratios and concentrations of sphingolipid reported as the total of dihydroceramides (DHCer), ceramides (Cer), dihydrosphingomyelins (DHSM), sphingomyelins (SM), and glycosylceramides (GlyCer) in the heart, gizzard, and breast muscle of chickens fed the control diet free of mycotoxins (C) and chickens fed 9 days with a diet containing 20.8 mg FB1 + FB2/kg (FB). * 0.05< p <0.01 and *** $p < 0.001$.

2.2. Effects of Fumonisins on Sphingolipids in Heart

Being fed a diet containing 20.8 mg FB1 + FB2/kg over a period of 9 days significantly increased the Sa level in heart (Table 1). A significant increase in 18:1/20:0, 18:1/22.1, 18:1/22:0, 18:1/23:0, 18:1/24:2, and 18:1/26:2 ceramides was also observed. The levels of most of the other ceramides and dihydroceramides were increased by fumonisins, but the effects were not significant. SM18:1/20:0 was the only sphingomyelin significantly increased in heart by fumonisins. The effects of the toxins on other sphingomyelins varied according to the fatty acid carbon chain length. SM14–16 decreased slightly, whereas a weak but not significant increase was observed for SM18–SM26. Concerning glycosylceramides, the levels of Hex18:1/22:0, Hex18:1/24:1, Lac18:1/24:1, Lac18:1/24:0, and ceramide sulfatide ST18:1/24:0 were significantly increased by fumonisins. The levels of all other hexosylceramides and lactosylceramides were also increased, but the effects were not significant (Table 1).

Partial least square-discriminant analysis was conducted to determine whether changes in the heart sphingolipidome were sufficient to separate chickens into two groups according to their exposure to fumonisins. As shown in Figure 3C, a good separation of chickens was observed in this analysis. The values of the R^2Y and R^2X indices were 0.608 and 0.632, respectively, indicating that the selected sphingolipids could predict the group that the chickens belonged to. The value of Q^2 with the two first components was 0.403, indicating that the model was of medium quality (Figure 3D). Not surprisingly, sphingolipids that were important in the projection corresponded to variables that differed significantly in chickens fed fumonisins versus the controls. It is also interesting to note that SM18:1/14:0 and SM18:1/16:0 were important in explaining the repartition of chickens into the two groups (Figure 3A,B), whereas only a very weak difference was noted in the mean concentration of these analytes in heart compared to the controls (Table 1). Other variables important in the projection corresponded to sphingolipids that were increased by fumonisins for which the changes measured were considered not significant by ANOVA.

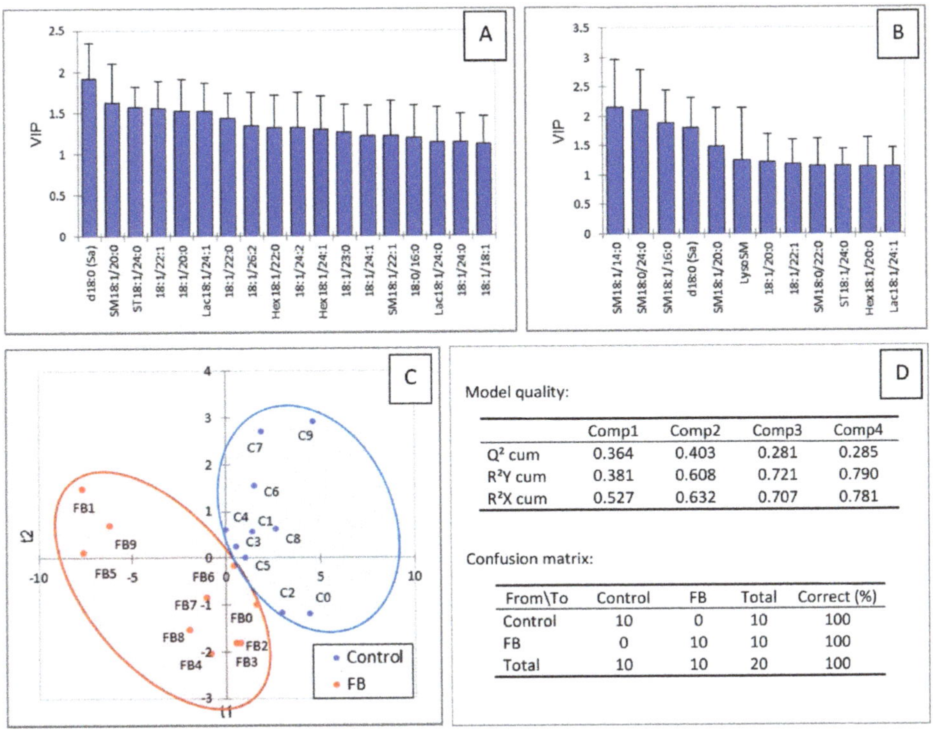

Figure 3. PLS-DA of sphingolipids measured in the heart of chickens fed 9 days with a control diet free of mycotoxins or a diet containing 20.8 mg FB1 + FB2/kg. (**A**) Scores of the VIP for the first component, and (**B**) the second component. (**C**) Discrimination on the factor axes extracted from the original explanatory variables. (**D**) Quality of the model and confusion matrix for the training sample (variable groups).

Figure 4 shows the correlation observed in heart for So, Sa, and sphingolipids with 14 to 24 carbon fatty acid saturated chain lengths. The numerical values of the Pearson coefficients of correlation are reported in Table S1.

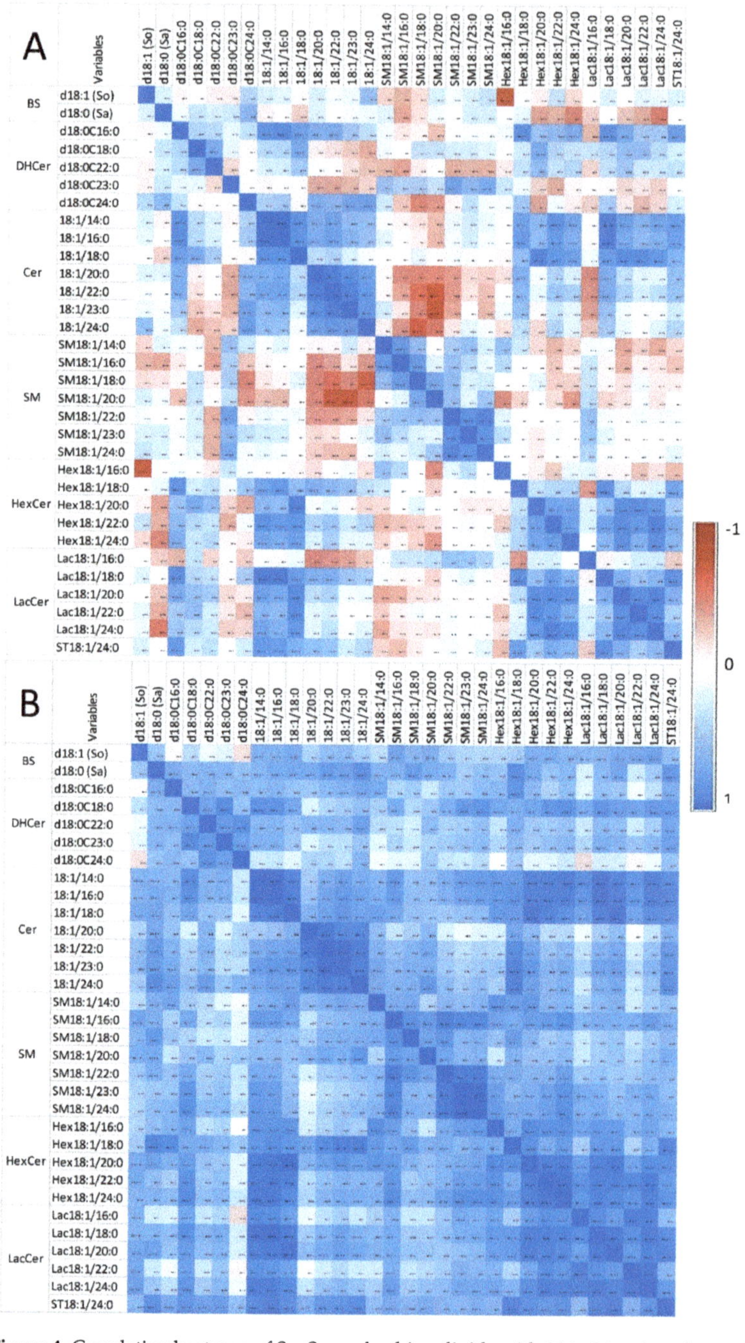

Figure 4. Correlation heatmap of So, Sa, and sphingolipids with 14 to 24 carbon fatty acid saturated chain lengths observed in the heart of chickens fed the control diet (**A**) and chickens fed 9 days with a diet containing 20.8 mg FB1 + FB2/kg (**B**). Numeric values of the Pearson coefficients of correlation observed among all sphingolipids assayed in this study are reported in Table S1.

Only weak correlations were observed between Sa and So, among the sphingoid bases and the other sphingolipids in control chickens (i.e., unexposed to fumonisins) (Figure 4A). Moreover, the correlations among the different classes of sphingolipids were weak, except for ceramides and glycosylceramides, which exhibited significant positive correlations, and ceramides and sphingomyelins, which exhibited significant negative correlations. By contrast, significant positive correlations were observed within the same class among dihydroceramides, ceramides, sphingomyelins, hexosylceramides, and lactosylceramides, which have similar fatty acid chain lengths. The feeding of fumonisins strongly increased the positive correlations measured among sphingolipids in the heart (Figure 4B). Notably, significant positive correlations among the different classes of sphingolipids were observed in chickens fed fumonisins, which was not the case in the controls. Moreover, the coefficients of correlation found among the same class of sphingolipids with similar fatty acid chain lengths were generally increased (Table S1B). The correlations among unsaturated sphingolipids and other analytes were close to those observed for the corresponding saturated forms. The correlations among C25 and C26 sphingolipids and other analytes were close to those observed for C24 sphingolipids (Table S1).

All of these results suggest a significant effect of fumonisins on sphingolipids in the heart. This effect was dominated by an increase in dihydroceramides and ceramides and, to a lesser extent, an increase in glycosylceramides. This was accompanied by strong positive correlations between the different classes and subclasses of sphingolipids. The effects of fumonisins also varied according to the fatty acid chain length. C20–C26 ceramides were significantly increased by the addition of the toxins to the feed ($p = 0.029$), whereas C14–C16 ceramides were not ($p = 0.112$). Additionally, Hex20–Hex26 hexosylceramides, Lac20–Lac26 lactosylceramides, and SM20–SM26 sphingomyelins tended to increase, whereas Hex16, Lac16, and SM14SM16 were unaffected.

2.3. *Effects of Fumonisins on Sphingolipids in Gizzard*

As shown in Table 1, the sphinganine concentration in the gizzards of chickens fed a diet containing 20.8 mg FB1 + FB2/kg over a period of 9 days was significantly increased. A significant increase in N-acetylsphinganine (d18:0/2:0) was also observed. By contrast, the levels of most of the dihydroceramides and most of the ceramides assayed were decreased by fumonisins, and this effect was significant for 18:1/14:0, 18:1/16.0, 18:0/16:0, 18:0/18:0, 18:1/22:2, 18:0/22:0, 18:1/23:0, 18:1/24:2, 18:1/26:2, 18:1/26:1, and 18:1/26:2. No significant difference between groups was observed in terms of the dihydrosphingomyelin and sphingomyelin levels in gizzard. Hex18:1/24:1 and Lac18:1/16:0 were significantly decreased by fumonisins, and most of the other glycosylceramides tended to decrease, but the effect was not significant (Table 1).

Figure 5 shows the results of the partial least square-discriminant analysis conducted on the sphingolipids in gizzard. A good separation of the chickens according to the feed was observed (Figure 5C). The values of the R^2Y, R^2X, and Q^2 indices, which were 0.755, 0.615, and 0.671, respectively, indicated the good predicted appurtenance of the chickens to the different groups and the good quality of the model. The confusion matrix confirmed the model was highly sensitive and specific (Figure 5D). The sphingolipids that were important in the projection mainly corresponded to the dihydroceramides, ceramides, and glycosylceramides that were significantly decreased in chickens fed fumonisins. SM18:1/16:0 and SM18:1/18:0 also had VIP scores above 1.1 in this analysis (Figure 5A,B), although their concentrations in gizzards did not differ from the controls (Table 1).

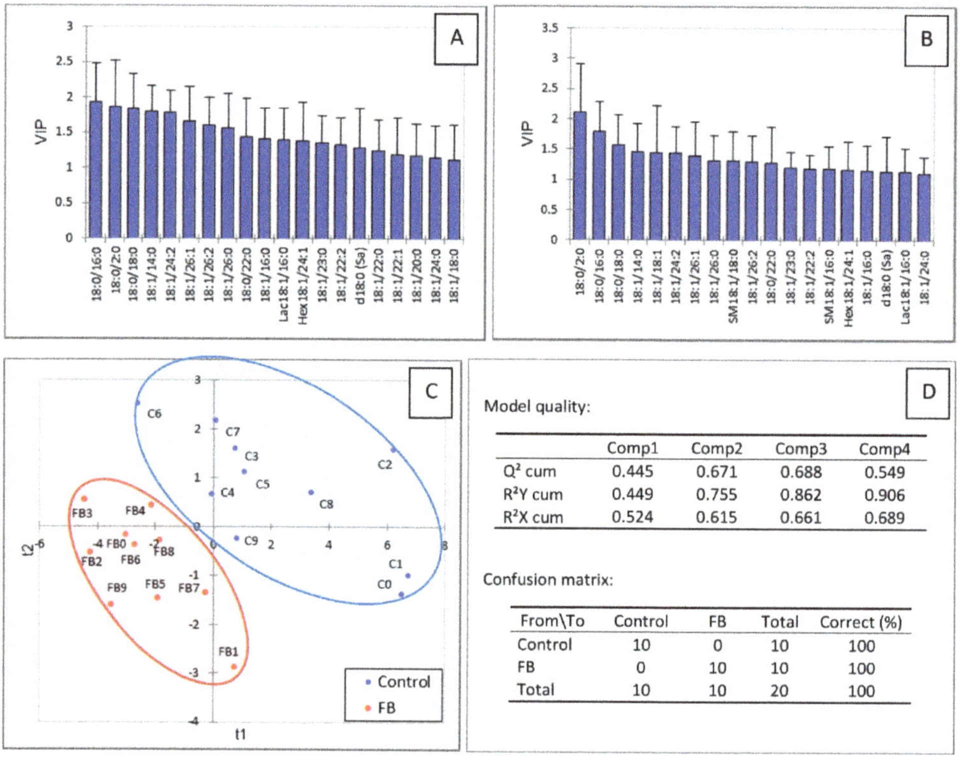

	Comp1	Comp2	Comp3	Comp4
Q^2 cum	0.445	0.671	0.688	0.549
R^2Y cum	0.449	0.755	0.862	0.906
R^2X cum	0.524	0.615	0.661	0.689

From\To	Control	FB	Total	Correct (%)
Control	10	0	10	100
FB	0	10	10	100
Total	10	10	20	100

Figure 5. PLS-DA of sphingolipids measured in the gizzards of chickens fed for 9 days with a control diet free of mycotoxins or a diet containing 20.8 mg FB1 + FB2/kg. (**A**) Scores of the VIP for the first component and (**B**) the second component. (**C**) Discrimination on the factor axes extracted from the original explanatory variables. (**D**) Quality of the model and confusion matrix for the training sample (variable groups).

The correlations measured among the assayed sphingolipids in gizzards in this study are reported in Figure 6 and Table S2. Strong significant positive correlations were observed among So and dihydroceramides and among So and ceramides, whereas weak negative correlations were observed among Sa and dihydroceramides in chickens unexposed to fumonisins (Figure 6A). Significant positive correlations were observed among sphingolipids of the same class with similar fatty acid chain lengths. Additionally, a significant positive correlation was observed between dihydroceramides and ceramides and between sphingomyelins and hexosylceramides. Most of the correlations between ceramides and sphingomyelins, ceramides and hexosylceramides, and between hexosylceramides and lactosylceramides were weak and not significant. Feeding fumonisins only resulted in weak effects on the correlations observed among sphingolipids (Figure 6B). The correlations among unsaturated sphingolipids and other analytes were close to those observed for the corresponding saturated forms. The correlations among C25 and C26 sphingolipids and other analytes were close to those observed for C24 sphingolipids (Table S2).

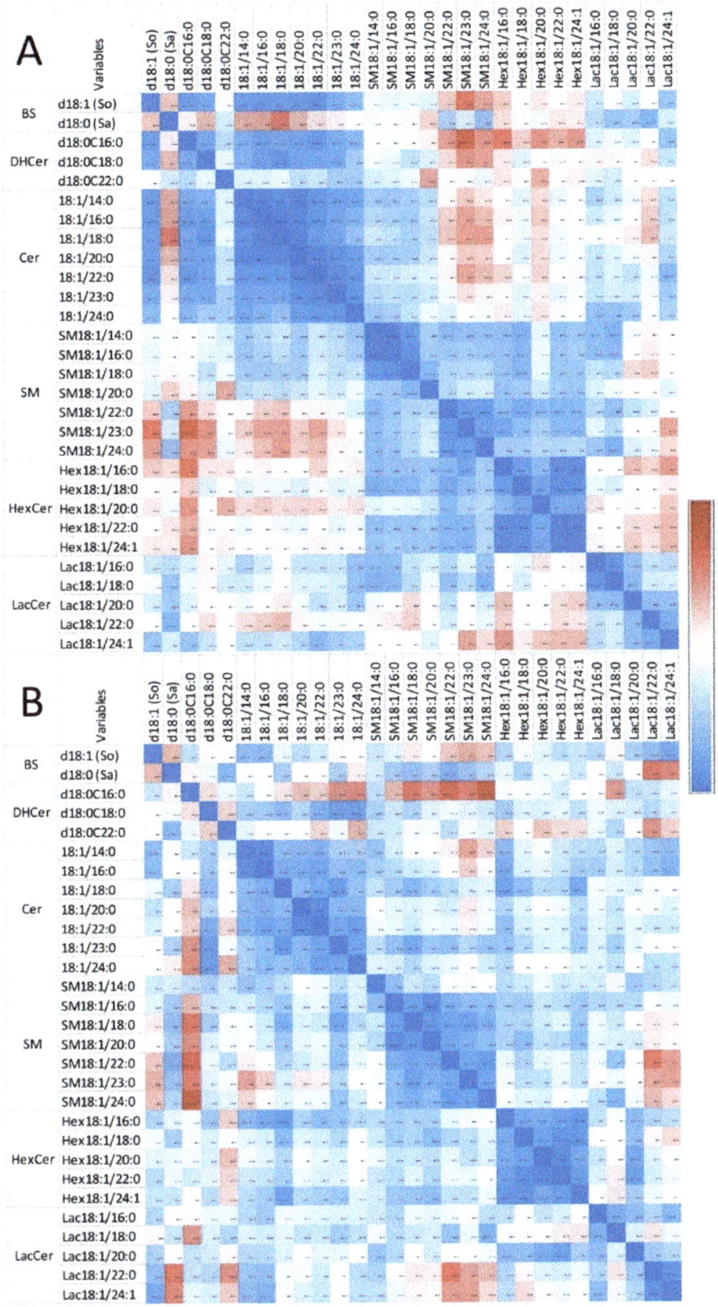

Figure 6. Correlation heatmap of So, Sa, and sphingolipids with 14 to 24 carbon fatty acid saturated chain lengths observed in the gizzard of chickens fed the control diet (**A**) and chickens fed 9 days with a diet containing 20.8 mg FB1 + FB2/kg (**B**). Numeric values of the Pearson coefficients of correlation observed among all sphingolipids assayed in this study are reported in Table S2.

These results suggest that there was a significant effect of fumonisins on sphingolipids in gizzards in this study. The effects observed were dominated by a decrease in dihydroceramides and ceramides. Sphingomyelins and glycosylceramides also tended to decrease in gizzards of chickens fed fumonisins, although the effects were not significant at the class level and varied with the analyte assayed. The feeding of fumonisins did not profoundly change the correlations among the sphingolipids in the gizzard.

2.4. Effects of Fumonisins on Sphingolipids in Breast Muscle

The concentrations of sphingolipids measured in breast muscle of chickens fed a control diet free of mycotoxins and chickens fed 9 days with a diet containing 20.8 mg FB1 + FB2/kg are reported in Table 1. Sphingosine was the only sphingolipid assayed for which the concentration in gizzard was significantly affected by fumonisins, and its concentration increased. There were only weak non-significant changes in the concentrations of dihydroceramides, ceramides, sphingomyelins, hexosylceramides, and lactosylceramides.

Partial least square-discriminant analysis was conducted to determine whether this analysis would be able to discriminate chickens according to the diet that they had been fed (Figure S1). As shown in Figure S1C, there was a good separation of chickens. The values of the R^2Y and the R^2X indices were satisfactory, and the confusion matrix revealed the model was highly sensitive and specific (Figure S1D). However, the value of Q^2 obtained with the two first components was only 0.185, indicating that the model can vary greatly depending on the sphingolipids selected for the modeling. Additionally, the standard deviations observed for the score of the variables important in the projection (VIP) were high in breast muscle (Figure S1A,B), confirming high variability in the modeling.

Finally, all together, the total amount of sphingolipids (Figure 2), the concentration of the various analytes assayed (Table 1), and the comparison of the repartition of the chickens into groups according to the diet that they had been fed using PLS-DA (Figure S1) revealed that the presence of fumonisins in the feed only had a minor effect on the sphingolipids in breast muscle in this study.

3. Discussion

Feeding 20.8 mg FB1 + FB2/kg over a period of 9 days did not lead to significant differences in the performances or the biochemistry in the chickens in this study [43]. This observation is consistent with the regulatory guidelines, which define a maximum tolerable level of fumonisins in feed of 20 mg FB1 + FB2/kg [1,4], whereas feeding 2.5 mg FB1 + FB2/kg of feed was reported to alter the length of intestinal crypts in chickens after 21 days of exposure, and feeding 5.3 mg FB1 + FB2/kg induced oxidative damage in the liver at 17 days, and histological damage to the liver and lungs at 21 days, effects on growth were only observed at 21 days [2,44].

Targeted analysis of sphingolipids in heart, gizzard, and breast muscle revealed that the effects of fumonisins varied greatly with the tissue studied. A significant increase in the Sa:So ratio was observed in the heart, and this increase was due to an increase in Sa, with a small increase in So also being observed. The change in sphingoid bases was accompanied by an increase in ceramides and, to a lesser extent, an increase in glycosylceramides and sphingomyelins. Most of the effects observed for dihydroceramides and dihydrosphingomyelins in the heart paralleled those found for ceramides and sphingomyelins. The effect of fumonisins on Sa is consistent with its well-known property of inhibiting CerS (Figure 1), and de novo synthesis of ceramides [10]. An increase in So concomitant with an increase in Sa has previously been reported to be the consequence of sphingomyelin hydrolysis occurring to maintain ceramide concentrations in cells [11], with this mechanism being known as the salvage synthesis pathway of ceramides (Figure 1). In this study, an increase rather than a decrease in sphingomyelins was observed, suggesting that sphingomyelin hydrolysis cannot explain the observed increase in So and ceramides. Moreover, glycosylceramides are derived from ceramides, and because glycosylceramides were increased in this study, the hypothesis of an increase in ceramides in heart secondary to hydrolysis of sphingomyelins appears unlikely (Figure 1). Other mechanisms that could explain an increase in ceramide concentrations in heart include: (1) change in the availability of the substrates used for their de novo synthesis [45,46]; (2) activation of de novo synthesis by different mechanisms linked with an inflammatory response such as Toll-like receptor (TLR) or tumor necrosis factor alpha (TNFα) activation [47]; and (3) increased delivery of sphingolipids from plasma, which includes recycling of sphingolipids from the gut microbiota [46,48].

Irrespective of the mechanism involved, sphingolipid concentrations were increased in the hearts of chickens in this study that had been fed fumonisins. PLS-DA revealed that the changes in sphingolipids were enough to distinguish chickens fed the control diet from chickens fed the fumonisin diet. Analysis of the correlations among sphingolipids also revealed that nearly all the sphingolipids positively correlated together in chickens fed fumonisins, which was not the case in the controls, all suggesting that pronounced changes in sphingolipid homeostasis occurred in the hearts of chickens fed fumonisins. Moreover, it is interesting to note that the increases in sphingolipids for which the fatty acids had 20–26 carbon chain lengths were more pronounced than for those with 14–16 carbon chain lengths. This observation is consistent with previous results in liver and kidney suggesting that fumonisins preferentially inhibit CerS5/6 activity in chickens [13,22]. Thus, it can be hypothesized that a general increase in sphingolipid synthesis occurred in the hearts of animals fed fumonisins, with this increase being less pronounced for sphingolipids for which the fatty acid chain length was 14 or 16 carbons due to partial inhibition in CerS5/6 activity.

Changes in sphingolipid concentrations in the hearts of chickens fed fumonisins could have negative consequences on health. Indeed, recent studies have revealed a prominent role of sphingolipids in cardiovascular diseases and heart failure [31]. Increased concentrations of ceramides in the myocardium have been reported in a rat model of ischemic reperfusion injury [49], and measurement of the expression of serine palmitoyl-transferase in mice suggested an increase in the de novo synthesis of sphingolipids in the infarct area and in the area at risk [50]. By contrast, sphingosine 1-phosphate has been reported to have a protective function in the myocardium [31]. Interestingly, a pronounced increase in So1P was observed in the plasma of chickens fed fumonisins, and a rapid increase in So1P has been reported in transient ischemia and reperfusion injury in humans [13,51]. Unfortunately, So1P was only found at trace levels in the myocardium in this study, and it was hence not possible to quantify its concentration. Only a few studies to date have reported cardiac alterations in animals fed fumonisins. A reduced heart rate has been reported to occur in pigs before the development of pulmonary edema and death [27]. Hypertrophy of the heart has been reported in pigs at very high doses of fumonisins in feed without histopathologic alterations despite an increase in the Sa:So ratio [28]. In contrast, a recent study of low doses of fumonisins revealed no alteration in heart weight, whereas histopathologic lesions characterized by hemorrhage and lymphocyte inflammatory infiltrate were observed [30]. Myocardial infiltration by lymphocytes leading to inflammatory damage is consistent with the hypothesis of activation of de novo synthesis of sphingolipids by TLR or TNFα [30,47]. Cardiac alterations characterized by macroscopic thinning of the heart and thinning of cardiomyocytes have been observed in Japanese quails fed fumonisins [29]. Transmission electron microscopy also revealed that the number of mitochondria was increased and that the mitochondria appeared swollen and pleomorphic in fumonisins-fed quails, although the outer membrane remained intact [29]. Alteration of the heart mitochondria of Japanese quails fed fumonisins is of considerable relevance because an in vitro study has revealed that ceramides cause mitochondrial dysfunction, oxidative stress, and cell death in cardiomyocytes [52]. Interestingly, overexpression of CerS2 induced oxidative stress, mitophagy, and apoptosis, which were prevented by depletion of CerS2. By contrast overexpression of CerS5 did not affect these processes, suggesting a chain-length dependent impact of ceramides on mitochondrial function [52].

A minor decrease in the Sa concentration was observed in the gizzards of chickens fed fumonisins, whereas fumonisins had no effect on the sphingolipid concentrations in breast muscle in this study. There are no data to compare the effects of fumonisins on sphingolipids in gizzard and breast muscle with our present results. Fumonisins have been reported to induce gizzard ulceration in chickens and have been reported to increase the relative weights of gizzards in chickens, turkeys, and ducks, but the mechanism underlying these alterations remains unknown [32–36]. In this study, the decrease in Sa observed in the gizzards of chickens receiving fumonisins was accompanied by a small decrease in So, so the Sa:So remained unchanged. Total ceramides in gizzards were decreased by fumonisins, and all of the ceramides appeared to decrease. The concentrations of dihydroceramides, sphingomyelins, dihydrosphingomyelins, hexosylceramides, and lactosylceramides paralleled those of ceramides. Interestingly, the effects of fumonisins on sphingolipids in gizzards appeared to be independent of the fatty acid carbon chain length. PLS-DA allowed the chickens to be separated into two groups according to the presence or absence of fumonisins in the feed, but the correlations measured among the sphingolipids in gizzards did not differ greatly in fumonisin-fed chickens and in controls. Finally, the effects of fumonisins on gizzards appeared to be consistent with the hypothesis of a decrease in the de novo synthesis of ceramides even though there was no accumulation of Sa. The decrease in sphingolipids in gizzards appeared to be independent of the fatty acid chain, which is different from what was observed in the liver and kidneys and to a lesser extent in the heart [22].

Sphingolipids have been shown to have important roles in skeletal muscle in obesity and aging. Accumulation of C16 ceramide occurs during the development of insulin resistance and CerS6 silencing may be a potential target for the treatment of insulin resistance, obesity, and type 2 diabetes [37]. C18 ceramide, produced mainly by CerS1, also accumulated in mice fed with a high-fat diet that promotes systemic insulin resistance [38,40]. By contrast, a decrease in C16 and C18 ceramides appears to be important in the skeletal muscle and myocardium of aged mice and humans [39]. Histological analysis of muscle in CerS1- and CerS5-deficient mice revealed reduced caliber sizes in slow (type 1) and fast (type 2) fibers of quadriceps femoris [39]. All of these results suggest that the decrease in Cer concentrations observed in gizzards could affect its contractility, even though no clinical consequences were observed.

4. Conclusions

In conclusion, this study revealed for the first time that feeding fumonisins at a concentration that did not alter performances in chickens nonetheless resulted in several changes in the sphingolipid composition of muscles that varied according to the type of muscle. Whereas no effects of fumonisins were noted in breast muscle, an increase in sphingoid bases, dihydroceramides, ceramides, and glycosylceramides, and to a lesser extent also sphingomyelins, was observed in the myocardium. These changes were more pronounced for sphingolipids with fatty acid chains of 20 to 26 carbons than for those with 14–16 carbon chain lengths. By contrast, a decrease in dihydroceramides and ceramides, and to a lesser extent also sphingomyelins, was observed in the gizzard. The decreases in sphingolipid levels in the gizzard appeared to be the same irrespective of the fatty acid chain length. Taken together, these results confirmed that the effects of fumonisins on sphingolipids at the scale of the organism are complex and probably involve cell- or metabolite-mediated effects in association with the inhibition of CerS activity. Further studies are necessary to understand these mechanisms and their consequences on health.

5. Materials and Methods

5.1. Analytes and Reagents

The analytes and reagents used in this study were obtained from Sharlab (Sharlab S.L., Sentmenat, Spain) or Sigma (Sigma Aldrich Chimie, Saint-Quentin-Fallavier, France). The separation of the analytes by UHPLC-MSMS was done with LC-MS grade solvents, whereas all other reagents were HPLC grade. The 42 sphingolipids used as standards were obtained from Sigma or Bertin (Bertin Technologies, Montigny-Le-Bretonneux, France) and corresponded to: deoxysphingosine (dSo = m18:1); deoxysphinganine (dSa = m18:0); sphingosine (So = d18:1); sphinganine (Sa = d18:0); sphingosine-1-P (d18:1P); sphinganine-1-P (d18:0P); glucosylsphingosine (GluSo); lysosphingomyelin (LysoSM); lactosylsphingosine (LacSo); N-acetylsphingosine (18:1/2:0); N-acetylsphinganine (18:0/2:0); ceramides: 18:1/14:0, 18:1/16:0, 18:1/18:0, 18:1/20:0, 18:1/22:0, 18:1/24:1, and 18:1/24:0; ceramide-1P: 18:1/16:0P; dihydroceramides: 18:0/16:0, and 18:0/24:0; deoxyceramides: m18:1/16:0, m18:1/22:0, m18:1/24:1, m18:1/24:0; deoxydihydroceramides: m18:0/22:0, m18:0/24:1, m18;0/24:0; glucosylceramides: Glu18:1/16:0 and Glu18:1/24:1; lactosylceramides: Lac18:1/16:0 and Lac18:1/24:1; ceramides sulfatides: ST18:1/24:1 and ST18:1/24:0; sphingomyelins: SM18:1/14:0, SM18:1/16:0, SM18:1/18:0, SM18:1/18:1, SM18:1/20:0, SM18:1/22:0, SM18:1/24:1, and SM18:1/24:0. The 12 sphingolipids used as IS corresponded to the 10 IS mixture "Ceramide/Sphingoid Internal Standard Mixture I" from Avanti Polar Lipids, which contains 25 μM C17 sphingosine (d17:1), C17 sphinganine (d17:0), C17 sphingosine-1-P (d17:1P), C17 sphinganine-1-P (d17:0P), lactosyl (ß) C12 ceramide (Lac18:1/12:0), C12 sphingomyelin (SM18:1/12:0), glucosyl (ß) C12 ceramide (Glu18:1/12:0), 12:0 ceramide (18:1/12:0), 12:0 ceramide-1-P (18:1/12:0P), and 25:0 ceramide (18:1/25:0) in ethanol solution. This mixture was completed by C12 deoxyceramide (m18:1/12:0) and C12 ceramide sulfatide (ST18:1/12:0) solubilized in ethanol at a concentration of 25 μM.

5.2. Tissue Samples

Heart, gizzard, and breast muscle were obtained from a Ross chicken study for which the animal maintenance conditions, feed formulation, and results of the effects of fumonisins on health and performance were detailed in [43]. This study was completed at Cebiphar (Cebiphar, ondettes, France) under number V9152 as a randomized, parallel, monocenter study under project number 2017062111426641 accepted by the French Ministry of Higher Education, Research, and Innovation (Paris, France) on 6 November 2017. Briefly, the experimental diets were formulated on a corn-soybean basis to best meet the nutritional needs of the chickens. Corn containing fumonisins was incorporated to a final concentration of FB1, FB2, and FB3 in the feed of 15.1, 5.6, and 0.9 mg/kg, respectively. Mycotoxin-free corn was used as the control diet. Mycotoxins in the raw materials and in the diets were measured by LC-MSMS according to AFNOR V03-110 [53]. Drinking water and feed were provided ad libitum. A diet containing fumonisins was provided to 10 chickens per group from the day of age 14 to the day of age 21. A control diet free of mycotoxins was provided to 10 other chickens until the day of age 21. On day 21, the feed was removed for eight hours before euthanasia and tissue collection. Heart, gizzard, and breast muscle were stored at −80 °C until analysis.

5.3. Sphingolipids in Tissues

The sphingolipids in heart, gizzard, and breast muscle were measured as previously described in [13] and completed in [22] for the determination of deoxyceramides, deoxydihydroceramides, and ceramides sulfatides. Briefly, 0.5 g of tissue was homogenized with a Potter grinder in 1.5 mL of phosphate buffer (0.1 M, pH 7.4) and centrifuged for 15 min at 3000× *g*. A 40 μL aliquot of the supernatant was collected and 120 μL of NaCl 0.9%, 600 μL of methanol/chloroform (2:1), and 10 μL of a solution containing the IS mixture were added to obtain a final concentration of each IS equivalent to 6250 pmol/g of tissue. The IS mixture was composed of the "Ceramide/Sphingoid Internal Standard Mixture I" that was completed with m18:1/12:0 and ST18:1/12:0. Samples were incubated overnight at 48 °C. After cooling to room temperature, 100 μL of KOH (1 M in methanol) was added and the samples were incubated for 2 h at 37 °C to hydrolyze glycerophospholipids, which could otherwise interfere with the determination of sphingolipids. A 10 μL aliquot of 50% acetic acid was added and the samples were centrifuged for 15 min at 4500× *g*. The supernatant was collected, and the residue was extracted again with 600 μL of methanol/chloroform (2:1). The second supernatant was added to the first and then evaporated to dryness. The dry residue was solubilized in 200 μL methanol, and a 5 μL aliquot was injected into the UHPLC-MSMS system comprising a 1260 binary pump, an autosampler, and an Agilent 6410 triple quadrupole spectrometer (Agilent, Santa Clara, CA, USA). The analytes were separated on a Poroshell 120 column (3.0 × 50 mm, 2.7 μm). The mobile phase was (A) methanol/acetonitrile/isopropanol (4/1/1) and (B) water, each containing 10 mM ammonium acetate and 0.2% formic acid. The mobile phase was delivered using a gradient of elution at a flow rate of 0.3 mL/min, as previously described [13]. Sphingolipids were detected after positive electrospray ionization under the following conditions: temperature 300 °C, flow rate of 10 L/min, pressure of 25 psi, capillary voltage 4000 V. Transitions, fragmentor voltages, and collision energies were reported in [22]. Agilent Mass-Hunter quantitative analysis software (B.05.00 SP03/Build 5.0.291.7) was used to analyze the chromatograms. The precision of the method expressed as a relative standard deviation (RSD) was considered acceptable for an RSD of 20%. As shown in Table S3, the linearity of the method of analysis was good over a relatively large range of concentrations, in agreement with previous results [13,22]. The recovery of the 12 IS in heart, gizzard, and breast muscle is presented in Table S4. The recovery varied according to the tissue and the analyte. The lowest recoveries of the IS were measured in breast muscle, whereas the recoveries in heart and gizzard were similar. A high recovery attributed to a positive matrix effect was observed for 18:1/12P and to a lesser extent for d17:1P and d17:0P, which is in agreement with previous results involving other tissues [13,22]. The lowest recovery was measured for 18:1:25:0, which is in agreement with previous results involving other tissues [13,22]. The repeatability of the method was considered good, with an RSD of 20%. Good repeatability was observed for all the IS in heart, gizzard, and breast muscle except for 18:1/25:0 in gizzard, for which an RSD of 25% was found. This value was considered acceptable, with a high RSD being already found for this analyte in other tissues [13,22].

The sphingolipid concentrations were determined from the calibration curves obtained with the standards. The final concentrations in tissues were corrected by the recovery measured for the corresponding IS. No correction was used for 18:1/2:0, 18:0/2:0, GluSo, LysoSM, and LacSo. The concentrations of sphingolipids not available as standards were calculated using the calibration curves obtained with standards of the same class with the closest mass and similar abundance. The final concentrations in tissues were corrected as carried out for the sphingolipids available as standards.

5.4. Analytes and Reagents

One-way ANOVA was carried out to compare the sphingolipid concentrations in the controls and in the chickens fed fumonisins after determination of the homogeneity of variance (Hartley's test). Significant differences were reported as follows: (*) $0.05 < p < 0.01$, (**) $0.01 < p < 0.001$, and (***) $p < 0.001$. Partial least squares-discriminant analysis (PLS-DA) was carried out to reveal the appurtenance of the chickens to the control and the fumonisins-fed group and to identify the sphingolipids that were important in the model. A score above 1.1 was retained to select the variables important in the projection (VIP) for heart and gizzard, whereas a score above 0.8 was retained for breast muscle. Pearson coefficients among the sphingolipids assayed in this study were measured and reported in bold font when significant in Figures S1 and S2. All the statistical analyses were carried out with XLSTAT Biomed software (Addinsoft, Bordeaux, France).

Supplementary Materials: The following supporting information can be downloaded at: https://www.mdpi.com/article/10.3390/toxins14120828/s1, Table S1: Pearson coefficients of correlation among sphingolipids dosed in heart; Table S2: Pearson coefficients of correlation among sphingolipids dosed in gizzard; Figure S1: Partial least squares discriminant analysis (PLS-DA) of sphingolipids measured in the breast muscle of chickens fed 9 days with a control diet free of mycotoxins or a diet containing 20.8 mg FB1 + FB2/kg. (A) Scores of variables important in the projection (VIP) for the first component. (B) Scores of variables important in the projection (VIP) for the second component.

(C) Discrimination on the factor axes extracted from the original explanatory variables. (D) Quality of the model and confusion matrix for the training sample (variable groups); Table S3: Linearity of the method measured for the sphingolipids available as standards; Table S4: Recovery of the internal standards of sphingolipids measured in the heart, gizzard, and breast muscle of chickens.

Author Contributions: Conceptualization, P.G. and P.N.C.; Methodology, P.G.; Formal analysis, P.G., C.G. and M.M.-M.; Investigation, P.G.; Resources, P.N.C.; Writing—original draft, P.G.; Writing—review & editing, P.G. All authors have read and agreed to the published version of the manuscript.

Funding: This research received no external funding.

Institutional Review Board Statement: All experimental procedures with animals were conducted in accordance with the French National Guidelines for the care and use of animals for research purposes. The animal study protocol was approved by the French Ministry of Higher Education, Research, and Innovation (protocol code 2017062111426641 accepted 6 November 2017).

Informed Consent Statement: Not applicable.

Data Availability Statement: Not applicable.

Conflicts of Interest: The authors declare no conflict of interest.

References

1. EFSA; Knutsen, H.; Alexander, J.; Barregård, L.; Bignami, M.; Brüschweiler, B.; Ceccatelli, S.; Cottrill, B.; Dinovi, M.; Edler, L.; et al. Risks for Animal Health Related to the Presence of Fumonisins, Their Modified Forms and Hidden Forms in Feed. *EFSA J.* **2018**, *16*, e05242. [CrossRef]
2. EFSA Panel on Contaminants in the Food Chain (CONTAM); Schrenk, D.; Bignami, M.; Bodin, L.; Chipman, J.K.; Del Mazo, J.; Grasl-Kraupp, B.; Hogstrand, C.; Leblanc, J.-C.; Nielsen, E.; et al. Assessment of Information as Regards the Toxicity of Fumonisins for Pigs, Poultry and Horses. *EFSA J. Eur. Food Saf. Auth.* **2022**, *20*, e07534. [CrossRef]
3. FUMONISINS (JECFA 47, 2001). Available online: http://www.inchem.org/documents/jecfa/jecmono/v47je03.htm (accessed on 18 March 2019).
4. FDA. 2011. Available online: https://www.ngfa.org/wp-content/uploads/NGFAComplianceGuide-FDARegulatoryGuidancef orMycotoxins8-2011.pdf (accessed on 11 August 2016).
5. IARC MOnograph 82, I.W.G. FUMONISIN B1. 2002. Available online: https://monographs.iarc.who.int/wp-content/uploads/2 018/06/mono82.pdf (accessed on 6 November 2022).
6. Liu, X.; Fan, L.; Yin, S.; Chen, H.; Hu, H. Molecular Mechanisms of Fumonisin B1-Induced Toxicities and Its Applications in the Mechanism-Based Interventions. *Toxicon* **2019**, *167*, 1–5. [CrossRef]
7. Arumugam, T.; Ghazi, T.; Chuturgoon, A.A. Molecular and Epigenetic Modes of Fumonisin B1 Mediated Toxicity and Carcinogenesis and Detoxification Strategies. *Crit. Rev. Toxicol.* **2021**, *51*, 76–94. [CrossRef]
8. Chen, J.; Wen, J.; Tang, Y.; Shi, J.; Mu, G.; Yan, R.; Cai, J.; Long, M. Research Progress on Fumonisin B1 Contamination and Toxicity: A Review. *Molecules* **2021**, *26*, 5238. [CrossRef] [PubMed]
9. Wangia-Dixon, R.N.; Nishimwe, K. Molecular Toxicology and Carcinogenesis of Fumonisins: A Review. *J. Environ. Sci. Health Part C Toxicol. Carcinog.* **2021**, *39*, 44–67. [CrossRef] [PubMed]
10. Wang, E.; Norred, W.P.; Bacon, C.W.; Riley, R.T.; Merrill, A.H. Inhibition of Sphingolipid Biosynthesis by Fumonisins. Implications for Diseases Associated with Fusarium Moniliforme. *J. Biol. Chem.* **1991**, *266*, 14486–14490. [CrossRef]
11. Riley, R.T.; Merrill, A.H. Ceramide Synthase Inhibition by Fumonisins: A Perfect Storm of Perturbed Sphingolipid Metabolism, Signaling, and Disease. *J. Lipid Res.* **2019**, *60*, 1183–1189. [CrossRef] [PubMed]
12. Tran, S.T.; Tardieu, D.; Auvergne, A.; Bailly, J.D.; Babilé, R.; Durand, S.; Benard, G.; Guerre, P. Serum Sphinganine and the Sphinganine to Sphingosine Ratio as a Biomarker of Dietary Fumonisins during Chronic Exposure in Ducks. *Chem. Biol. Interact.* **2006**, *160*, 41–50. [CrossRef] [PubMed]
13. Tardieu, D.; Matard-Mann, M.; Collén, P.N.; Guerre, P. Strong Alterations in the Sphingolipid Profile of Chickens Fed a Dose of Fumonisins Considered Safe. *Toxins* **2021**, *13*, 770. [CrossRef]
14. Guerre, P.; Travel, A.; Tardieu, D. Targeted Analysis of Sphingolipids in Turkeys Fed Fusariotoxins: First Evidence of Key Changes That Could Help Explain Their Relative Resistance to Fumonisin Toxicity. *Int. J. Mol. Sci.* **2022**, *23*, 2512. [CrossRef]
15. Loiseau, N.; Polizzi, A.; Dupuy, A.; Therville, N.; Rakotonirainy, M.; Loy, J.; Viadere, J.-L.; Cossalter, A.-M.; Bailly, J.-D.; Puel, O.; et al. New Insights into the Organ-Specific Adverse Effects of Fumonisin B1: Comparison between Lung and Liver. *Arch. Toxicol.* **2015**, *89*, 1619–1629. [CrossRef] [PubMed]
16. Pewzner-Jung, Y.; Park, H.; Laviad, E.L.; Silva, L.C.; Lahiri, S.; Stiban, J.; Erez-Roman, R.; Brügger, B.; Sachsenheimer, T.; Wieland, F.; et al. A Critical Role for Ceramide Synthase 2 in Liver Homeostasis: I. Alterations in Lipid Metabolic Pathways. *J. Biol. Chem.* **2010**, *285*, 10902–10910. [CrossRef] [PubMed]

17. Pewzner-Jung, Y.; Brenner, O.; Braun, S.; Laviad, E.L.; Ben-Dor, S.; Feldmesser, E.; Horn-Saban, S.; Amann-Zalcenstein, D.; Raanan, C.; Berkutzki, T.; et al. A Critical Role for Ceramide Synthase 2 in Liver Homeostasis: II. Insights into Molecular Changes Leading to Hepatopathy. *J. Biol. Chem.* **2010**, *285*, 10911–10923. [CrossRef] [PubMed]
18. Riedel, S.; Abel, S.; Burger, H.-M.; van der Westhuizen, L.; Swanevelder, S.; Gelderblom, W.C.A. Differential Modulation of the Lipid Metabolism as a Model for Cellular Resistance to Fumonisin B1–Induced Cytotoxic Effects in Vitro. *Prostaglandins Leukot. Essent. Fat. Acids* **2016**, *109*, 39–51. [CrossRef]
19. Song, Y.; Liu, W.; Zhao, Y.; Zang, J.; Gao, H. Fumonisin B1 Exposure Induces Apoptosis of Human Kidney Tubular Epithelial Cells through Regulating PTEN/PI3K/AKT Signaling Pathway via Disrupting Lipid Raft Formation. *Toxicon* **2021**, *204*, 31–36. [CrossRef]
20. Lumsangkul, C.; Tso, K.-H.; Fan, Y.-K.; Chiang, H.-I.; Ju, J.-C. Mycotoxin Fumonisin B1 Interferes Sphingolipid Metabolisms and Neural Tube Closure during Early Embryogenesis in Brown Tsaiya Ducks. *Toxins* **2021**, *13*, 743. [CrossRef]
21. Mignard, V.; Dubois, N.; Lanoé, D.; Joalland, M.-P.; Oliver, L.; Pecqueur, C.; Heymann, D.; Paris, F.; Vallette, F.M.; Lalier, L. Sphingolipid Distribution at Mitochondria-Associated Membranes (MAMs) upon Induction of Apoptosis. *J. Lipid Res.* **2020**, *61*, 1025–1037. [CrossRef]
22. Guerre, P.; Matard-Mann, M.; Nyvall Collén, P. Targeted Sphingolipid Analysis in Chickens Suggests Different Mechanisms of Fumonisin Toxicity in Kidney, Lung, and Brain. *Food Chem. Toxicol. Int. J. Publ. Br. Ind. Biol. Res. Assoc.* **2022**, *170*, 113467. [CrossRef]
23. Bowler, R.P.; Jacobson, S.; Cruickshank, C.; Hughes, G.J.; Siska, C.; Ory, D.S.; Petrache, I.; Schaffer, J.E.; Reisdorph, N.; Kechris, K. Plasma Sphingolipids Associated with Chronic Obstructive Pulmonary Disease Phenotypes. *Am. J. Respir. Crit. Care Med.* **2015**, *191*, 275–284. [CrossRef]
24. Kovacic, B.; Sehl, C.; Wilker, B.; Kamler, M.; Gulbins, E.; Becker, K.A. Glucosylceramide Critically Contributes to the Host Defense of Cystic Fibrosis Lungs. *Cell Physiol. Biochem. Int. J. Exp. Cell Physiol. Biochem. Pharmacol.* **2017**, *41*, 1208–1218. [CrossRef] [PubMed]
25. Koike, K.; Berdyshev, E.V.; Mikosz, A.M.; Bronova, I.A.; Bronoff, A.S.; Jung, J.P.; Beatman, E.L.; Ni, K.; Cao, D.; Scruggs, A.K.; et al. Role of Glucosylceramide in Lung Endothelial Cell Fate and Emphysema. *Am. J. Respir. Crit. Care Med.* **2019**, *200*, 1113–1125. [CrossRef] [PubMed]
26. Alessenko, A.V.; Albi, E. Exploring Sphingolipid Implications in Neurodegeneration. *Front. Neurol.* **2020**, *11*, 437. [CrossRef]
27. Smith, G.W.; Constable, P.D.; Tumbleson, M.E.; Rottinghaus, G.E.; Haschek, W.M. Sequence of Cardiovascular Changes Leading to Pulmonary Edema in Swine Fed Culture Material Containing Fumonisin. *Am. J. Vet. Res.* **1999**, *60*, 1292–1300.
28. Haschek, W.M.; Gumprecht, L.A.; Smith, G.; Tumbleson, M.E.; Constable, P.D. Fumonisin Toxicosis in Swine: An Overview of Porcine Pulmonary Edema and Current Perspectives. *Environ. Health Perspect.* **2001**, *109* (Suppl. 2), 251–257. [CrossRef]
29. Sharma, D.; Asrani, R.K.; Ledoux, D.R.; Rottinghaus, G.E.; Gupta, V.K. Toxic Interaction between Fumonisin B1 and Moniliformin for Cardiac Lesions in Japanese Quail. *Avian Dis.* **2012**, *56*, 545–554. [CrossRef]
30. Terciolo, C.; Bracarense, A.P.; Souto, P.C.M.C.; Cossalter, A.-M.; Dopavogui, L.; Loiseau, N.; Oliveira, C.A.F.; Pinton, P.; Oswald, I.P. Fumonisins at Doses below EU Regulatory Limits Induce Histological Alterations in Piglets. *Toxins* **2019**, *11*, 548. [CrossRef]
31. Borodzicz-Jażdżyk, S.; Jażdżyk, P.; Łysik, W.; Cudnoch-Jędrzejewska, A.; Czarzasta, K. Sphingolipid Metabolism and Signaling in Cardiovascular Diseases. *Front. Cardiovasc. Med.* **2022**, *9*, 915961. [CrossRef]
32. Ledoux, D.R.; Brown, T.P.; Weibking, T.S.; Rottinghaus, G.E. Fumonisin Toxicity in Broiler Chicks. *J. Vet. Diagn. Investig. Off. Publ. Am. Assoc. Vet. Lab. Diagn. Inc.* **1992**, *4*, 330–333. [CrossRef] [PubMed]
33. Kubena, L.F.; Edrington, T.S.; Kamps-Holtzapple, C.; Harvey, R.B.; Elissalde, M.H.; Rottinghaus, G.E. Influence of Fumonisin B1, Present in Fusarium Moniliforme Culture Material, and T-2 Toxin on Turkey Poults. *Poult. Sci.* **1995**, *74*, 306–313. [CrossRef] [PubMed]
34. Javed, T.; Bunte, R.M.; Dombrink-Kurtzman, M.A.; Richard, J.L.; Bennett, G.A.; Côté, L.M.; Buck, W.B. Comparative Pathologic Changes in Broiler Chicks on Feed Amended with Fusarium Proliferatum Culture Material or Purified Fumonisin B1 and Moniliformin. *Mycopathologia* **2005**, *159*, 553–564. [CrossRef] [PubMed]
35. Tran, S.T.; Auvergne, A.; Benard, G.; Bailly, J.D.; Tardieu, D.; Babilé, R.; Guerre, P. Chronic Effects of Fumonisin B1 on Ducks. *Poult. Sci.* **2005**, *84*, 22–28. [CrossRef] [PubMed]
36. Wang, F.; Chen, Y.; Hu, H.; Liu, X.; Wang, Y.; Saleemi, M.K.; He, C.; Haque, M.A. Protocatechuic Acid: A Novel Detoxication Agent of Fumonisin B1 for Poultry Industry. *Front. Vet. Sci.* **2022**, *9*, 923238. [CrossRef] [PubMed]
37. Raichur, S.; Brunner, B.; Bielohuby, M.; Hansen, G.; Pfenninger, A.; Wang, B.; Bruning, J.C.; Larsen, P.J.; Tennagels, N. The Role of C16:0 Ceramide in the Development of Obesity and Type 2 Diabetes: CerS6 Inhibition as a Novel Therapeutic Approach. *Mol. Metab.* **2019**, *21*, 36–50. [CrossRef] [PubMed]
38. Turpin-Nolan, S.M.; Hammerschmidt, P.; Chen, W.; Jais, A.; Timper, K.; Awazawa, M.; Brodesser, S.; Brüning, J.C. CerS1-Derived C18:0 Ceramide in Skeletal Muscle Promotes Obesity-Induced Insulin Resistance. *Cell Rep.* **2019**, *26*, 1–10.e7. [CrossRef]
39. Tosetti, B.; Brodesser, S.; Brunn, A.; Deckert, M.; Blüher, M.; Doehner, W.; Anker, S.D.; Wenzel, D.; Fleischmann, B.; Pongratz, C.; et al. A Tissue-Specific Screen of Ceramide Expression in Aged Mice Identifies Ceramide Synthase-1 and Ceramide Synthase-5 as Potential Regulators of Fiber Size and Strength in Skeletal Muscle. *Aging Cell* **2020**, *19*, e13049. [CrossRef] [PubMed]
40. Błachnio-Zabielska, A.U.; Roszczyc-Owsiejczuk, K.; Imierska, M.; Pogodzińska, K.; Rogalski, P.; Daniluk, J.; Zabielski, P. CerS1 but Not CerS5 Gene Silencing, Improves Insulin Sensitivity and Glucose Uptake in Skeletal Muscle. *Cells* **2022**, *11*, 206. [CrossRef]

41. Cingolani, F.; Futerman, A.H.; Casas, J. Ceramide Synthases in Biomedical Research. *Chem. Phys. Lipids* **2016**, *197*, 25–32. [CrossRef]
42. Truman, J.-P.; Ruiz, C.F.; Montal, E.; Garcia-Barros, M.; Mileva, I.; Snider, A.J.; Hannun, Y.A.; Obeid, L.M.; Mao, C. 1-Deoxysphinganine Initiates Adaptive Responses to Serine and Glycine Starvation in Cancer Cells via Proteolysis of Sphingosine Kinase. *J. Lipid Res.* **2022**, *63*, 100154. [CrossRef]
43. Laurain, J.; Tardieu, D.; Matard-Mann, M.; Rodriguez, M.A.; Guerre, P. Fumonisin B1 Accumulates in Chicken Tissues over Time and This Accumulation Was Reduced by Feeding Algo-Clay. *Toxins* **2021**, *13*, 701. [CrossRef]
44. Sousa, M.C.S.; Galli, G.M.; Alba, D.F.; Griss, L.G.; Gebert, R.R.; Souza, C.F.; Baldissera, M.D.; Gloria, E.M.; Mendes, R.E.; Zanelato, G.O.; et al. Pathogenetic Effects of Feed Intake Containing of Fumonisin (Fusarium Verticillioides) in Early Broiler Chicks and Consequences on Weight Gain. *Microb. Pathog.* **2020**, *147*, 104247. [CrossRef] [PubMed]
45. Bergouignan, A.; Trudel, G.; Simon, C.; Chopard, A.; Schoeller, D.A.; Momken, I.; Votruba, S.B.; Desage, M.; Burdge, G.C.; Gauquelin-Koch, G.; et al. Physical Inactivity Differentially Alters Dietary Oleate and Palmitate Trafficking. *Diabetes* **2009**, *58*, 367–376. [CrossRef] [PubMed]
46. Nicholson, R.J.; Norris, M.K.; Poss, A.M.; Holland, W.L.; Summers, S.A. The Lard Works in Mysterious Ways: Ceramides in Nutrition-Linked Chronic Disease. *Annu. Rev. Nutr.* **2022**, *42*, 115–144. [CrossRef] [PubMed]
47. Zietzer, A.; Düsing, P.; Reese, L.; Nickenig, G.; Jansen, F. Ceramide Metabolism in Cardiovascular Disease: A Network with High Therapeutic Potential. *Arterioscler. Thromb. Vasc. Biol.* **2022**, *42*, 1220–1228. [CrossRef] [PubMed]
48. Guerre, P. Mycotoxin and Gut Microbiota Interactions. *Toxins* **2020**, *12*, 769. [CrossRef]
49. Beresewicz, A.; Dobrzyń, A.; Górski, J. Accumulation of Specific Ceramides in Ischemic/Reperfused Rat Heart; Effect of Ischemic Preconditioning. *J. Physiol. Pharmacol. Off. J. Pol. Physiol. Soc.* **2002**, *53*, 371–382.
50. Reforgiato, M.R.; Milano, G.; Fabriàs, G.; Casas, J.; Gasco, P.; Paroni, R.; Samaja, M.; Ghidoni, R.; Caretti, A.; Signorelli, P. Inhibition of Ceramide de Novo Synthesis as a Postischemic Strategy to Reduce Myocardial Reperfusion Injury. *Basic Res. Cardiol.* **2016**, *111*, 12. [CrossRef]
51. Egom, E.E.; Mamas, M.A.; Chacko, S.; Stringer, S.E.; Charlton-Menys, V.; El-Omar, M.; Chirico, D.; Clarke, B.; Neyses, L.; Cruickshank, J.K.; et al. Serum Sphingolipids Level as a Novel Potential Marker for Early Detection of Human Myocardial Ischaemic Injury. *Front. Physiol.* **2013**, *4*, 130. [CrossRef]
52. Law, B.A.; Liao, X.; Moore, K.S.; Southard, A.; Roddy, P.; Ji, R.; Szulc, Z.; Bielawska, A.; Schulze, P.C.; Cowart, L.A. Lipotoxic Very-Long-Chain Ceramides Cause Mitochondrial Dysfunction, Oxidative Stress, and Cell Death in Cardiomyocytes. *FASEB J. Off. Publ. Fed. Am. Soc. Exp. Biol.* **2018**, *32*, 1403–1416. [CrossRef]
53. ANSES_GuideValidation.Pdf. Available online: https://www.anses.fr/fr/system/files/ANSES_GuideValidation.pdf (accessed on 19 December 2018).

toxins

Article

A Novel Cost-Effective Nanobody against Fumonisin B1 Contaminations: Efficacy Test in Dairy Milk and Chickens

Yi Chen [1,†], Guanggang Qu [2,†], Hongkun Quan [1], Yihui Wang [1], Changjiang Wang [2], Md Atiqul Haque [1,3] and Cheng He [1,*]

1 Key Laboratory of Animal Epidemiology and Zoonoses of Ministry of Agriculture, College of Veterinary Medicine, China Agricultural University, Beijing 100019, China
2 Shandong Binzhou Animal Science and Veterinary Medicine Academy, Binzhou 256600, China
3 Department of Microbiology, Faculty of Veterinary and Animal Science, Hajee Mohammad Danesh Science and Technology University, Dinajpur 5200, Bangladesh
* Correspondence: hecheng@cau.edu.cn
† These authors contributed equally to this work.

Abstract: AbstractBackground: Fumonisin B1 (FB1) is a secondary metabolite produced mainly by *Fusarium verticillioides* or *Fusarium proliferatum*. It poses a huge threat to the sustainable animal industry and human health as well via food chains (egg, meat and milk). Although *E. coli*-expressed nanobodies are documented for diagnostic applications, nanobodies remain elusive as FB1 detoxifiers in feed and food. Results: In the present study, the *E. coli*-expressed nanobody was assessed to remove FB1 in fresh milk, embryonated eggs and broilers. Firstly, 2 alpacas received intramuscularly FB1-adjuvanted BSA 6 times, and then the variable domain of the heavy-chain antibody (VHH) of fb1 genes were amplified to clone into the pCANTAB 5 E vector in order to generate a VHH-FB1 phage antibody display library, yielding 3.4×10^{10} capacity with 96.7% positivity. Afterwards, 5 anti-FB1 nanobodies were expressed and identified. Furthermore, maximal 43.2% FB1 was removed from milk by 1:2000 concentration of nanobody 5 (Nb5). Furthermore, SPF-embryonated eggs were inoculated into albumens with nanobody-treated FB1. The Nb5 group yielded an 83.3% hatching rate, higher body weight, lower gizzard ulceration and fewer FB1 residuals. In order to warrant the above results, 50 broilers aged 10 days were received orally with 20 ppm of FB1 for 20 days. At the same time, birds were fed orally with 50 μg of Nb5 or bivalent nanobody 11 (BiNb11). Finally, the Nb5 group showed a higher relative body weight gain and lower gastric ulcerations and fewer inflammations in the thymus and bursa. Conclusions: Based on the above evidence, the Nb5 nanobody may be considered as an additional FB1 detoxifier, contributing to FB1 decontamination.

Keywords: fumonisin B1; nanobody; phage-display technology; detoxification; efficacy test

Key Contribution: 1. Five nanobodies were firstly expressed and bound specifically to FB1 mycotoxin. 2. Monomer nanobody Nb5 is a promising food detoxifier by degrading FB1-contaminated fresh milk. 3. Engineered nanobody Nb5 is a potential feed detoxifier for poultry industry.

Citation: Chen, Y.; Qu, G.; Quan, H.; Wang, Y.; Wang, C.; Haque, M.A.; He, C. A Novel Cost-Effective Nanobody against Fumonisin B1 Contaminations: Efficacy Test in Dairy Milk and Chickens. *Toxins* **2022**, *14*, 821. https://doi.org/10.3390/toxins14120821

Received: 16 September 2022
Accepted: 17 November 2022
Published: 23 November 2022

Publisher's Note: MDPI stays neutral with regard to jurisdictional claims in published maps and institutional affiliations.

Copyright: © 2022 by the authors. Licensee MDPI, Basel, Switzerland. This article is an open access article distributed under the terms and conditions of the Creative Commons Attribution (CC BY) license (https://creativecommons.org/licenses/by/4.0/).

1. Introduction

Mycotoxins are toxic byproducts of fungi that have the ability to cause cancer, mutagenesis, teratogenicity, cytotoxicity, nephrotoxicity, neurotoxicity, immunotoxicity, dermotoxicity, and estrogenic potential. Their pollution has raised worries about the safety of feed and food throughout the world [1–3]. More than 500 mycotoxins, notably, aflatoxins, ochratoxins, trichothecenes, fumonisins, zearalenone, patulin, citrinin, and ergot alkaloids, have been identified in recent years [4]. Fumonisins (FBs) are a group of hydrophilic mycotoxins produced largely by the fungi Fusarium (*verticillioides, proliferatum, moniliforme, anthophilum, dlamini, globosum, fujikuroi, napiforme, nygamai, oxysporum*) and Aspergillus (*awamori, niger*) [2,4,5]. They frequently contaminate corn (maize), corn-based foodstuffs, asparagus, sorghum, soybeans, rice, pineapple, banana, sugarcane, beer, and animal feed worldwide, with FB1 being the predominant (>70%) cause of toxicity and posing a potential threat to human health and animal production [2,5]. FB1 has diverse serious impacts on various organs (brain, lungs, liver, kidneys, etc.) [6]. For example, it has been linked to equine leukoencephalomalacia in horse, porcine pulmonary edema syndrome in pigs, liver cancer in rat, stunted growth, developmental disorder, and neural tube defect in Brown Tsaiya Duck embryos, negative structural bone modifications in young chickens, and decreased hatchability and gizzard ulceration in chicken progenies [5,7,8]. Epidemiological data revealed that FB1 pollution in the human diet had a certain correlation with the high incidence of esophageal cancer, primary liver cancer, neural tube defects (NTDs), growth problem, idiopathic congestive cardiopathy (ICC) and other diseases in humans [5–7,9]. Furthermore, FB1 has been listed as a possible Group 2B human carcinogen by the International Agency for Research on Cancer (IARC) [10], with a tolerable daily intake (TDI) of 2 μg/kg BW/day set by the joint Food and Agricultural Organization (FAO) [11] and World Health Organization (WHO) [11].

A recent study on female BALB/c mice revealed that FB1 could cause significant hepatotoxicity, nephrotoxicity, and hematological toxicity, indicating that the foregoing maximum TDI of FB1did not appear to provide adequate protection [1]. In lower doses, FB1 triggers cell death in all body parts of plants, both at the cellular and organ levels, and has similar adverse effects on animals and humans [2]. Confronting high FB1 contaminations in food chains, it is still a big challenge for scientists to innovate novel FB1 detoxifier. Therefore, attempts have been made for the discovery of an effective FB1 detoxification method. Physical, chemical, and biological methods of detoxification are currently available. Furthermore, both organic and inorganic mycotoxin binders have been used to limit toxicity in animal feed. Feed and food by-products contaminated with FB1 can be destroyed by heat (>150–200 °C) and alkali treatment to reduce FB1 damage to the body [12]. In vitro, tri-octahedral bentonites, an inorganic binder, have been shown to adsorb >90% of Zearalenon (ZEN) and FB1 [13]. Isothiocyanate (ITC), which contains electrophilic carbon and can react with the free amino groups of mycotoxins, reduced the levels of FB1 by 53–96%, FB2 by 29–91%, and FB3 by 29–96% through ITC fumigation. These results suggest that the primary amine group of FB is critical to its toxicity, as the naturally occurring acetylfumonisin B is not considered toxic. FB can also react with reducing sugars (glucose or fructose) to block primary amine groups and can undergo the Maillard reaction to form N-carboxymethyl-FB1, which is less toxic, and detoxify FB1 contamination [14,15]. Traditional physical and chemical methods can only achieve partial FB reduction while depleting nutrients and having undesired effects. With the development of modern biotechnology, metabolic detoxification has gradually supplanted physical and chemical detoxification as the primary method for mycotoxin reduction. Metabolic detoxification has mild action and does not reduce the nutritional value or the palatability of feed [4]. Some effective enzyme preparations and strains can reduce FB1 toxicity in feed. The two genes *fumD* and *fumI* of *Sphingopyxis* spp. MTA144 were found to produce enzymes that catalyze the de-esterification and deamination of FB1 in a sequential manner, detoxifying FB1 through their continuous action. The *Pseudomonas* genus was identified as a significant FB1 degradation member using 16 SrDNA sequencing and antibiotic-driven selection of a bacterial consortium (SAAS79) and its crude enzymes with strong FB1 degradation activity (90%) isolated from waste water mushroom [9]. Moreover, the hydrolase and transferase enzymes of Serratia marcescens are capable of degrading FB1 at a rate of 37% [15]. Although many studies have been focused on the degradation of FB, there is still a lack of effective methods for FB1 degradation in food and feed.

Nanobodies (Nbs), also known as single-domain antibodies, are derived from the variable domain of heavy-chain antibodies (VHH) in camels and are thought to be the smallest intact antigen-binding fragment currently available [8,9]. Nbs have many unique antibody characteristics, including small molecular weight (15 kDa), high solubility, high specificity, high affinity, strong stability, and easy cloning. Importantly, industrial production of recombinant nanobodies by microorganisms is highly cost-effective, and nanobodies can be easily utilized as building blocks for multi-domain constructs [16,17]. Nbs have high stability to organic solvents, which is useful for mycotoxin immunoassays. He et al. assessed Nbs' solvent stability against AFB1 and found that it was more stable to methanol, acetone, and acetonitrile than monoclonal antibody (mAb) [18]. Kunz et al. examined the thermostabilities of 78 pure nanobody binders and determined that the stability of eight modified nanobodies varied by a mean of 2.3 °C and a maximum of 6.1 °C [19]. Alsulami et al. reported the FNanoBiT assay for the detection of FB1 in the maize extract, and the relative standard deviation (RSD) observed suggested high stability allowed it to be better suited for field application [6]. Due to the aforementioned advantages of nanobodies, their application in the detection of mycotoxins in agricultural products has gained increasingly attention in recent years. The anti-FB1-idiotypic nanobody and alkaline phosphatase were fused to express, and a one-step competitive enzyme-linked immunosorbent assay (ELISA) method was established for detecting FB1 level in grains. Likewise, for FB1 detection, a noncompetitive idiometric immunoassay using a combination of β-type anti-idiotypic Nb and phage-displayed α-type anti-idiotypic Nb revealed a 17-fold increase in sensitivity compared to the competitive ELISA (LOD = 3.41 ng/mL), implying that this approach has broad utility for checking small molecules in foods [17]. In addition to the ELISA detection method, the application of an immunosensor for aflatoxin B1 (AFB1) detection also been established by using an Nb. For instance, an Nb, conjugated by a horseradish peroxidase, is combined with a hybridization chain reaction signal amplification system to achieve rapid and ultra-sensitive AFB1 detection. Under ideal conditions, the LOD of the immunosensor was 68 fg/mL, and the linear range was 0.5–10 ng/mL with a sensitivity of 2.7 μA • (mL/ng) [20]. So far, scarce reports have been published on toxin neutralization with nanobodies. In 2016, Andersen et al. immunized llamas with toxin B and obtained four Nbs capable of neutralizing toxin B in an in vitro cell-based assay [21]. Further, the in vivo protective effect was validated in a hamster model. When challenged with *Clostridium difficile*, half of the hamsters survived after being treated with Nbs expressed in *Lactobacillus* and showed no damage or only limited inflammation of the intestinal mucosa. In another investigation, Harmsen et al. showed that oral administration of high doses of nanobodies against *Escherichia coli* F4 pili reduced *E. coli*-induced diarrhea in piglets [22]. However, no report has been published until now on the detoxifying effect of Nbs on FB1.

In this study, FB1-specific nanobodies were developed by phage-display technology and expressed in an *E. coli* expression system. The binding ability of the Nbs was analyzed by ELISA and their antagonistic effect on FB1 was verified via chicken embryo model and broiler model. Our results showed that Nb5 was identified to reduce FB1 toxicity on embryonated eggs and it also improved body weight gain and reduced gizzard ulceration in broilers, suggesting that the nanobody Nb5 had the effect of antagonizing FB1 and potential application for human food additive and poultry industry.

2. Results

2.1. Screening and Identification of FB1 Nanobodies

The alpaca was immunized with FB1-adjuvanted-BSA whole antigen. Post the final inoculation on day 4, the titers of the FB1 nanobody were detected to be 1:5000 in alpaca sera by an indirect ELISA. The lymphocytes were isolated from a total volume of 200 mL of alpaca peripheral anticoagulant. After extracting RNA, it was reversely transcribed into cDNA. The *VHH* fragment was amplified by nested PCR, and the target band of approximately 700 bp was recovered in the 1st round and the approximately 400 bp was produced in the 2nd round. The *VHH* fragment was inserted into a pCANTAB 5 E vector and electrically converted to TG1-competent cells to obtain a phage-display library. The library capacity was roughly 3.4×10^{10} and the positive rate was approximately 96.7%.

After the 3rd round of panning, the enrichment rate was identified to be 16. After the 3rd rounds of panning, the recombinant bacteriophage was infected into logarithmic TG1 cells, LB/AMP-Glu plates were coated, and 60 single colonies were randomly selected. The reactivity of the crude extract with FB1-OVA was identified by ELISA assay. Out of the 60 selected monoclonal clones chosen, 58 positive clones were sequenced and analyzed, where 5 nanobody phages coding different sequences were identified and named as Nb5, Nb11, Nb12, Nb13, and Nb16 in sequence.

2.2. Prokaryotic Expression and Identification of Anti-FB1 Nanobody

The target fragment was amplified by specific primers and the PCR amplification produced a roughly 400 bp band, which was compatible with the expected target band size (Figure 1A). For PCR identification, a single colony was picked and incubated overnight at 37 °C. The prokaryotic expression vector of *VHH*-FB1-PSF was constructed successfully (Figure 1B). Afterwards, the *VHH*-FB1-PSF plasmid was transformed into *E. coli* BL21(DE3) competent cells, and IPTG was added to a final concentration of 1 mM before incubation overnight at 37 °C. The proteins were expressed as soluble proteins and then identified by SDS-PAGE (Figure 1C). The recombinant Nbs were confirmed to be approximately 35 KDa in size. The supernatants were purified with Ni-NTA columns at room temperature and then identified by SDS-PAGE (Figure 1D).

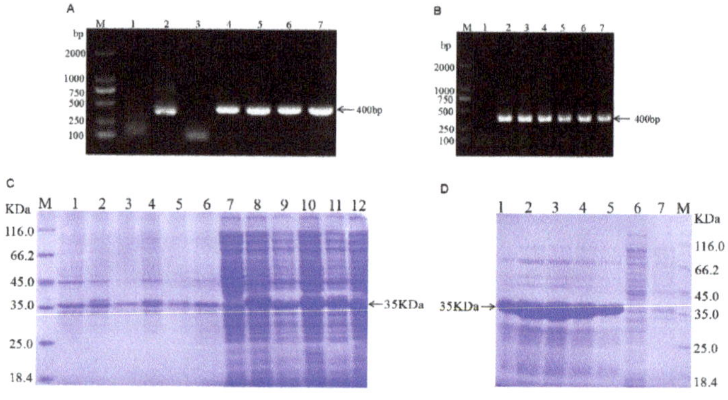

Figure 1. Construction of the *VHH*-FB1 prokaryotic expression vector and expression of VHH recombinant protein. (**A**) Cloning *VHH*-FB1 gene. Lane M, DNA marker (TAKARA); Lane 1, Negative control of *VHH*-FB1 gene (400 bp); Lane 2–7, PCR products of *VHH*-FB1 gene; (**B**) PCR colony. (**C**) Expression of VHH recombinant protein (M. Protein marker; 1. Precipitation of bacteria of PSF without induction; 2–6. Precipitation of bacteria of Nb1–Nb5 with induction; 7. Supernatants of bacteria of PSF without induction; 8–10. Supernatants of bacteria of Nb1–Nb5 with induction). (**D**) Purification of VHH recombinant protein (1–5. Recombinant protein eluted with 250 mM imidazole concentration; 6. Supernatant of flow-through; 7. Precipitation of flow-through).

2.3. Nb5 Detoxification on FB1-Contaminated Milk

Serial concentrations of Nb5 were incubated with FB1-contaminated milk sample at 25 °C and significant difference of toxin removal was determined post treatment with the initial solution, 1:2000, or 1:3000, or 1:5000, or 1:10,000, while a dose-dependent manner of detoxification was 43.3%, 21.98%, 19.62%, and 3.07%, respectively ($p < 0.01$). Obviously, the optimal detoxification was determined to be 1:2000 solution and FB1 concentration was reduced to 27.04 ppb with a degradation rate of 43.26% (Figure 2).

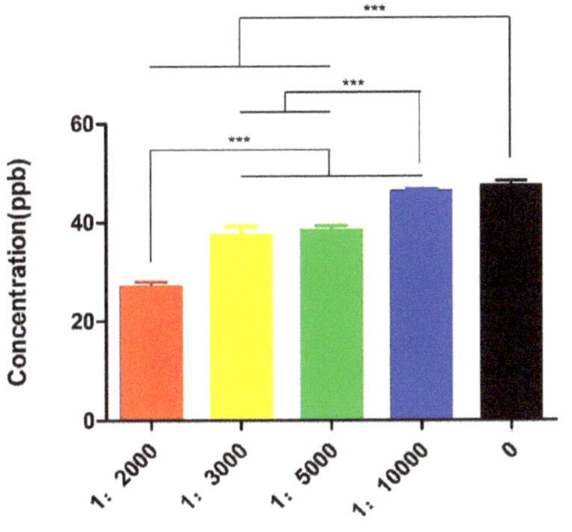

Figure 2. Detoxification of Nb5 at different concentrations. A dose-dependent manner of decontamination was found among 1:2000, 1:3000, 1:5000 and initial solutions post treatment at 25 °C for 2 h. The maximal detoxification was determined to be 1:2000 solution. The data were expressed as mean ± SD.; ***: $p < 0.01$.

2.4. *Effect of Nanobodies on Growth and Gizzard Ulceration of the Embryonated Eggs*

On day 21, the hatchability of the embryonated eggs exposed to 64 µg of FB1 was significantly lower than that of the control groups or the nanobody-treated groups ($p < 0.01$). The hatchability was 83.33% and 80%, respectively, in the Nb5 + FB1 group and the D-glucose + FB1 group. However, lower hatchability was found in the Nb13 + FB1, the BiNb11 + FB1, and the BiNb13 + FB1 compared to the Nb5 + FB1 group (Figure 3A).

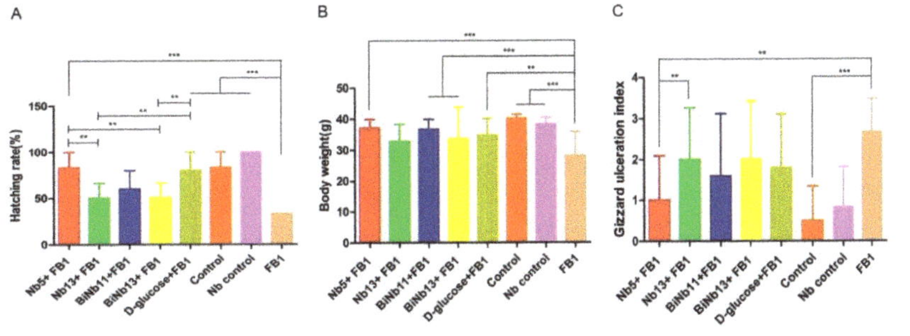

Figure 3. Effects of nanobodies on FB1-contaminated chicken embryos. (**A**): Hatching rate post inoculation with FB1 or in combination with monomer nanobodies, bivalent nanobodies, and D-glucose. Lower hatchability was observed in the FB1 control group compared to other groups ($p < 0.01$). Both the Nb5 + FB1 group and the D-glucose group yielded higher hatching abilities than other Nb groups did in the experiment. No statistical difference was determined between the Nb5 + FB1 group and the D-glucose + FB1 group either ($p = 0.795$). (**B**): Hatching body weight after treatment with monomer nanobodies, bivalent nanobodies, and D-glucose. Slower growth rate was observed in the FB1 control group compared to the other groups ($p < 0.01$). However, no statistical difference of body weight was found among the Nb5 + FB1 group, or the Nb13 + FB1, or the BiNb11 + FB1 group, or the BiNb13 + FB1 group, or the D-glucose + FB1 group ($p > 0.05$). (**C**): Lesion scores of gizzard ulceration of chicken post FB1 treatment with monomer nanobodies, bivalent nanobodies, and D-glucose. Postmortem, the FB1 group induced higher lesions of gizzard

ulcerations compared to the Nb5 + FB1 group, the BiNb11 + FB1 group ($p < 0.05$), and the control group ($p < 0.01$). No significant difference was found in the Nb13 + FB1 group, BiNb13 + FB1 group, and the D-glucose + FB1 group ($p > 0.05$) and no significant difference was found between the Nb5 + FB1 group and the D-glucose + FB1 group either ($p = 0.255$). The data were expressed as mean ± SD. **: $p < 0.05$, ***: $p < 0.01$. Control: PBS; Nb control: Bi-Nb-11 nanobody.

As for body weight of the new-borne chickens, the FB1 group induced slow growth compared to the control groups and the Nb control group ($p < 0.01$). However, no statistical difference was found among the Nb5 + FB1 group, or the BiNb11 + FB1 group, the BiNb13 + FB1 group and the D-glucose + FB1 group (Figure 3B). Postmortem, FB1 induced a highly gastric ulceration index while lower lesions were determined in the Nb5 + FB1 group ($p < 0.05$) and all the control groups ($p < 0.01$). In addition to the Nb5 + FB1 group, no statistical difference was found among the Nb13+FB1 group, the BiNb11 + FB1 group, the BiNb13 + FB1 group, and the D-glucose + FB1 group (Figure 3C).

As for FB1 residues in the lungs of hatching chickens, lower FB1 contamination was determined in the Nb5 + FB1 group and the BiNb13 + FB1 group compared to that of the FB1 control ($p < 0.05$). However, no significant difference was found among the Nb13 + FB1 group, the BiNb11 + FB1 group, and the D-glucose + FB1 group ($p > 0.05$). No residual FB1 was detected in the gizzards of all groups (Figure 4).

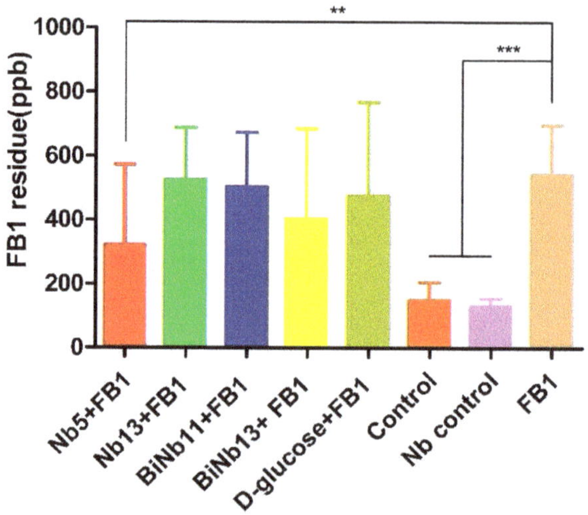

Figure 4. FB1 residuals in the lungs and gizzard of new-borne chickens. Lower FB1 contamination was observed in the Nb5 + FB1 group compared to that of the FB1 control ($p < 0.05$). However, no significant difference was found among the Nb13 + FB1 group, the BiNb11 + FB1 group, and the D-glucose + FB1 group ($p > 0.05$) and no significant difference was found between the Nb5 + FB1 group and the D-glucose + FB1 group either ($p = 0.448$). No residual FB1 was detected in the gizzards of all groups. The data were expressed as mean ± SD.**: $p < 0.05$; ***: $p < 0.01$. Note: Control: PBS; Nb control: Bi-Nb-11 nanobody.

2.5. *Effect of Nanobodies on Growth, Gizzard Ulceration, and Antibody Levels of the Broilers*

Regarding the body weight of the broilers (Figure 5A), FB1 greatly reduced the body weight of broilers on day 7 and day 14 ($p < 0.01$). Significant increasing body weight was determined in the Nb5 + FB1 group and the BiNb11 + FB1 group in the first two weeks ($p < 0.01$). However, no significant difference was observed among all the groups in the final week ($p > 0.05$). Post-mortem, lesion score index was used to access the gizzard ulceration. Obviously, FB1 aggravated gizzard lesions in broilers and gastric ulceration index amounted to 3.0 ± 0.0. On the contrary, lower gizzard ulceration index

was observed in the Nb5 + FB1 group ($p < 0.01$), the BiNb11 + FB1 group ($p < 0.05$), and the Qinankang + FB1 group ($p < 0.01$), compared to the FB1 group. However, no significant difference was found among the Nb5 + FB1 group, the BiNb11 + FB1 group, and the Qinankang + FB1 group ($p > 0.05$) (Figure 5B).

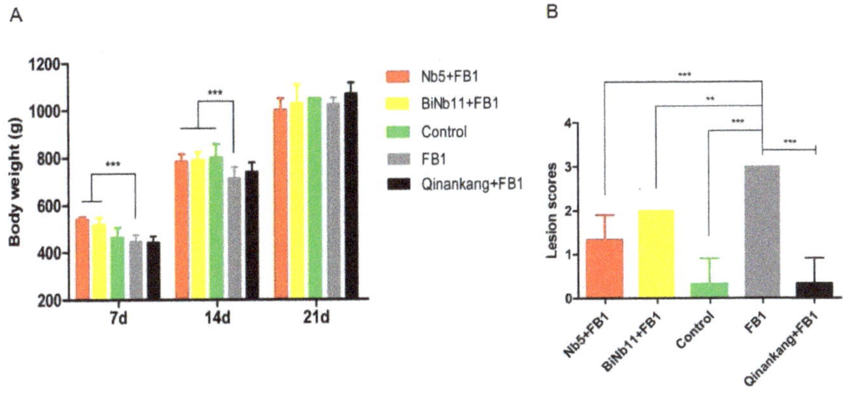

Figure 5. Effects of nanobodies on the FB1-contaminated broiler's development. (**A**) Both Nb5 + FB1 and BiNb11 + FB1 improved body weight on day 7 and on day 14 compared to the FB1 control ($p < 0.01$). No discernible difference was found among all the groups on day 21 ($p > 0.05$). (**B**) Lesion scores of gizzard ulceration of broiler chickens post treatment with Nb5 and BiNb11. Post mortem observation, both the Nb5 + FB1 group ($p < 0.01$) and BiNb11 + FB1 ($p < 0.05$) group developed lower lesions of gizzard ulcerations compared to the FB1 control group. However, no statistical difference was found between the Nb5 + FB1 group and the commercial detoxifier, Qingankang ($p > 0.05$). The data were expressed as mean ± SD. **: $p < 0.05$; ***: $p < 0.01$.

Post treatment, antibody titers against Newcastle disease virus (NDV), Infectious bronchitis virus (IBV), and Infectious bursal disease virus (IBDV) were determined from initial day to day 21. Increasing antibody titers against IBDV were determined in the Nb5 + FB1 group compared to other groups on day 14 and day 21 ($p < 0.05$); no significant difference was found between the Nb5 + FB1 group and the BiNb11 + FB1 group during all the observation ($p > 0.05$) (Figure 6A).

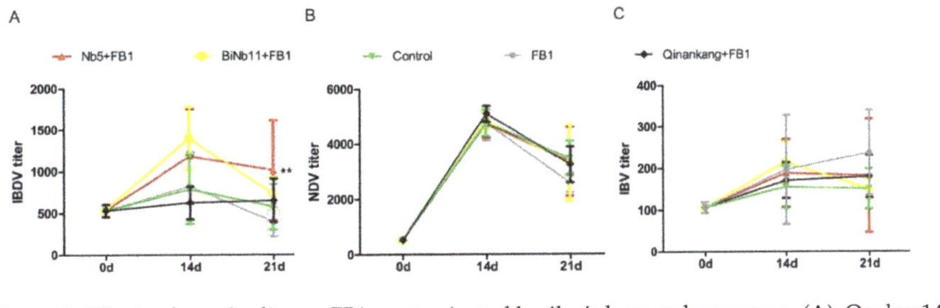

Figure 6. Effects of nanobodies on FB1-contaminated broiler's humoral responses. (**A**): On day 14 and 21, the Nb5 + FB1 group induced higher IBDV antibody titers than the Control group or the Qinankang + FB1 group or the FB1 group ($p < 0.05$), whereas no significant difference was found between the Nb5 + FB1 group and the BiNb11 + FB1 group ($p > 0.05$) throughout the study. (**B**): On day 21, both the Nb5 + FB1 and BiNb11 + FB1 groups yielded higher NDV antibody levels than the FB1 control group, whereas no statistical difference was found on day 14 ($p > 0.05$). (**C**): Regarding the IBV antibody, both the Nb5 + FB1 and BiNb11 + FB1 groups yielded a low antibody compared to the FB1 control group, whereas no significant difference was determined among all the groups ($p > 0.05$). The data were expressed as mean ± SD. **: $p < 0.05$.

As for NDV antibody response, the increasing antibody levels were determined both in the Nb5 + FB1 group and the BiNb11 + FB1 group compared to the FB1 control on day 21, but no statistical difference was found in the study ($p > 0.05$) (Figure 6B). Regarding the IBV antibody, on day 21, compared to the FB1 control group, the IBV antibody level of the Nb5 + FB1, BiNb11 + FB1 and Qinankang groups were lower and moreover, the Nb5 + FB1 and BiNb11 + FB1 showed a declining trend of antibody level. However, there is no significant difference of IBV antibodies level among all groups ($p > 0.05$) (Figure 6C).

3. Discussion

In this study, five FB1 toxin-specific Nbs were obtained post immunization in the alpaca. Embryonated eggs are recommended as a toxicity bioassay due to FB1 and DON contaminations [23]. We used the embryonated eggs to test the efficacy of the five obtained nanobodies against FB1 embryo toxicity. The primary amino group of FB plays an important role in the toxicity of FB1, because N-acetyl-FB1 was not considered toxic. Lu et al. found that D- glucose reacted with the primary amino group of FB1 at 60 °C and induced its detoxification [15]. Our results showed that Nb5 not only reduced FB1 in the contaminated milk at 1:2000 dilution, but also increased egg hatching ability and body weight and reduced gizzard ulcerations of new-borne chickens. Moreover, the Nb5 and BiNb11 improved the broiler's body weight and the Nb5 also reduced gizzard ulceration, which was comparable to the popular commercial detoxifier, Qiangankang, as the positive control. Our study revealed that Nb5 exerted decontamination by detoxifying FB1 toxins.

Regarding detoxification of mycotoxin contamination, physical measure and chemical decontamination techniques may be quite efficient. However, the more sustainable and restricted use of fungicides require new approaches to control this hazard. Food safety demands permanent research efforts for exploring new strategies to reduce mycotoxin contamination. In the present study, Nb5 was identified to remove the FB1 toxin in dairy milk, with a suggestion of a potential detoxifying agent due to an efficient and robust method for the generation of antibodies against a wide range of targets with highly specific binding properties. FB1 was documented to be transmitted from diet into animal milk and exerted toxic effects on human and animal health. The interaction of different mycotoxins may be additive or synergetic. Given that milk is a source of nutrients, especially in childhood, a thorough investigation of the occurrence of mycotoxins as well the adoption of measures to minimize their contamination of milk is urgently needed. Moreover, FB1 and FB2 in milk samples are stable to pasteurization (62 °C/30 min) and storage at 4 °C for 11 days [8], and the incidence of these contaminants is a major issue for human health. The carry-over of fumonisin B1 from contaminated feed into dairy milk also suggests its carry-over from contaminated food into breast milk in Northern Tanzania; 10.3% had fumonisin B1 levels above the EU limit of 200 ppb for fumonisins in infants' food that leads to unacceptable exposures in infants [24]. In the present study, a 43.26% detoxifying rate was determined in fresh milk post treatment with 1:2000 concentration of Nb5 concentration. Compared to the anti-FB1 monoclonal antibody, a 25% degradation rate of FB1 was determined post incubation at 25 °C for 2 h in our previous study [25]. Regarding cost-effective detoxification, Nb5 might be a promising detoxifier for combating mycotoxin contamination in the daily milk consumption. The nanobody is the smallest specific binding entity compared to the monoclonal antibody and polyclonal antibody. In addition, the nanobody will avoid random chemical conjugation and the use of secondary antibodies, contributing to mycotoxin removal in the study.

Nanobodies have benefits over standard monoclonal and polyclonal antibodies in terms of size, stability, and expression level when used as antibodies. There have been some successful examples regarding the use of Nbs to effectively recognize small molecules, such as the mycotoxins AFB1, OTA, and 15-acetyl-DON, and some environmental contaminants. The isolated anti-idiotypic-Nb was subjected to an ELISA for the detection of FB1 contaminated in cereals and feedstuffs [26]. The developed assay showed an IC50 value of 0.95 ng/mL, with a limit of detection of 0.15 ng/mL, linear range of 0.27–5.92 ng/mL, and low cross-reactivity toward FB2 (4.93%). The sensitivity of anti-idiotypic ELISA was enhanced approximately 20-fold compared with that of the chemosynthetic FB1-BSA conjugates-based ELISA (IC50 = 21.14 ng/mL). The established anti-idiotypic ELISA was validated to be suitable for monitoring the total fumonisin concentration under the current regulatory limits of fumonisins in most countries [18]. In the present study, we tested the ELISA titers after the 4th immunization of alpaca; the titer reached only 1:100, which might be related to the low dose of the FB1 compound. Our phage-display library yielded a capacity of 3.4×10^{10} after identification, and 46 single colonies were randomly selected, with a positive rate of 96.7%. The library showed good diversity once the positive monoclonal antibodies were sequenced. The library was then subjected to rescue, including 3 rounds of selective panning and enrichment, and 5 nanobodies with different amino acid sequences were finally screened out.

The detoxification was further verified by embryonated eggs and the broiler model. A previous study confirmed that FB1 had moderate toxicity in chick embryos, causing pathological changes, such as hydrocephalus, beak enlargement, neck elongation, as well as heart, lung, liver, kidney, and small intestine abnormalities [27]. In a prior study, chick embryos were inoculated with 16 μg FB1 per embryo and examined for the presence of any serious developmental defects, characterized as severe hemorrhagical inflammations evident in the head, neck, and chest of the deceased embryo [28]. In a recent study, 24 μg FB1 was inoculated into 11-day-old chicken embryos, resulting in lung hemorrhage and gastric ulcer in the new-borne chickens, as well as FB1 residuals in gizzard and lung [8]. In the current study, the average body weight was significantly reduced, indicating that FB1 had certain effects on the development of chicken embryos. In the present study, Nb5 not only improved the egg hatching ability and body weight. It also alleviated the gizzard ulcers, while the D-glucose did not reduce the gizzard ulceration of new-born chickens. More interestingly, broilers' gizzards health was improved greatly post Nb5 ($p < 0.01$) and the BiNb11 ($p < 0.05$) treatment, without a significant difference comparable to the treatment of the commercial detoxifier product, the Qingankang group ($p > 0.05$) (Figure 5B). In terms of clinical efficacy, Qinankang with 20% protocatechuic acid (PCA) is commercialized against mycotoxin contamination in the broiler industry, characterized by fewer gizzard ulcerations and enhanced body weight in the study (Figure 5A). However, no stimulating IgG responses against IBDV, NDV, and IBV were observed in the Qinankang group compared to those of the Nb5 + FB1 group. As for material supplying, both PCA and nanobody are environmentally friendly products using *E. coli*-engineered strains in the fermentation process. The nanobody is cheaper to produce for large scale production than PCA in the manufacturer due to complicated purification. In this sense, the nanobody might be a low-cost detoxifier in the poultry industry.

FB1 has diverse effects on the immune system, causing both stimulation and suppression of the response to foreign antigens, and apparently inducing an antigenic response to FB1 [29]. Our previous study indicated that the combination of FB1 and DON was associated with a low hatching rate and gizzard ulcerations in chicken progenies [8]. The previous report confirmed that FB1 was an immunosuppressive to chickens when present in the diet with a ratio of 200 mg FB1/kg [30]. In the present study, our data confirmed that limited immune suppression against IBDV was induced in the birds that received 10 mg FB1/kg of diet for 14 days. No significant suppressive impact on NDV and IBV was observed in the study. However, Nb5, as a monomer nanobody, could enhance the IBDV humoral response of chickens exposed to daily FB1 diet.

Nanobodies termed as VHH antibody are usually isolated from the library constructed by primary antibody immunization and eluted with the free antigen. Although the nanobodies are a particularly useful tool for monitoring mycotoxins in food and feedstuffs, as they are easily genetic engineered and have superior stability [31], the potential removal of mycotoxin has not been fully investigated. Nanobodies are reported to recognize the active site of the antigen, and also serve as a surrogate for the original antigen and compete with the original antigen for the primary antibody [32]. In the present study, both monomer VHH and bivalent VHH were produced in large quantities, excellent solubility, and resilience to gastric pH value. Post oral administration with Nb5 and BiNb11, nanobodies might bind to FB1 and block the specific site of the toxin, leading to an improvement of body weight and gastric lesions. Compared to the bivalent VHH of BiNb11, the monomer Nb5 showed the advantages of chicken health and embryo development, a suggestion of highly affinity to FB1 and efficient binding activities. Initially, bivalent and monomer nanobodies were used for toxin quantitative, and our presented data indicated novel detoxifiers in poultry industry and diary consumption. Further investigation is needed to verify the commercial administration. On the other hand, more understanding is needed about the conversion procedures, the toxicological characteristics of the products obtained by transformation, the effect of the conversion on the nutrition of feed and on animal safety. Such feed additive must be harmless and stable in the digestive tract of animals. Therefore, future work is required to elucidate wide application as feed or food detoxifier.

4. Conclusions

In the present study, 5 nanobodies that specifically bound to FB1 were successfully screened out using phage-display technology. Based on detoxifying efficacy in fresh milk, chicken embryos, and broilers, it is the first report in which the monomer nanobody termed Nb5 has been identified to be a putative novel detoxifier by degrading FB1 mycotoxins in fresh milk, embryonated chickens and broiler, contributing to sustainable poultry industry and food safety for human beings.

5. Materials and Methods

5.1. Alpaca Immunization

Two adult male alpaca received subcutaneously 0.2 mL of the FB1-adjuvanted bovine serum albumin (BSA) 6 times at 14-day intervals. The immunogen complex contained a total of 1 mg of for each immunization prepared with Freund's adjuvant in equal volume (complete Freund's adjuvant was used only once; subsequent immunizations were done using incomplete Freund's adjuvant). Blood was taken before the first injection, and two week later, after the sixth injection. ELISA assay was used to analyze serum samples for FB1-adjuvanted-BSA-specific antibody response. Peripheral blood mononuclear cells (PBMCs) were isolated from blood samples using standardized density gradient technique (Ficoll–Paque) (Solarbio, Beijing, China).

5.2. Construction of VHH Phage-Display Library

Total RNA was extracted from lymphocytes according to the instructions on the RNA extraction kit (QIAGEN, Shanghai, China) and reverse-transcribed into cDNA using Oligo (dT) 20 primers. With two cycles of PCR, the variable domain of heavy chain antibody (VHH) gene was amplified. The *VHH* up-forward gene and *VHH* up-reverse gene were used for the first round and *VHH* down-forward gene and *VHH* down-reverse gene were used for the second amplification reaction (Table 1) and the target fragments were then cloned into the pCANTAB 5 E vector and electrically transferred to TG1-competent cells. After three rounds of acid elutions, the bacteriophages bound to FB1-BSA were enriched successfully. In addition, the positive clones were screened by phage-ELISA and the titer of output phage determination was determined. The library diversity and capacity were also identified.

Table 1. Primers for amplifying *VHH* genes.

Primers	Sequences (5′→3′)
VHH-up-F	GTCCTGGCTGCTCTTCTACAAGG
VHH-up-R	GGTACGTGCTGTTGAACTGTTCC
VHH-down-F	TTTCTATTACTAGGCCCAGCCGGCCATGGCTCAGGTGTGGCTCGTGGAGTC
VHH-down-R	AAGGAAAAAAGCGGCCGCGCCATAATGGCCTGGTTGTG
pCANTAB5-R1	CCATGATTACGCCAAGCTTTGGAGCC
pCANTAB5-R2	CGATCTAAAGTTTTGTCGTCTTTCC

5.3. Prokaryotic Expression and Identification of FB1 Nanobody Protein

Based on the target gene sequence, the upstream and downstream primers were designed using Primer 5.0 software (Table 2). Using the above elutriation, the target sequence of *VHH*-FB1-5 was amplified by the *VHH* up-forward gene and *VHH* up-reverse gene, then cloned into a pCold-SUMO vector at Bam HI and Hind III restriction sites (Solarbio, Beijing, China) and identified by bacterial liquid PCR. Positive plasmids were extracted by using OMEGA Plasmid DNA Kit (Omega Bio-tek, Shanghai, China), and the concentration was measured by NanoDrop 2000 (Thermo Fisher Scientific, Waltham, MA, USA) and then sent for sequencing (Sangong Biotech, Shanghai, China) and stored at −20 °C for further use. The *VHH*-FB1-PSF-positive plasmid was transferred to *E. coli* BL21 (DE3) competent cells and induced using 1 M of isopropyl-β-D-thiogalactopyranoside (IPTG, Merck, Beijing, China) at 25 °C for 5 h. The bacterial precipitate was collected by centrifugation and re-suspended with PBS. After ultrasound, the supernatant and precipitate were collected, and SDS-PAGE was used to determine the solubility of the protein. Nanobody 5 (termed Nb5), Nanobody11 (termed Nb11), Nanobody 12 (termed Nb12), Nanobody (termed Nb13) and Nanobody (termed Nb16) were purified individually by Ni-NTA (GenScript, Nanjing, China). Further, the titer of the nanobodies was determined by an indirect ELISA assay. The OD measurement was conducted at OD450 nm.

Table 2. Primer sequences for constructing recombinant FB1 nanobody prokaryotic expression vector.

Primers	Sequences (5′→3′)
VHH-up-F	CGAGCTCATGGAGGTGCAGCTCCTGGTG
VHH-up-R	CGGGATCCCGAGACGGTGACCAGGGTC

5.4. Detoxification of Nb5 on FB1 in Milk

Milk samples were purchased from the local shop and extracted according to the method previously described [24]. In brief, a skimming process was carried out by centrifugation (4500× *g* for 10 min at 4 °C), and the 9 milk samples were thoroughly vortex mixed. Afterwards, an aliquot of 2 mL of each sample was added to 6 mL of acetonitrile (Aladdin Biochemical Technology Co., Ltd., Shanghai, China), mixed by vortexing for 2 min and processed with ultrasounds for 5 min. The samples were centrifuged to remove large biomolecules (such as proteins) at 12,000× *g* for 15 min. The supernatants were then filtered through 0.22 μm cellulose syringe filters in amber vials until the further HPLC analysis.

Afterwards, FB1 was added to 900 μL of milk to the final concentration of 47.66 ppb and then incubated with 100 μL of 4 dilutions (1:2000, 1:3000, 1:5000, and 1:10,000 of Nb5) at 25 °C for 2 h. Subsequently, the samples were extracted according to the aforementioned method.

5.5. Antagonistic Effect of FB1-Specific Nanobody against the Chicken Embryotoxicity of FB1

To analyze the antagonistic effect of nanobodies on FB1 detoxification of embryonated chickens, 64 μg of FB1 was treated separately with a different FB1-specific nanobody (2 μg/egg) at 25 °C, or 0.1 M D-glucose at 70 °C for 2 h while PBS, FB1, and Nanobody were regarded as the control groups.

A total of 48 SPF embryonated chickens, aged 11 days, were purchased from a commercial company (Boehringer Ingelheim Inc, Beijing, China) and divided into 8 groups, 6 embryos per group. The experimental protocols were approved by an Ethical Reviewing Board at the China Agricultural University (approved code: IACUC 20190802, date of approval 2 August 2019), based on guidelines from the Institutional Animal Care and Use Committee of China Agricultural University (IACUC). This follows humane protocols that minimize pain in the animals. The experimental groups were inoculated into albumen with 100 μL of the treated mixtures, and then hatched at 37 °C until day 21 (Table 3). Upon hatching day, chickens were euthanized in a CO_2 chamber using 100% CO_2 at a flow rate of 10–30% of the chamber volume per minute, and the birds were observed for the absence of breathing activities and loss of the heartbeat. The CO_2 flow lasted for at least 1 min after breathing arrest. The body weight and lesions of the lungs and gizzards were observed and scored. The FB1 concentrations from the lungs and gizzards were determined by HPLC (SPD-M20A, Shimadzu Corporation, Japan) as previously described [24].

Table 3. Experimental protocol of detoxification in the embryonated eggs.

Groups	Treatment	Incubation Conditions	Eggs
Control	PBS	25 °C; 2 h	6
FB1	FB1 (64 μg)	25 °C; 2 h	6
Nb control	1: 50 Nb	25 °C; 2 h	6
Nb5 + FB1	FB1 (64 μg)) + 1: 50 Nb5	25 °C; 2 h	6
Nb13 + FB1	FB1 (64 μg) + 1: 50 Nb13	25 °C; 2 h	6
BiNb11 + FB1	FB1 (64 μg) + 1: 50 BiNb11	25 °C; 2 h	6
BiNb13 + FB1	FB1 (64 μg) + 1: 50 BiNb13	25 °C; 2 h	6
D-glucose + FB1	FB1(64 μg) + 0.1 M D-glucose	70 °C; 2 h	6

Forty-eight embryonated eggs aged 11 days were randomly divided into 8 groups, including 5 experimental groups (Nb5, Nb13, BiNb13, BiNb11, and D-glucose groups, 3 replicates per group) and 3 control groups (PBS control, FB1 control, and Nb control, 6 chicken embryos per group). Before inoculation, poorly growing embryonated eggs were eliminated from the experiment.

5.6. *Antagonistic Effect of Nanobody on Growth, Gizzard Ulceration, and Immune Response of Broilers*

A total of 50 ten-day-old Elvin broilers were purchased from a commercial company (Meikeduo Food Group Co. Ltd., Hebei, China). The broilers were randomly divided into 5 groups: 2 experimental groups (Nb5 group and BiNb11 group) and 3 control groups (Control group, FB1 group, and Qinankang + FB1 group) (Table 4). The experimental protocols were approved by an Ethical Reviewing Board at China Agricultural University (Approved code: IACUC 20190803). Qinankang is comprised of 20% protocatechuic acid and reported to degrade 71.43% FB1 at 80 °C for 2 h [26]. It is registered as a commercial detoxifier (Genten Biotech, Beijing, China) against FB1 toxin in poultry. From day 10 onwards, the control group was fed with basic diet while the other groups were given feed with corn gluten meal naturally contaminated with 10 mg FB1 per kg. Simultaneously, broilers orally received 50 μg of Nb5, or 50 μg of BiNb11 and 75 mg/kg of commercial product of Qiangankang for 21 days, respectively. Before treatment, all the birds received an oral dose of attenuated vaccines against NDV and IBV on day 10, as well as one dose of live vaccine against IBDV on day 14. During observation, broilers were weighed weekly and monitored activities. Sera were collected on days 0, 14, and 21. Subsequently, the specific antibody levels for IBV or IBDV, and NDV were determined using commercial ELISA kits (IDEXX Laboratories. Inc, Beijing, China). In addition, lymphocyte proliferation index was determined on day 21 as previously described [33]. At the end of the experiment, the chickens were euthanized humanely using cervical dislocation, which caused death by breaking the blood vessels so that the brains run out of oxygen. Afterwards, gizzard lesions were determined while immune organs were observed, including spleen, thymus, and bursa of Fabricius.

Table 4. Effect of monomer nanobodies and bivalent nanobodies on hatchability and development of chicken embryos.

Groups	Hatchability (%)	Body Weight (g)
Control	83.33	40.11 ± 1.29 [a]
FB1	33.33	28.14 ± 7.70 [b]
Nb5 + FB1	83.33	37.10 ± 2.74 [a]
Nb13 + FB1	50	32.87 ± 5.47 [b]
BiNb11 + FB1	60	36.69 ± 3.17 [a]
BiNb13 +FB1	50	33.61 ± 10.22 [a]
D-glucose + FB1	80	34.74 ± 5.27 [a]
Nb control	100	38.35 ± 2.11 [a]

a indicated $p > 0.05$ when compared the body weight of the Nb5 + FB1 group, the BiNb11 + FB1 group, the BiNb13 +FB1 group, the D-glucose + FB1 group and the Nb control group with that of the control group. b indicated $p < 0.01$ when compared the body weight of the FB1 group and the Nb13 + FB1 group with that of the control group. The data are expressed as the mean ± SD.

Fifty broilers were randomly divided into 5 groups, including two experimental (Nb5 and BiNb11 groups, ten birds per group) and 3 control groups (Control group, FB1 group, and Qinankang + FB1 group, 10 birds per group). Except for the control group, the other 4 groups received corn gluten meal containing 10 ppm of FB1 per day, and then treated with 50 μg of Nb5, 50 μg of BiNb11, and 75 mg/kg of commercial detoxifier, Qinankang, respectively for 21 days.

As the Table 4 showed, the hatchability was 83.33% and 80%, respectively, in the Nb5 + FB1 group and the D-glucose + FB1 group compared to 33.3% in the FB1 group. Moreover, decreasing hatchability was observed in the Nb13 + FB1 group, the BiNb11 + FB1 group, and the BiNb13 + FB1 group compared to the Nb5 + FB1 group. As for body weight, the FB1 group induced slow growth compared to the control groups and the Nb control group ($p < 0.01$). No statistical difference was found among the Nb5 + FB1 group, or the BiNb11 + FB1 group, the BiNb13 + FB1 group and the D-glucose + FB1 group. Postmortem, FB1 induced a highly gastric ulceration index while lower lesions were determined in the Nb5 + FB1 group and all the control groups.No statistical difference was found among the Nb13 + FB1 group, the BiNb11 + FB1 group, the BiNb13 + FB1 group, and the D-glucose + FB1 group.

As for FB1 residuals, lung and gizzard were collected aseptically, the organs were homogenized, and 1 g of tissue was blended with 5 mL of 80% methanol aqueous solution containing 0.1% acetic acid. Subsequently, the supernatant solutions were filtered by glass fiber filter paper, collected and mixed with 5 mL n-hexane. Afterwards, the samples were centrifugated to discard n-hexane at 12,000 rpm for 5 min. Finally, the samples were collected and diluted with PBS 5 times and passed in column purification. The FB1 residuals were analyzed using high-performance liquid chromatography (Agilent Technologies Inc., Santa Clara, CA, USA). A ZORBAX SB-C18 Column (150 mm × 4.6 mm) was employed with a mobile phase of methanol: sodium biphosphate (77:23). The sample was detected at a flow rate of 1 mL/minute using a 335 nm/440 nm wavelength.

5.7. Data Analysis

FB1 concentrations and lesion scores were statistically analyzed using SPSS 17.0 version to perform a one-way ANOVA with the LSD post hoc test on at least three independent replicates. p-values of <0.05 were considered statistically significant for each test, and when $p < 0.01$, the results were highly significant. The hatching rate were statistically analyzed using SPSS 17.0 version to perform a Chi-square test with a categorical variable. A p-value of <0.05 was considered to be a significant difference for each test and a p-value of <0.01 was considered to be highly significant.

Author Contributions: Y.C.: Animal experiments and data analysis and manuscript writing. G.Q.: Provided the techniques for nanobody development and manuscript writing. H.Q.: Developed recombinant FB1 nanobodies. Y.W.: Data analysis and manuscript reviewing. C.W.: Development of FB1 Nanobody phage library. M.A.H.: Manuscript reviewing and English improvement. C.H.: Designed the experiment and did the final approval of the manuscript. All authors have read and agreed to the published version of the manuscript.

Funding: This work was supported by grant from Asia Cooperation Foundation (Grant No.12200111) and Natural Science Foundation of Shandong Province (No. ZR2020MC186).

Institutional Review Board Statement: The experimental protocols were approved by an Ethical Reviewing Board at the China Agricultural University (approved code: IACUC 20190802, date of approval 2 August 2019), based on guidelines from the Institutional Animal Care and Use Committee of China Agricultural University (IACUC).

Informed Consent Statement: Not applicable.

Data Availability Statement: The data presented in this study are available on request from the corresponding author via email address: hecheng@cau.edu.cn.

Acknowledgments: This work was supported by a grant from the Asia Cooperation Foundation (Grant No.12200111) and Natural Science Foundation of Shandong Province (No. ZR2020MC186).

Conflicts of Interest: The authors declare no conflict of interest.

Ethics Approval: All the procedures involving animals were conducted in strict accordance with the Guidelines for the Ethical Review of Laboratory Animal Welfare of China National Standard GB/T 35892-2018. Furthermore, the procedures for animal inoculation as well as other animal procedures were approved by the Institutional Animal Care and Use Committee (IACUC) at the China Agricultural University (Approved code: IACUC 20190802; date of approval 2 August 2019).

References

1. Chen, Z.; Zhang, F.; Jiang, L.; Chen, Z.; Sun, H. Toxic Effects of Mycotoxin Fumonisin B1 at Six Different Doses on Female BALB/c Mice. *Toxins* **2022**, *14*, 21. [CrossRef] [PubMed]
2. Iqbal, N.; Czékus, Z.; Poór, P.; Ördög, A. Plant defence mechanisms against mycotoxin Fumonisin B1. *Chem. Biol. Interact.* **2021**, *343*, 109494. [CrossRef] [PubMed]
3. Rocchetti, G.; Ghilardelli, F.; Masoero, F.; Gallo, A. Screening of Regulated and Emerging Mycotoxins in Bulk Milk Samples by High-Resolution Mass Spectrometry. *Foods* **2021**, *10*, 2025. [CrossRef] [PubMed]
4. Haque, M.A.; Wang, Y.; Shen, Z.; Li, X.; Saleemi, M.K.; He, C. Mycotoxin contamination and control strategy in human, domestic animal and poultry: A review. *Microb. Pathog.* **2020**, *142*, 104095. [CrossRef] [PubMed]
5. Lumsangkul, C.; Tso, K.H.; Fan, Y.K.; Chiang, H.I.; Ju, J.C. Mycotoxin Fumonisin B1 Interferes Sphingolipid Metabolisms and Neural Tube Closure during Early Embryogenesis in Brown Tsaiya Ducks. *Toxins* **2021**, *13*, 743. [CrossRef] [PubMed]
6. Alsulami, T.; Nath, N.; Flemming, R.; Wang, H.; Zhou, W.; Yu, J.H. Development of a novel homogeneous immunoassay using the engineered luminescent enzyme NanoLuc for the quantification of the mycotoxin fumonisin B1. *Biosens. Bioelectron.* **2021**, *177*, 112939. [CrossRef]
7. Tomaszewska, E.; Rudyk, H.; Dobrowolski, P.; Donaldson, J.; Swietlicka, I.; Puzio, I.; Kamiński, D.; Wiącek, D.; Kushnir, V.; Brezvyn, O.; et al. Changes in the Intestinal Histomorphometry, the Expression of Intestinal Tight Junction Proteins, and the Bone Structure and Liver of Pre-Laying Hens Following Oral Administration of Fumonisins for 21 Days. *Toxins* **2021**, *13*, 375. [CrossRef]
8. Wang, Y.; Quan, H.; Li, X.; Li, Q.; Haque, M.A.; Shi, Q.; Fu, Q.; He, C. Contamination with Fumonisin B and Deoxynivalenol Is a Threat to Egg Safety and Contributes to Gizzard Ulcerations of Newborn Chickens. *Front. Microbiol.* **2021**, *12*, 676671. [CrossRef]
9. Zhao, Z.; Zhang, Y.; Gong, A.; Liu, N.; Chen, S.; Zhao, X.; Li, X.; Chen, L.; Zhou, C.; Wang, J. Biodegradation of mycotoxin fumonisin B1 by a novel bacterial consortium SAAS79. *Appl. Microbiol. Biotechnol.* **2019**, *103*, 7129–7140. [CrossRef]
10. IARC. *Monographs on the Evaluation of Carcinogenic Risks to Humans: Some Naturally Occurring Substances: Food Items and Constituents, Heterocyclic Aromatic Amines and Mycotoxins*; International Agency for Research on Cancer: Lyon, France, 1993; Volume 56, pp. 1–599.
11. Joint FAO/WHO Expert Committee on Food Additives. *Evaluation of Certain Contaminants in Food: Eighty-Third Report of the Joint FAO/WHO Expert Committee on Food Additives*, 83rd ed.; World Health Organization: Geneva, Switzerland, 2017.
12. Humpf, H.U.; Voss, K.A. Effects of thermal food processing on the chemical structure and toxicity of fumonisin mycotoxins. *Mol. Nutr. Food Res.* **2004**, *48*, 255–269. [CrossRef]
13. Vila-Donat, P.; Marín, S.; Sanchis, V.; Ramos, A.J. Tri-octahedral bentonites as potential technological feed additive for *Fusarium* mycotoxin reduction. *Food Addit. Contaminants. Part A Chem. Anal. Control Expo. Risk Assess.* **2020**, *37*, 1374–1387. [CrossRef] [PubMed]

14. Murphy, P.A.; Hendrich, S.; Hopmans, E.C.; Hauck, C.C.; Lu, Z.; Buseman, G.; Munkvold, G. Effect of processing on fumonisin content of corn. *Adv. Exp. Med. Biol.* **1996**, *392*, 323–334. [CrossRef] [PubMed]
15. Lu, Y.; Clifford, L.; Hauck, C.C.; Hendrich, S.; Osweiler, G.; Murphy, P.A. Characterization of fumonisin B (1)-glucose reaction kinetics and products. *J. Agric. Food Chem.* **2002**, *50*, 4726–4733. [CrossRef] [PubMed]
16. Muyldermans, S. Nanobodies: Natural single-domain antibodies. *Annu. Rev. Biochem.* **2013**, *82*, 775–797. [CrossRef]
17. Shu, M.; Xu, Y.; Wang, D.; Liu, X.; Li, Y.; He, Q.; Tu, Z.; Qiu, Y.; Ji, Y.; Wang, X. Anti-idiotypic nanobody: A strategy for development of sensitive and green immunoassay for Fumonisin B_1. *Talanta* **2015**, *143*, 388–393. [CrossRef] [PubMed]
18. He, T.; Zhu, J.; Nie, Y.; Hu, R.; Wang, T.; Li, P.; Zhang, Q.; Yang, Y. Nanobody Technology for Mycotoxin Detection in the Field of Food Safety: Current Status and Prospects. *Toxins* **2018**, *10*, 180. [CrossRef]
19. Kunz, P.; Flock, T.; Soler, N.; Zaiss, M.; Vincke, C.; Sterckx, Y.; Kastelic, D.; Muyldermans, S.; Hoheisel, J.D. Exploiting sequence and stability information for directing nanobody stability engineering. *Biochim. Biophys. Acta Gen. Subj.* **2017**, *1861*, 2196–2205. [CrossRef]
20. Liu, X.; Wen, Y.; Wang, W.; Zhao, Z.; Han, Y.; Tang, K.; Wang, D. Nanobody-based electrochemical competitive immunosensor for the detection of AFB1 through AFB1-HCR as signal amplifier. *Mikrochim. Acta* **2020**, *187*, 352. [CrossRef]
21. Andersen, K.K.; Strokappe, N.M.; Hultberg, A.; Truusalu, K.; Smidt, I.; Mikelsaar, R.H.; Mikelsaar, M.; Verrip, T.; Hammarström, L.; Marcotte, H. Neutralization of Clostridium difficile Toxin B Mediated by Engineered Lactobacilli That Produce Single-Domain Antibodies. *Infect. Immun.* **2015**, *84*, 395–406. [CrossRef]
22. Harmsen, M.M.; van Solt, C.B.; Hoogendoorn, A.; van Zijderveld, F.G.; Niewold, T.A.; van der Meulen, J. Escherichia coli F4 fimbriae specific llama single-domain antibody fragments effectively inhibit bacterial adhesion in vitro but poorly protect against diarrhoea. *Vet. Microbiol.* **2005**, *111*, 89–98. [CrossRef]
23. Driehuis, F.; Spanjer, M.C.; Scholten, J.M.; Te Giffel, M.C. Occurrence of mycotoxins in maize, grass and wheat silage for dairy cattle in The Netherlands. *Food Addit. Contaminants. Part B Surveill.* **2008**, *1*, 41–50. [CrossRef] [PubMed]
24. Magoha, H.; De Meulenaer, B.; Kimanya, M.; Hipolite, D.; Lachat, C.; Kolsteren, P. Fumonisin B1 contamination in breast milk and its exposure in infants under 6 months of age in Rombo, Northern Tanzania. *Food Chem. Toxicol. Int. J. Publ. Br. Ind. Biol. Res. Assoc.* **2014**, *74*, 112–116. [CrossRef] [PubMed]
25. Wang, F.; Chen, Y.; Hu, H.; Liu, X.; Wang, Y.; Saleemi, M.K.; He, C.; Haque, M.A. Protocatechuic acid: A novel detoxication agent of fumonisin B1 for poultry industry. *Front. Vet. Sci.* **2022**, *9*, 923238. [CrossRef] [PubMed]
26. Yang, S.; Luo, Y.; Mu, L.; Yang, Y.; Yang, Y. Risk screening of mycotoxins and their derivatives in dairy products using a stable isotope dilution assay and LC-MS/MS. *J. Sep. Sci.* **2021**, *44*, 782–792. [CrossRef]
27. Javed, T.; Richard, J.L.; Bennett, G.A.; Dombrink-Kurtzman, M.A.; Bunte, R.M.; Koelkebeck, K.W.; Côté, L.M.; Leeper, R.W.; Buck, W.B. Embryopathic and embryocidal effects of purified fumonisin B1 or Fusarium proliferatum culture material extract on chicken embryos. *Mycopathologia* **1993**, *123*, 185–193. [CrossRef]
28. Henry, M.H.; Wyatt, R.D. The toxicity of fumonisin B1, B2, and B3, individually and in combination, in chicken embryos. *Poult. Sci.* **2001**, *80*, 401–407. [CrossRef]
29. Keck, B.B.; Bodine, A.B. The effect of fumonisin B1 on viability and mitogenic response of avian immune cells. *Poult. Sci.* **2006**, *85*, 1020–1024. [CrossRef]
30. Li, Y.C.; Ledoux, D.R.; Bermudez, A.J.; Fritsche, K.L.; Rottinghaus, G.E. Effect of fumonisin B1 on selected immune responses in broiler chicks. *Poult. Sci.* **1999**, *78*, 1275–1282. [CrossRef]
31. Leivo, J.; Vehniäinen, M.; Lamminmäki, U. Phage Display Selection of an Anti-Idiotype-Antibody with Broad-Specificity to Deoxynivalenol Mycotoxins. *Toxins* **2020**, *13*, 18. [CrossRef]
32. Wang, J.; Mukhta, H.; Ma, L.; Pang, Q.; Wang, X. VHH Antibodies: Reagents for Mycotoxin Detection in Food Products. *Sensors* **2018**, *18*, 485. [CrossRef]
33. Chu, J.; Zhang, Q.; Zuo, Z.; El-Ashram, S.; Guo, Y.; Zhao, P.; Huang, S.; He, C.; Khan, A. Co-infection of Chlamydia psittaci with H9N2, ORT and Aspergillus fumigatus contributes to severe pneumonia and high mortality in SPF chickens. *Sci. Rep.* **2017**, *7*, 13997. [CrossRef] [PubMed]

MDPI
St. Alban-Anlage 66
4052 Basel
Switzerland
Tel. +41 61 683 77 34
Fax +41 61 302 89 18
www.mdpi.com

Toxins Editorial Office
E-mail: toxins@mdpi.com
www.mdpi.com/journal/toxins

MDPI

www.ingramcontent.com/pod-product-compliance
Lightning Source LLC
Chambersburg PA
CBHW061445210926
51850CB00008B/1619
9783036579238